Genetic Management of Fragmented Animal and Plant Populations

Genetic Management of Fragmented Animal and Plant Populations

Richard Frankham
Macquarie University and Australian Museum, Sydney, NSW, Australia

Jonathan D. Ballou & Katherine Ralls
Smithsonian Institution, Washington, DC, USA

Mark D. B. Eldridge
Australian Museum, Sydney, NSW, Australia

Michele R. Dudash & Charles B. Fenster
South Dakota State University, Brookings, SD, USA

Robert C. Lacy
Chicago Zoological Society, Brookfield, IL, USA

Paul Sunnucks
Monash University, Clayton, Vic, Australia

Line drawings by Karina McInnes, Melbourne, Vic, Australia

OXFORD
UNIVERSITY PRESS

OXFORD
UNIVERSITY PRESS

Great Clarendon Street, Oxford, OX2 6DP,
United Kingdom

Oxford University Press is a department of the University of Oxford. It furthers the University's objective of excellence in research, scholarship, and education by publishing worldwide. Oxford is a registered trade mark of Oxford University Press in the UK and in certain other countries

First Edition published in 2017

Published in the United States of America by Oxford University Press
198 Madison Avenue, New York, NY 10016, United States of America

British Library Cataloguing in Publication Data
Data available

Library of Congress Control Number: 2017933536

ISBN 978–0–19–878339–8

Contents

Preface

Importance of the topic

One of the greatest unmet issues in conservation biology is the genetic management of fragmented animal and plant populations. Many species across the planet have fragmented distributions with some small isolated populations that are potentially suffering from inbreeding, loss of genetic diversity, and elevated extinction risk (genetic erosion). Fortunately, these effects can usually be remedied by augmenting gene flow (crossing populations within species), but this is rarely conducted. Disturbingly, evidence of any genetic differentiation among populations typically leads managers to conclude that the populations should be kept isolated, thereby dooming many to eventual extinction. We are particularly concerned with populations fragmented by human actions in the last 500 years, encompassing most of the impacts arising from a 13-fold increase in the human population.

A **paradigm shift** is urgently required where the existence of genetic differentiation among populations acts as a trigger to ask if any populations are suffering from genetic problems. If so, are there populations to which they can be crossed to reduce inbreeding and augment genetic diversity, and will this gene flow be beneficial or harmful?

Whether to maintain genetic isolation is a major issue for managers of wild animal and plant populations. It is critical that managers, and those who advise them, become aware of the issues relating to restoration of genetic diversity and contribute to the reversal of genetic deterioration associated with small population size.

Genesis of this book

The events that triggered this book began in February 2007 during a book writing session on the 2nd edition of *Introduction to Conservation Genetics* involving Jon Ballou and Dick Frankham. When revising material on population fragmentation, we were deeply concerned about over-emphasis in the literature on the risks of outbreeding depression (where crossing between populations is harmful), while the benefits of crossing between populations within species to rescue small inbred populations were largely neglected. We subsequently outlined means to predict the risk of outbreeding depression (thus allowing the benefits of genetic rescue to be utilized widely), recruited Mark Eldridge, Kathy Ralls, Bob Lacy, Michele Dudash, and Charlie Fenster as collaborators, and subsequently published the Frankham et al. (2011) paper on the topic. That work influenced a paper on the use of genetic translocations to cope with global climate change (Weeks et al. 2011) co-authored by Mark Eldridge, Dick Frankham, and Paul

Sunnucks among others. Our paper on implications of different species concepts for biodiversity conservation (Frankham et al. 2012) was provoked by practical concerns that Ingrid Porton of St Louis Zoo had about genetic management of ruffed lemurs in light of taxonomic revisions and competing taxonomies, and also drew upon our work on predicting outbreeding depression.

This book arose as an extension of the above studies and our previous work on inbreeding depression, evolutionary potential, outbreeding depression, genetic rescue, and genetic management of fragmented populations, plus our extensive involvement in threatened species recovery. While drafting Chapter 6 on genetic rescue, Dick Frankham identified the lack of a quantitative overview and this provoked him to complete two meta-analyses that inform the book chapter (Frankham 2015, 2016), as did the content of a symposium on the topic at the 2014 meeting of the Society for Conservation Biology in Missoula, Montana.

The material on managing gene flow among isolated population fragments (optimizing gene flow) was initiated largely as an extension of the seminal Ballou & Lacy (1995) work on genetic management by minimizing mean kinship—the recommended management strategy for threatened species in captivity. The content of this book was also informed by a symposium on genetic management of fragmented populations at the 2011 meeting of the Society for Conservation Biology in New Zealand, organized by authors of this book.

Purpose

This book provides the conceptual background for understanding genetic issues in the management of fragmented wild animal and plant populations, and offers means for making practical decisions on genetic management of population fragments. We focus particularly on the harmful consequences of mating among relatives on offspring fitness (inbreeding depression) that is universal among sexually reproducing non-haploid organisms. A second major focus is on the maintenance of genetic diversity, so that species and populations can continue to adapt to changing environments. We also define the criteria that must be satisfied by definitions of species for conservation purposes, and make recommendations regarding definitions that are appropriate and inappropriate. Our goal of preserving endangered species through genetic management can provide additional benefits beyond single species preservation, because species interact within communities that are subject to abiotic and biotic environments, contributing to ecosystem function and overall global biodiversity.

Intended audience

This book is designed for senior undergraduate and graduate students and professional scientists in conservation and wildlife biology, and evolutionary ecology, as well as for managers of wild animal and plant populations. We will subsequently write a shorter, simpler Practical Guide for managers who are less familiar with conservation genetics.

To make the book accessible to a wide audience, we have assumed knowledge only of introductory genetics and basic statistics. For the same reason, we have restricted use of mathematics to simple algebra, even though genetic management requires a quantitative approach. Readers wishing for a simple concise introduction to conservation genetics are referred to *A Primer of Conservation Genetics* by Frankham et al. (2004) or the Italian, Portuguese, or Korean translations of it.

Précis of contents

The first eight chapters of the book consist of background material and concepts, the subsequent six chapters are concerned with genetic management (including objectives, questions, and decision trees), and the final chapter summarizes the messages from the prior chapters. Genetic issues in single populations are addressed prior to the more complex issues of multiple populations.

Chapter 1 provides a general outline of the issues involved in genetic management of fragmented animal and plant populations and its importance.

Section I provides detailed background on genetic problems in fragmented populations. This begins with a brief introduction to the evolutionary genetics of small populations (Chapter 2) where the roles of mutation, migration, selection, and chance (genetic drift) are described, inbreeding and its measurement considered, and population size in the evolutionary genetics context (effective population size) defined.

Threats to small isolated populations from inbreeding depression and mutational accumulation are covered in Chapter 3, and those from reduced ability to adapt genetically in Chapter 4. We next address genetic problems caused by inadequate gene flow in fragmented populations (Chapter 5).

Section II addresses risk/benefit analyses of remedies for genetic problems in fragmented populations. Genetic rescue by augmenting gene flow is addressed in Chapter 6. Chapter 7 delineates the causes of outbreeding depression and considers the potential of populations to recover from outbreeding depression through natural selection. Chapter 8 evaluates how expectations for remedies and risk from crossing populations need to be modified for species with diverse mating systems and modes of inheritance (self-incompatible, self-fertilizing, asexual, polypoid, haploid, haplodiploid, etc.).

Section III of the book deals with genetic management of fragmented populations. Chapter 9 defines the criteria that must be satisfied by a species definition used for conservation purposes, provides recommendations on definitions that are and are not appropriate, identifies the many pitfalls in species delineation, and recommends remedies. Chapter 10 covers methods for delineating the number and locations of population fragments. Chapter 11 presents practical guidelines (including decision trees) for determining whether populations are suffering genetic erosion and whether they would benefit from augmented gene flow. Chapters 12 and 13 address management of gene flow among isolated population fragments, the former chapter based on limited information, and the latter on extensive genetic information (using mean kinship). Chapter 14 explores genetic management under the moving target created by global climate change. Finally, Chapter 15 draws the prior threads together to present the **Take home messages** from the book.

Format

To make the books easy to read and follow, we use a textbook format similar to that of the Frankham, Ballou, and Briscoe textbooks *Introduction to Conservation Genetics* (2nd edn, 2010) and *A Primer of Conservation Genetics* (2004). In particular, we use:

- main messages boxes at the start of each chapter
- list of terms for each chapter
- frequent headings and sub-headings
- small main point boxes and dot point lists throughout chapters
- ample use of figures, maps, and tables
- boxes with case studies
- example boxes
- decision trees
- line drawings by Karina McInnes
- links to software for genetic analysis and management
- List of symbols
- Glossary
- Appendices
- References and Index.

We use questions for most of the major headings within chapters, as science seeks to answer questions. This approach was successfully trialed in undergraduate teaching, presentations to professional scientists, and for communicating genetic concepts to conservation practitioners.

References

Given length constraints and issues of readability, we restrict referencing to examples and a few recent or key citations (especially meta-analyses and reviews) to allow access to the relevant literature, except for contentious issues where we present more extensive referencing.

Acknowledgments

We thank our partners Annette Lindsay (RF), Vanessa Ballou (JDB), Robert L. Brownell Jr. (KR), Anne Baker (RCL), and Andrea Taylor (PS) for their forbearance and support during the preparation of this book.

We are grateful for the support of our home institutions Macquarie University (RF), the Australian Museum (ME and RF), the Smithsonian Institution (JDB and KR), University of Maryland and South Dakota State University (MRD and CBF), Chicago Zoological Society (RCL), and Monash University (PS). The Smithsonian's National Zoological Park and the University of Maryland kindly hosted book writing sessions and RF thanks Jon and Vanessa Ballou and Michele Dudash and Charlie Fenster for their hospitality during several of these sessions.

The authors thank the following for financial research support for projects relevant to the development of ideas presented in the book during the writing period 2011–2016, as follows:

M. Dudash: her work on inbreeding depression and its genetic basis and how mating system evolution may be impacted was funded through the US National Science Foundation and USDA (USA).

M. Eldridge: his group's work has been supported by the Australian Research Council, the Hermon Slade Foundation, the Foundation for National Parks and Wildlife, and the Holsworth Wildlife Research Endowment.

C. Fenster: his work on the genetic and ecological consequences of population fragmentation has been supported by the US National Science Foundation and DOE (USA), while that on mating system evolution and its genetic consequences has been backed by grants from NSERC (Canada) and NFR (Norway).

R. Lacy: his work on population management has been supported by various grants from the US National Science Foundation, the Institute of Museum and Library Services, and the Association of Zoos and Aquariums. Much of his work to apply genetic management to species conservation arose from collaborations within the IUCN (International Union for Conservation of Nature) Species Survival Commission, and its Conservation Breeding Specialist Group.

P. Sunnucks: his group's work has been supported by diverse sources including the Australian Research Council and the Holsworth Wildlife Research Endowment.

We thank Karina McInnes for preparing the excellent line drawings. Harry Crosby skilfully revised figures that were based on published material and Jon Ballou prepared the new figures in the book and revised many others.

We thank M. Culver for providing us with the raw data on puma genotypes used for the calculations in Box 13.2, J. Deitz, T. King, J. Mickelberg, and I. Porton for supplying information, I. Porton for checking Box 9.3, and T. King for checking Box 11.2, and to N. Hancock and P. Mahoney for commenting on Chapter 14. We thank J. Howard and B. Pukazhenthi for supplying material for illustrations.

The staff of Oxford University Press and their outsourced production company turned our material into this book. At Oxford University Press, we thank Ian Sherman for championing our book proposal and for overall supervision of the project, Lucy Nash for preparing the book contract, and Bethany Kershaw for her efficient and cheerful work as commissioning editor. The CEO Nigel Portwood and Helen Tyas facilitated improvements in production, liaising with the team at SPi Global led by Cheryl Brant. We thank Cheryl and her team for the effort they put into the production of the book and for accommodating our numerous suggestions and revisions. We are grateful to Julie Musk for the outstanding job she did proofreading the book.

Copyright acknowledgments

We thank Cambridge University Press for permission to reproduce the following figures: 2.3, 2.4, 2.6, 2.7, 3.1, 3.2, 3.3, 3.4, 3.8, 3.9, 3.10, 3.11, 4.1, 4.2, 4.3, 4.4, 4.5, 4.6, 5.2, 5.3, 5.4, 5.5, 5.6, 5.8, 5.10, 5.11, 5.12, 7.2, and A1, and figures within Box 2.3, Example 2.2, Box 3.1, Chapter 4 frontispiece, Box 4.1, Box 4.2, Box 4.4, Chapter 5 frontispiece, Chapter 6 frontispiece, Box 6.3 (two), Box 8.2, Box 2.4, Chapter 9 frontispiece, Box 10.1, and Box 14.3 which first appeared in Frankham, Ballou, & Briscoe (2010) *Introduction to Conservation Genetics*, 2nd edition, and Fig. 13.3 from the first edition (Frankham et al. 2002).

We are grateful to the following for kind permission to reproduce copyright material:

Chapter 1 frontispiece reprinted from Oedekoven (1960), Page 184, Copyright (1960), with permission from IOS Press; figures in Box 1.1 from figs 1 and 3 in Zhang et al. 2007 by permission of Oxford University Press; Chapter 3 frontispiece from fig 3 in Szulkin et al. 2007, © 2014 The Authors, Journal of Evolutionary Biology © 2014 European Society for Evolutionary Biology; Chapter 3 frontispiece (c) republished with permission of John Wiley and Sons Inc., from fig 2 in Nason and Ellstrand 1995; permission conveyed through Copyright Clearance Center, Inc.; Fig 3.5 from fig 1a in Losdat et al. 2014, © 2014 The Authors, Journal of Evolutionary Biology © 2014 European Society for Evolutionary Biology; Fig 3.6 from fig 4 in Szulkin et al. 2007, © 2014 The Authors, Journal of Evolutionary Biology © 2014 European Society for Evolutionary Biology; Fig 4.2 republished with permission of John Wiley and Sons Inc., from fig 11.3 in Weber 2004, permission conveyed through Copyright Clearance Center, Inc.; Fig 4.4 reprinted from fig 3 in Montgomery et al. 2010, Copyright (2010), with permission from Elsevier; Chapter 5 frontispiece from fig 2 in James 1995; Fig 5.1 from fig 1 in Jump and Peñuelas 2006, Copyright (2006) National Academy of Sciences, U.S.A.; Fig 5.7 from fig 5B in Jousimo et al. 2014, reprinted with permission from AAAS, permission conveyed through Copyright Clearance Center, Inc.; Fig 5.9 from fig 5 in Aguilar et al. 2008,

© 2008 Blackwell Publishing Ltd.; figures in Box 5.2 from fig 1 and 3 in Phillipsen et al. 2015, © 2014 John Wiley & Sons Ltd.; Fig 5.12 from fig 4 in Forbes and Hogg 1999, Copyright © 1999–2016 John Wiley & Sons, Inc. All Rights Reserved; Box 5.3 from figs 1 and 3 in Major et al. 2014, © 2014 The Zoological Society of London; Fig 6.1a republished with permission of John Wiley & Sons Inc., from fig 2 in Newman and Tallmon 2001, permission conveyed through Copyright Clearance Center, Inc.; Fig 6.2 from fig 2 in Frankham 2015, Copyright © 1999–2016 John Wiley & Sons, Inc.; Chapter 7 frontispiece from fig 3 in Yang et al. 2004, Copyright © 2004, Kluwer Academic Publishers; Fig 7.1 republished with permission of Oxford University Press—Journals, from figs 3a and b in Murray and Young 2001, permission conveyed through Copyright Clearance Center, Inc.; Fig 7.3 reprinted from fig 1 in Trends in Ecology & Evolution, 27 (4), Cutter 2012, with permission from Elsevier; Box 8.2 from fig 1 in Young and Brown 1999, © Cambridge University Press; Fig 8.2 republished with permission of John Wiley & Sons Inc., from fig 1 in Jarne and Auld 2006, permission conveyed through Copyright Clearance Center, Inc.; Fig 8.3 from fig 1 in Bousalem et al. 2010, © 2008 Canadian Science Publishing or its licensors, reproduced with permission; Fig 9.3 republished with permission of John Wiley & Sons Inc., from figs 1 and 3 in Bouzat et al. 1998, permission conveyed through Copyright Clearance Center, Inc.; Box 9.1 reproduced from fig 1 from Eldridge and Close 1992, with permission from CSIRO Publishing; Box 9.4 from fig 1 in Degnan 1993, Copyright © 1999–2016 John Wiley & Sons, Inc.; Chapter 10 frontispiece from figs 1 and 3 in Bergl and Vigilant 2007, Copyright © 1999–2016 John Wiley & Sons, Inc.; Box 10.2 republished with permission of John Wiley & Sons Inc., from figs 1, 4, and 5 from Coleman et al. 2013, © 2013 Blackwell Publishing Ltd, permission conveyed through Copyright Clearance Center, Inc.; Fig 10.3 republished with permission of Genetics Society of America, from fig 1a in Gao et al. 2007, permission conveyed through Copyright Clearance Center, Inc.; Fig 10.4 from fig 1 in Coulon et al. 2008, © 2008 Blackwell Publishing Ltd.; Fig 10.5 from figs 1a and 1c from Bittner and King 2003, Copyright © 1999–2016 John Wiley & Sons, Inc.; Fig 10.6 from fig 1 in Eldridge et al. 2001, Copyright © 1999–2016 John Wiley & Sons, Inc.; Box 11.4 republished with permission of Oxford University Press, from fig 3B in Stephen et al. 2005, permission conveyed through Copyright Clearance Center, Inc.; Chapter 11 frontispiece from fig 1 in Dolan et al. 1999, Copyright © 2016 by American Journal of Botany, Inc.; Chapter 11 frontispiece histogram from fig 2 in Oakley and Winn 2012, © 2012 The Authors, New Phytologist © 2012 New Phytologist Trust; Box 11.1 image of kinked tail reprinted from Roelke et al. 1993, Copyright 1993, with permission from Elsevier; Example 13.3 from fig 1 in Finger et al. 2011, © 2008 Blackwell Publishing Ltd.; Box 14.1 figure republished with permission of John Wiley & Sons Inc., from figs 1 and 2 in Miller-Rushing and Primack 2008, permission conveyed through Copyright Clearance Center, Inc.

List of Symbols

The following table contains the list of symbols used in the book, along with their meaning and the chapters where they are mainly considered. Where the same symbol is used for more than one purpose in the published literature, we have added an additional letter to avoid ambiguity. Different symbols for the same variable are sometimes used in diverse publications.

Symbol	Definition	Chapter
a	Arbitrary genotypic value assigned to the effect on the phenotype of genotype A_1A_1, where genotype A_2A_2 has a value of –a, and the two homozygotes differ in mean by $2a$	3, 4
A	Allelic diversity, mean number of alleles per locus	2
B	Haploid lethal equivalents	3, 11
C	A parameter used in analyses of the impact of inbreeding on fitness, where e^{-C} is the survival of an outbred population (often referred to as A, but we use that for another purpose)	3
d	Dominance deviation for the effect of the heterozygote on a quantitative trait: deviation of heterozygote effect from the mean of the two homozygotes	3, 4
D_J	Jost's measure of population differentiation	5
D_M	Absolute genetic differentiation between population fragments, also referred to as the average minimum genetic distance	13
D_N	Nei's genetic distance	5, 13
D_{ST}	Average genetic diversity between population fragments, the numerator for G_{ST}	13
E	Proportion of adaptation that is due to non-shared features of the environments of two populations	7
ER	Evolutionary rescue, ratio of evolutionary potential in a crossed population compared to that in its inbred parents	6

Symbol	Definition	Chapter
f	Relative frequency	2, 11, 12
F	Inbreeding coefficient of individuals, or the mean for a population	2, 3, 5, 6, 8, 11, 12
F_{IS}	Inbreeding within population fragments (sub-populations)	5, 11
F_{IT}	Total inbreeding over a number of population fragments	5
F_{ST}	Inbreeding due to population differentiation	5, 10, 11, 13
GA	Genetic adaptation over a generation, where ΣGA_t is cumulative genetic adaptation over t generations	4, 7
G_{ST}	Analogue of F_{ST} for multiallelic loci	5
G'_{ST}	G_{ST} corrected so that it scales 0–1 (G_{ST} divided by G_{STmax})	5, 11, 13
G_{STmax}	Maximum value of G_{ST} for a particular locus (or set of loci) with particular allele frequencies	5
H	Heterozygosity (see also $\hat{H}$, H_e, H_o, H_S, and H_T)	2, 4, 5, 6, 11
$\hat{H}$	Equilibrium heterozygosity, for example due to mutation-drift equilibrium	8
h^2	Heritability of a quantitative character	4, 6, 11, 14
ht	Haplotype diversity	2
H_e	Hardy–Weinberg equilibrium heterozygosity	2, 5, 8, 11
H_o	Observed heterozygosity	2, 8, 11
hs	Dominance effects of an allele on fitness, where the two homozygotes have fitnesses of 1 and 1 – s, respectively, and the heterozygote 1 – hs. Thus $h = 0$ is dominant, 1 is recessive, and 0.5 additive, etc.	2, 8
H_S	Expected heterozygosities averaged across population fragments	5
H_T	Expected heterozygosity across the totality of a group of population fragments	5
I	Normalized identity of genes between two populations	13
ID	Inbreeding depression, difference in mean between outbred and inbred offspring for a quantitative trait (e.g. fitness)	3
J_{XY}	Gene identity between population fragments X and Y (Nei 1987), a measure of population similarity, related to mk_{AB}	5, 13
k	Number of items (e.g. population fragments, combinations, etc.)	5, 10, 13

k_{ij}	Kinship (or coancestry) between two individuals: the probability that two randomly chosen alleles at a locus, one from each individual, are identical by descent	5, 13
K_{ST}	Analogue of F_{ST} for DNA sequence data, a measure of population differentiation	5
ln	Natural logarithm, $\log_e$	3
m	Migration rate, proportion of migrant alleles introduced over a single generation, where Nm is the number of migrants	5, 10, 12
$\overline{mk}$	Average mean kinship of individuals in a population	13
mk_{AB}	Mean kinships between populations A and B, average of pairwise kinships of all individuals in population A with all in population B	5, 13
mk_i	Mean kinship of individual i, average of its kinships with all individuals in the population	5, 13
MU	Contribution to genetic adaptation from selection acting on new mutations	7
n	Sample size	2, 6
N	Census population size (usually potentially reproducing adults)	2, 8, 9, 10, 12, 14
n_A	Effective number of alleles	2, 5
N_e	Effective population size. This determines the genetic impacts of small population size	2, 3, 4, 5, 8, 9, 11, 14
p	Frequency of allele A_1 (or A)	2, 3, 4, 5, 8, 12, 13
PD	Contribution to genetic adaptation from selection operating on initial (pre-existing) genetic diversity	7
$P(E)$	Probability of extinction of a population over a defined number of generations	10
q	Frequency of allele A_2 (or a)	2, 3, 4, 5, 8, 12
$\hat{q}$	Equilibrium frequency of allele A_2 (a), as for example from mutation-selection equilibrium	2, 8
Q_{ST}	Proportion of variation for a quantitative character due to differences among population fragments (quantitative trait analogue of F_{ST})	5
r	Correlation coefficient	8, 10
r_e	Relatedness	13

Symbol	Definition	Chapter
R_{ST}	Analogue of F_{ST} that measures the proportion of differences in length of microsatellite alleles among population fragments (to incorporate evolutionary relationships among alleles)	5
s	Selection coefficient against a genotype at a locus	2, 8, 12
S	Probability of survival	3
Sd	Selection differential for a quantitative trait	4, 6
Sf	Selfing rate	8
S_P	Number of polymorphic sites at a locus in a segment of DNA	2
t	Number of generations	3, 4, 5, 9, 11
T	Outcrossing rate (1 – selfing rate)	8
u	Mutation rate per generation for a locus, or nucleotide	2, 8
V_A	Additive genetic variation for a quantitative character	4, 11
V_m	Increase per generation in additive genetic variation due to mutation	4, 11
V_P	Phenotypic (total) variance for a quantitative character	4
W	Mean fitness of a population or group	3
δ	Measure of inbreeding depression: proportionate decrease in trait mean due to inbreeding, compared to outbreeding (usually from the impact of selfing)	3
Δ	Change in value, usually from one generation to the next, e.g. ΔF, Δq	3, 4, 8
ΔF	Differences in inbreeding coefficient between individuals in different generations	3, 6, 8, 11
ΔF_m	Differences in inbreeding coefficient between mothers of individuals in different generations	6, 11
ΔF_z	Differences in inbreeding coefficient between zygotes in different generations	6, 11
ΔGR	Genetic rescue effect, measured as the proportionate benefit in fitness of a crossed population compared to the inbred parent	6
Δq	Change in allele frequency over one generation	8, 12
ΔW	Change in fitness, for example due to inbreeding	11
π	Nucleotide diversity	2
Π	Product of the quantities specified	12

σ_p^2 and σ_q^2	Variance in allele frequencies among population fragments for the A_1 and A_2 alleles, respectively. With only two alleles at a locus, the quantities are equal	2, 5
σ_P	Phenotypic standard deviation for a quantitative character (square root of V_P)	4
Σ	Sum of the following terms	2, 3, 4, 13
Φ_{ST}	Analogue of F_{ST} that incorporates genealogies	5

Introduction

CHAPTER 1

Genetic management of fragmented populations is one of the major, largely unaddressed issues in biodiversity conservation. Many species across the planet have fragmented distributions with small isolated populations that are potentially suffering from inbreeding and loss of genetic diversity (genetic erosion), leading to elevated extinction risk. Fortunately, genetic deterioration can usually be remedied by augmenting gene flow (crossing between populations within species), yet this is rarely done, in part because crossing is sometimes harmful (but it is possible to predict when this will occur). Benefits and risks of genetic problems are sometimes altered in species with diverse mating systems and modes of inheritance. Adequate genetic management depends on appropriate delineation of species. We address management of gene flow between previously isolated populations and genetic management under global climate change.

TERMS

Critically endangered, endangered, genetic diversity, genetic erosion, genetic rescue, inbreeding, inbreeding depression, outbreeding depression, population, population fragmentation, threatened

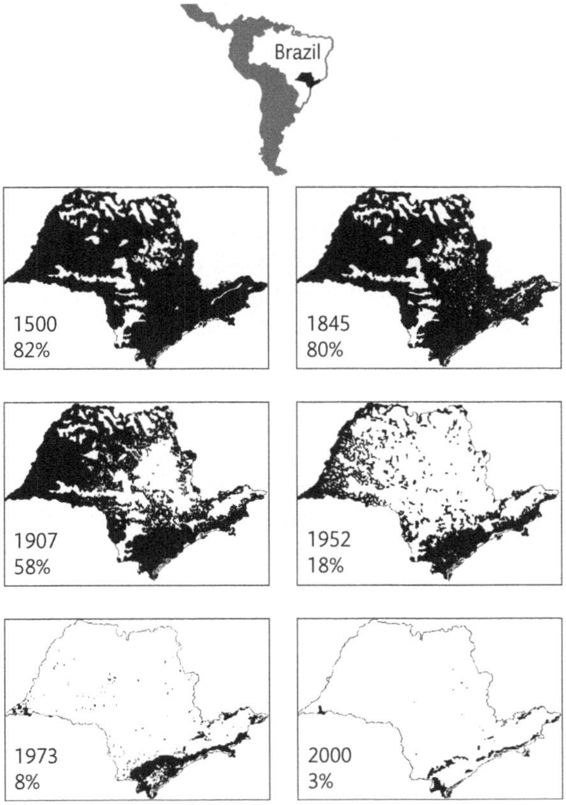

Loss and fragmentation of the Atlantic coastal forest of Brazil from 1500 to 2000 (from Oedekoven 1980) indicating that the distributions of many species have becoming increasingly fragmented. The percentages are extent of forest cover.

Genetic Management of Fragmented Animal and Plant Populations. R. Frankham, J. D. Ballou, K. Ralls, M. D. B. Eldridge, M. R. Dudash, C. B. Fenster, R. C. Lacy & P. Sunnucks. Oxford University Press (2017) © R. Frankham, J. D. Ballou, K. Ralls, M. D. B. Eldridge, M. R. Dudash, C. B. Fenster, R. C. Lacy & P. Sunnucks 2017. DOI 10.1093/oso/9780198783398.001.0001

Why should we conserve genetic diversity?

Conservation of genetic diversity is one of three levels of biodiversity recommended by the IUCN for conservation, but there has been limited activity to implement this recommendation for wild populations

IUCN (International Union for Conservation of Nature), the premier international conservation agency, recommends conservation of biodiversity at three levels (McNeely et al. 1990):

- genetic diversity
- species diversity
- ecosystem diversity.

Genetics is involved in all three levels. First, genetic diversity is worthy of conservation in its own right, because it controls the variety of life—the shape, size, color, biochemistry, behavior, and all other characters that define species and underpin their place in the world. Second, adequate genetic diversity favors species persistence by defending against harmful effects of inbreeding, maintaining fitness, and providing the ability to evolve. Through these influences on species diversity, genetic diversity enhances ecosystem viability (Bangert et al. 2005; Lankau & Strauss 2007). More directly, genetic diversity within species can have profound effects on the species composition of ecological communities and higher-order ecological processes, such as nutrient cycling (Reusch et al. 2005; Whitham et al. 2006; Hughes et al. 2008). Conservation at the species level has received most attention, while ecosystem conservation is now receiving increasing attention.

Genetics has received little attention in the management of threatened species in natural habitats, despite genetic diversity being protected under legislation, such as the United Nations Convention on Biological Diversity (UNEP 1992) and national laws in many countries (Moyle et al. 2003; Laikre et al. 2009). For example, genetic threats to persistence and recovery are mentioned in 63% of endangered species recovery plans in the USA, 55% in Australian plans, and 33% in European plans (Pierson et al. 2016). Inbreeding was commented upon only in ~7% of plans and even fewer discussed inbreeding depression. Few plans mentioned active genetic management. Examples of the consequences of ignoring genetic issues and poor handling of them are common (Frankham et al. 2010). In contrast, genetics has played a large role in the management of captive populations over the past 30 years (Ballou et al. 2010).

Efforts have recently been made to incorporate genetic considerations into the management of wild populations, such as actions in Sweden (Laikre et al. 2008), a Summit on Conservation of Australia's Genetic Diversity in 2012, and creation of the Conservation Genetics Specialist Group within the IUCN Species Survival Commission. This book is designed to aid in this undertaking by addressing arguably the most important issue in the conservation of genetic biodiversity, the genetic management of fragmented animal and plant populations (Frankham 2010a, 2010b).

Ubiquity of fragmented populations

Most species on the planet have fragmented distributions, with some population fragments genetically isolated and suffering genetic problems

Many species on the planet have fragmented distributions, much of it due to human conversion of natural habitats over the last few hundred years. This includes many threatened animal and plant species listed in the IUCN Red List (IUCN 2016). Many isolated populations have reduced genetic diversity relative to historic levels (Tallmon et al. 2004; Aguilar et al. 2008). For example, 26% of invertebrate, 29% of vertebrate, and 55% of plant species show clear evidence of inadequate gene flow among populations (Frankham et al. 2014a).

Harmful effects of inbreeding (inbreeding depression) and loss of genetic diversity will ultimately contribute to the extirpation of many small populations of sexually reproducing organisms (Fenster & Dudash 1994; Moritz 1999; Frankham et al. 2010) (see later, and Chapters 3 and 4). Box 1.1 illustrates the fragmented distribution of the giant panda (*Ailuropoda melanoleuca*) and some of its genetic consequences.

Box 1.1 Fragmented distribution of the giant panda in China and its genetic consequences

(after Lu et al. 2001; Zhang et al. 2007; Hu et al. 2010; Zhu et al. 2011)

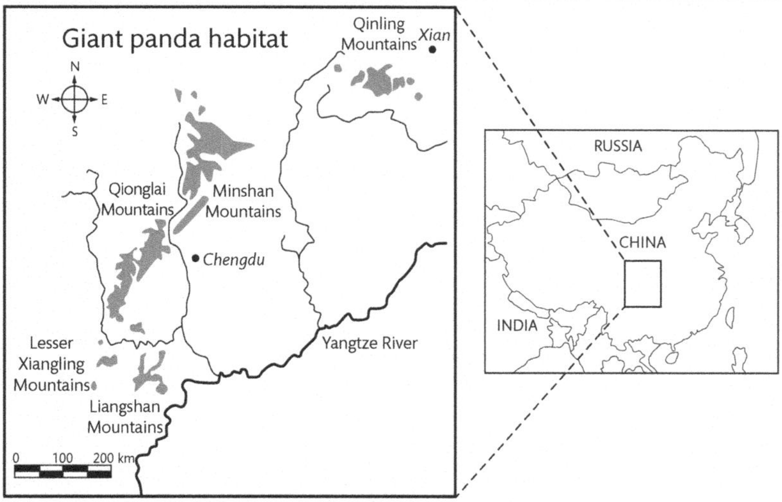

Giant panda (China)

Giant pandas in China have a geographically fragmented distribution (left map). Genetic analyses have shown that the most geographically isolated populations in the Qinling and Liangshan Mountains have lower genetic diversity than other larger, less physically isolated populations, such as those in the Minshan and Qionglai Mountains, based on data for 10 microsatellite loci (see Box Table below). Overall, the giant panda has levels of genetic diversity that are similar to other bears (Zhang et al. 2007).

Population	Heterozygosity	Allelic richness
Qinling (QIN)	0.486	2.58
Minshan (MIN)	0.559	3.03
Qionglai (QIO)	0.610	3.18
Liangshan (LIS)	0.366	2.30
Lesser Xiangling (XL)	0.635	3.29

The populations fall into four different genetic clusters, QIN, MIN, QIO, and LIS plus XL as shown in the STRUCTURE plot below (Zhang et al. 2007). Bars represent the genotypes of individual pandas. Individuals with genotypes typical of populations other than their own are immigrants, while those with genotypes intermediate between populations indicate gene flow. The pandas seem able to migrate considerable distances, even across some areas that are not their primary habitat. Consequently, there is some gene flow between the populations, but very little into Qinling.

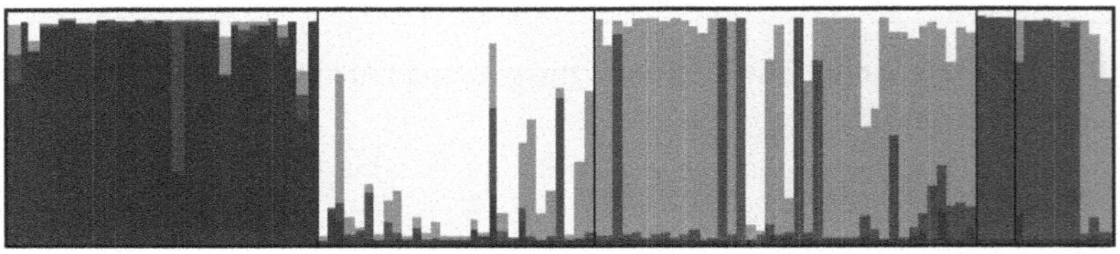

As the term "population" is used widely in this book, we pause to define it in a conservation genetics context.

What is a population?

A population is a group of individuals living in sufficiently close proximity that any member of the group can potentially reproduce with any member of the other sex

There are at least 16 definitions of population (Waples & Gaggiotti 2006), but we follow the evolutionary definition, as given above, as it is most appropriate for the issues in this book.

There is a continuum between completely isolated populations and completely connected ones (random mating), depending on levels of gene flow and divergence time in generations (Fig. 1.1). We elaborate on this in Chapter 5 and subsequent chapters.

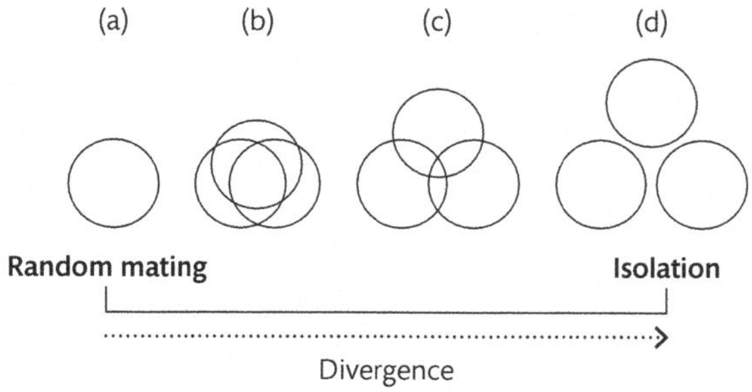

Fig. 1.1 The continuum of genetic differentiation among populations (after Waples & Gaggiotti 2006). Each group of circles represents a group of sub-populations with varying degrees of connectivity (geographic overlap or gene flow). (a) Random mating where "sub-populations" are completely congruent. (b) Substantial connectivity. (c) Modest connectivity. (d) Complete genetic isolation.

What genetic problems occur in fragmented populations?

Fragmentation with no gene flow leads to greater inbreeding and loss of genetic diversity than for a species consisting of a single population of the same total population size, and those population fragments with severely limited gene flow typically experience harmful genetic consequences

Fragmented populations with inadequate gene flow among population segments suffer from the following genetic problems:

- inbreeding and associated reduced reproductive fitness (inbreeding depression) and elevated extinction risks
- loss of genetic diversity and reduced ability to adapt evolutionarily to environmental change.

Small isolated populations of outbreeding diploid species suffer unavoidable loss of genetic diversity and inbreeding over generations with consequent inbreeding depression and reduction in the ability to evolve in response to environmental change (Frankham et al. 2010). The adverse genetic changes occur at a rate that is inversely proportional to the population size (Chapter 2). When populations are fragmented into genetically isolated units, the rate of genetic deterioration accelerates because it is determined by the population size of the unit, not that of the whole species (Fig. 1.2). Over generations, these genetic problems accumulate in closed populations and may cause extinctions, often in concert with other threats such as human impacts, and demographic and environmental stochasticity (Frankham 2005). These problems need to be considered in the context of environmental change, especially global climate change.

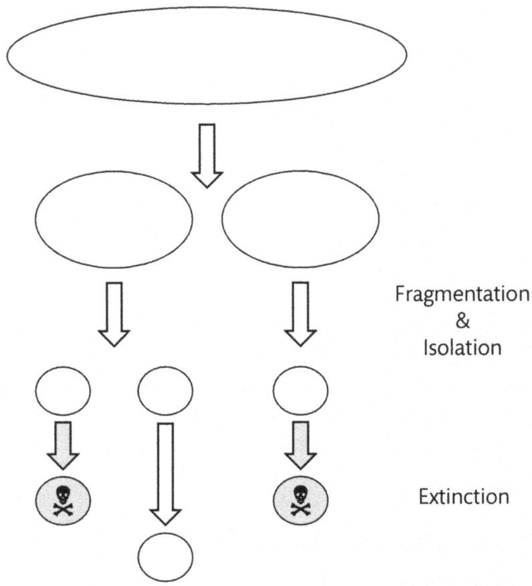

Fig. 1.2 Stages over time (top to bottom) from one large continuous population, through fragmentations into two substantial populations, to three small populations, followed by two extinctions, resulting in one small surviving population.

Why conserve isolated populations?

Conservation of geographically separate populations is important for conserving species genetic diversity, hedging against catastrophes, maintaining suitable habitat, and minimizing species extinction risk

Genetic diversity is distributed among species, populations within species, and individuals within populations. Genetic diversity is lost through random sampling in finite populations (Chapter 2), through extinction of populations, and due to selection favoring some alleles over others. Extinction of populations is a step towards species extinction (Ceballos & Ehrlich 2002). Geographically dispersed populations represent a buffer against extinction risk from catastrophes, help retain genetic diversity, and provide a sound justification for conserving habitat in the face of pressure from human activities.

Can we reverse the adverse effects of population fragmentation?

The adverse effects of genetic isolation can often be reversed by re-establishing gene flow among populations

Inbreeding depression and loss of genetic diversity in small genetically isolated populations can often be reversed by re-establishing gene flow between populations, referred to as **genetic rescue** (Tallmon et al. 2004; Frankham 2015; Whiteley et al. 2015). Gene flow into small inbred populations with low genetic diversity and reduced fitness often has large benefits for wild and domestic species of animals and plants (Chapter 6). For example, gene flow substantially improved the fitness of small inbred populations of African lions (*Panthera leo*), Glanville fritillary butterflies (*Melitaea cinxia*), and the critically endangered jellyfish tree (*Medusagyne oppositifolia*) (Fig. 1.3). Gene flow sometimes has harmful fitness consequences (Chapter 7), but such cases are uncommon and generally predictable (Chapter 10).

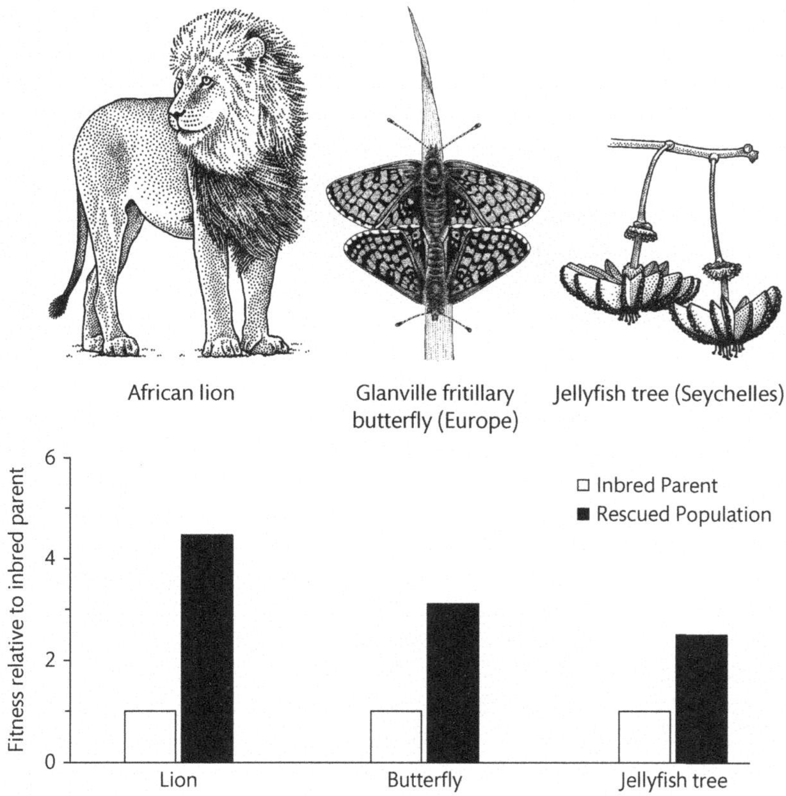

Fig. 1.3 Genetic rescue due to gene flow into small inbred populations of African lions, Glanville fritillary butterflies on the island of Pikku Tytärsaari in the northern Baltic Sea, and critically endangered jellyfish trees from the Seychelles. For each species, the black right-hand column represents the fitness of the genetically rescued population after gene flow relative to that of the inbred parent population (white left column) scaled as 1. Composite fitness (reproduction plus survival) was measured in the lions and trees and egg hatchability in the butterflies (Trinkel et al. 2008; Finger et al. 2011; Mattila et al. 2012).

How important is genetic management of fragmented populations?

Effective genetic management of fragmented populations is one of the major, largely unaddressed issues in conservation biology

Most species have fragmented distributions, many with isolated population fragments suffering genetic problems, but such genetic problems are usually neglected and the populations left to go extinct. However, most isolated inbred populations could be saved from extinction by crossing between populations, but this is very rarely done. For example, the Isle Royale population of the gray wolf (*Canis lupus*) is highly inbred and suffering severe skeletal problems, but the US National Parks Service in 2014 refused to augment the population from the mainland population (without giving a reason) and the wolf population declined to a single highly inbred pair by 2016 (Anonymous 2014; Mlot 2015, 2016). Consequently, genetic management of fragmented populations has been described as one of the major, largely unaddressed issues in all of conservation biology (Frankham 2010a, 2010b).

How many populations would potentially benefit from augmented gene flow?

Many species have isolated populations that would probably benefit from augmented gene flow

All species with fragmented distributions that have at least one totally isolated population (no gene flow) and at least one other population can potentially benefit genetically from augmented gene flow (Frankham et al. 2011). The smaller the population sizes of isolated fragments and the longer they have been isolated (in generations), the greater the need to augment gene flow. In sexual species such gene flow increases genetic diversity and reduces inbreeding, while asexual species benefit from augmented clonal genetic diversity. Island populations typically have reduced genetic diversity and are inbred compared to mainland populations of the species and would often benefit from re-establishment of gene flow (Frankham 1997, 1998; Eldridge et al. 2004).

Examples of species that have isolated populations and would likely benefit from augmented gene flow include greater one-horned rhinoceroses (*Rhinoceros unicornis*) in India and Nepal; black-footed rock-wallabies (*Petrogale lateralis*), koalas (*Phascolarctos cinereus*), ghost bats (*Macroderma gigas*), Cunningham's skinks (*Egernia cunninghami*), orchids, and matchstick banksias (*Banksia cuneata*) in Australia; giant pandas in China; black rhinoceros (*Diceros bicornis*) in Kenya; leopards (*Panthera pardus*) in South Africa; black-footed ferrets (*Mustela nigripes*), grizzly bears (*Ursus arctos horribilis*), gray wolves, desert topminnow fish (*Poeciliopsis monacha*), scarlet gilia (*Ipomopsis aggregata*), spreading avens (*Geum radiatum*), and swamp pink (*Helonias bullata*) plants in North America; tuataras (*Sphenodon punctatus*) and several species of birds in New Zealand; and Glanville fritillary butterflies in Finland and Russia (Frankham et al. 2011).

A rough estimate of the number of fragmented populations of threatened species that would benefit from augmenting gene flow by crossing between populations is given as follows: total number of eukaryotic species (~ 8.7 million; Mora et al. 2011) × proportion threatened (~ 20%; IUCN 2016) × proportion with fragmented populations (~ 40%; Frankham et al. 2014a) × mean number of isolated populations for each of these (genetic populations per species 220 [Hughes et al. 1997], but discount as fewer per threatened species: say 10) × proportion suffering genetic problems (say 20%). Thus, genetic problems due to population fragmentation may be affecting ~ 1.4 million isolated populations of threatened species. If we include non-threatened species, this number climbs to ~ 150 million isolated populations with genetic problems.

We are not aware of any other scientific issue in conservation biology that could have such beneficial practical effects on species conservation in the wild if resolved, and furthermore, at a relatively modest cost.

How frequently is genetic rescue attempted?

Augmentation of gene flow is rarely used to reverse inbreeding depression and loss of genetic diversity in small isolated populations of conservation concern

Despite the large number of populations that would probably benefit from crossing between populations, we are aware of only ~ 30 cases where augmentation of gene flow from non-locally derived populations may have been implemented for conservation purposes in threatened and near threatened populations worldwide (Table 1.1). These cases represent a miniscule proportion of the small fragmented populations that could benefit from management of gene flow.

Table 1.1 Threatened species and populations where gene flow has been augmented potentially for conservation purposes to alleviate genetic problems (known or suspected) (updated from Frankham et al. 2011).

Common name	Taxon	References
Mammals		
African elephant	*Loxodonta africana*	Frankham (2009a)
African lion	*Panthera leo*	Trinkel et al. (2008); Frankham (2009a)
African wild dog	*Lycaon pictus*	Davies-Mostert et al. (2009)
Black rhinoceros	*Diceros bicornis*	Frankham (2009a)
Bighorn sheep	*Ovis canadensis*	Hogg et al. (2006)
Black-footed rock-wallaby	*Petrogale lateralis*	Pearson (2013)
Brush-tailed rock-wallaby	*Petrogale penicillata*	Soderquist (2011)
Burrowing bettong	*Bettongia lesueur*	Finlayson et al. (2010)
Columbia Basin pygmy rabbit	*Brachylagus idahoensis*	Goodall et al. (2009); Elias et al. (2013)
Florida panther	*Puma concolor coryi*	Johnson et al. (2010)

Common name	Taxon	References
Golden lion tamarin	*Leontopithecus rosalia*	Frankham et al. (2010)
Mexican wolf	*Canis lupus baileyi*	Hedrick & Fredrickson (2010)
Mountain pygmy possum	*Burramys parvus*	Weeks et al. (2011); Mansergh et al. (2013)
Northern quoll	*Dasyurus hallucatus*	Cardoso et al. (2009)
Proserpine rock-wallaby	*Petrogale persephone*	Johnson et al. (2003)
Western barred bandicoot	*Perameles bougainville*	Smith & Hughes (2008)
Birds		
Greater prairie chicken in Illinois	*Tympanuchus cupido pinnatus*	Westemeier et al. (1998); Bouzat et al. (2009)
Red-cockaded woodpecker	*Picoides borealis*	US Fish and Wildlife Service (2003)
Reptiles		
Swedish adder	*Vipera berus*	Madsen et al. (2004)
Plants		
Beach clustervine	*Jacquemontia reclinata*	Maschinski et al. (2013)
Brown's banksia	*Banksia brownii*	Barrett & Jackson (2007)
Button wrinklewort	*Rutidosis leptorrhynchoides*	Pickup & Young (2008)
Florida ziziphus	*Ziziphus celata*	Florida Ziziphus Recovery Team Report (2012)
Lakeside daisy population in Illinois	*Hymenoxys acaulis* var. *glabra*	Demauro (1993)
Mauna Kea silversword	*Argyroxiphium sandwicense* ssp. *sandwicense*	Robichaux et al. (1997)
Marsh grass of Parnassus	*Parnassia palustris*	Bossuyt (2007)
Round-leafed honeysuckle	*Lambertia orbifolia*	Coates et al. (1998)
Spiny daisy	*Acanthocladium dockeri*	Clarke et al. (2013)
Twinflower	*Linnaea borealis*	Scobie & Wilcock (2009)

Why are there so few cases?

The limited use of genetic rescue (augmented gene flow between populations reversing inbreeding and leading to recovery of genetic diversity and reproductive fitness) is due to:

- fears of outbreeding depression
- uncertainty in predicting the probable outcomes of augmenting gene flow
- loss of genetic purity, identity, and local characteristics
- overly stringent guidelines for attempting genetic rescue.

Outbreeding depression is a reduction in reproductive fitness (reduced ability to mate/pollinate, fertilize, produce offspring, and survive) in the first or later generations following attempted crossing of populations (Templeton 1986; Fenster 1991; Thornhill 1993; Edmands 2007; Frankham et al. 2011). Fear of outbreeding depression is widely viewed as a major impediment to genetic rescue attempts (Frankham et al. 2011; Frankham 2015).

Thus, it is critical to predict the risk of outbreeding depression in crosses between fragmented populations of species that previously had continuous distributions. Outbreeding depression occurs in only some crosses (Whitlock et al. 2013; Chapter 7), and its occurrence has, until recently, been poorly predicted (Edmands 2002; McClelland & Naish 2007), but we have developed practical means to predict this risk (Frankham et al. 2011; Chapters 7 and 11).

How consistent and large are the benefits of crossing inbred populations to other populations? Recent meta-analyses have revealed that gene flow from other populations into inbred populations had large and consistently beneficial effects on fitness across F_1, F_2, and F_3 and later generations, provided that crosses with an elevated risk of outbreeding depression were excluded (Frankham 2015, 2016).

Until recently, overly stringent guidelines for genetic rescue may have discouraged managers of wild populations from attempting it (Edmands 2007; Hedrick & Fredrickson 2010). We provide appropriate, updated guidelines based on recent scientific advances including Frankham et al. (2011, 2012) and Frankham (2015) (Chapter 11).

How should we genetically manage fragmented populations?

In this book, we provide practical guidelines for genetically managing fragmented populations. These involve (a) determining if populations are suffering genetic erosion, and if so (b) whether they would potentially benefit from augmented gene flow, (c) whether gene flow would be beneficial or harmful, and (d) how much gene flow is appropriate

What questions do we need to answer to genetically manage fragmented animal and plant populations? A flowchart of these questions is given in Fig. 1.4. We first ask whether the taxonomy is appropriate and if not, recommend appropriate means for resolving it (Chapter 9). The second group of critical questions relates to population size and fragmentation and how they impact the genetic characteristics of the population. We have already indicated that the species' population size and the extent of fragmentation are central to an assessment of extinction risk through genetic factors.

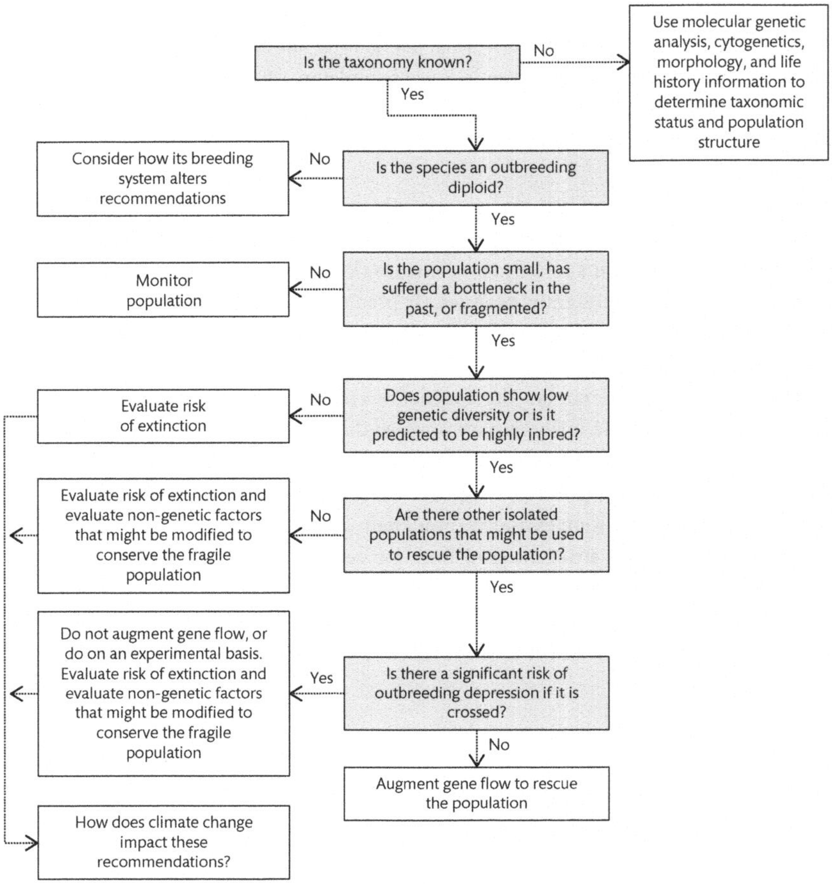

Fig. 1.4 Decision tree with the main questions that need to be asked when genetically managing fragmented populations.

These questions are addressed in Chapters 2–5. The next group of chapters (6–8) deals with the impacts of augmenting gene flow, whether they are likely to be beneficial (genetic rescue) or harmful (outbreeding depression), and how these expectations of risk and rescue need to be adjusted for species with mating systems and modes of inheritance other than outbreeding diploids (Chapter 8). Chapters 10–14 deal with developing management decisions. Chapter 10 addresses means for determining the number and location of population fragments, while Chapter 11 considers whether any of these are isolated fragments suffering genetic erosion that would benefit from augmented gene flow. Managing gene flow by moving individuals or gametes (Chapters 12 and 13) involves answering the questions:

- From where?
- How many?

- What population genetic statistics should we manage?
- Which individuals?
- How often?
- When should we cease?

Chapter 14 considers how global climate change increases the need for genetic management, and the book concludes with Chapter 15, the "Take home messages". We emphasize that genetic management issues must be integrated with those of other disciplines into a single comprehensive plan that includes all stakeholders and relevant political jurisdictions, as in the One Plan approach (Byers et al. 2013).

This book is designed to guide people addressing the critical problem of genetic management of fragmented populations.

In the next Section we consider genetic problems in small isolated populations. The first of these chapters provides a brief introduction to the evolutionary genetics of small populations, as a prelude to dealing in more detail with the genetic problems associated with small populations.

Summary

1. Instituting appropriate genetic management of fragmented populations is one of the major, largely unaddressed issues in conservation biology.
2. Many species have isolated populations that are suffering loss of genetic diversity, inbreeding, and elevated extinction risks: many of these would benefit from augmented gene flow.
3. There are very few cases where genetic rescue has been attempted.
4. Fear of outbreeding depression, lack of quantitative information on the magnitude of effects of augmenting gene flow in inbred populations, the desire to maintain genetic purity, and overly stringent guidelines are impeding appropriate genetic management of fragmented populations.
5. We present the genetic background for understanding inbreeding and outbreeding depression, and provide practical guidelines for identifying genetic problems in fragmented populations. Further, we consider means for determining whether genetic problems might be alleviated by augmented gene flow, evaluating the risks of outbreeding depression, evaluating the potential benefits of augmenting gene flow, and for optimizing gene flow regimes.
6. Modified recommendations may apply to species with reproduction and breeding systems other than outbreeding diploids.
7. Genetic management of fragmented populations under global climate change is also addressed.
8. Genetic issues need to be integrated with those from other disciplines.

FURTHER READING

Frankham (2015) A meta-analysis on the consistency and magnitude of effects of genetic rescues on fitness and evolutionary potential, and the variables affecting them.

Frankham et al. (2004) *A Primer of Conservation Genetics*: A simple concise introductory textbook on conservation genetics.

Frankham et al. (2010) *Introduction to Conservation Genetics*: A comprehensive textbook on conservation genetics, briefly encompassing the topics introduced within this chapter.

Frankham et al. (2011) Presents means for predicting the risk of outbreeding depression and evaluates its predictions.

Pierson et al. (2016) An evaluation of how often genetic factors were considered in endangered species recovery plans from Australia, Europe, and the USA.

SECTION I

Genetic problems in small isolated populations

This book applies theory and observations from evolutionary and conservation genetics to the genetic management of fragmented animal and plant populations. We begin this section with a brief introduction to evolutionary genetics of small populations (Chapter 2), especially for readers who are not familiar with entry-level conservation or population genetics. In this chapter, we consider the roles of mutation, migration, selection, and chance in population evolution. Furthermore, we describe the standing genetic diversity in populations that results from these factors and how it is measured. Subsequently, we define genetically effective population size and inbreeding (the mating of individuals related by descent), and how they are measured.

Several genetic problems occur within species as a consequence of small populations, namely:

- inbreeding depression (reduced reproductive fitness as a consequence of inbreeding)
- genetic drift (due to sampling of gametes in reproduction), leading to loss of genetic diversity
- reduced ability to evolve to cope genetically with environmental change (as a consequence of loss of genetic diversity).

In Chapter 3 we address the harmful fitness consequences of inbreeding (inbreeding depression). Inbreeding is unavoidable in small closed populations (no immigration), and has been known since at least Darwin's (1876) book to have overwhelmingly harmful effects on reproductive fitness across essentially all studied sexually reproducing species.

Genetic diversity is lost in small isolated populations, reducing their ability to evolve in response to environmental change, whether global climate change, pests, parasites and diseases, or human-associated pollution. We deal with these issues broadly in Chapter 4, and focus specifically on evolutionary adaptation to cope with global climate change in Chapter 14.

Fragmentation of populations is a major problem when it leads to inadequate gene flow, because problems due to inbreeding and loss of genetic diversity are worsened, and extinction risk elevated (Chapter 5). If population fragments are totally isolated, then their genetic fate is determined by their population size, rather than that of the whole species.

As conservation genetics is a quantitative science, we use equations to predict genetic changes in populations. These equations are restricted to simple algebra. The "**List of Symbols**", located after the "Acknowledgments" details the symbols used in this book, along with their definitions and the chapter(s) in which they are mainly considered.

Evolutionary genetics of small populations

CHAPTER 2

Genetic management of fragmented populations involves the application of evolutionary genetic theory and knowledge to alleviate problems due to inbreeding and loss of genetic diversity in small population fragments. Populations evolve through the effects of mutation, selection, chance (genetic drift), and gene flow (migration). Large outbreeding, sexually reproducing populations typically contain substantial genetic diversity, while small populations typically contain reduced levels. Genetic impacts of small population size on inbreeding, loss of genetic diversity, and population differentiation are determined by the genetically effective population size, which is usually much smaller than the number of individuals.

TERMS

Adaptive evolution, allozyme, balancing selection, bottleneck, census population size, coalescence, effective population size, electrophoresis, endangered, equilibrium, fitness, fixed, gene flow, genetic drift, genetic load, genotype × environment interaction, haplotype, heterozygosity, idealized population, identity by descent, inbreeding coefficient, lethal, microsatellite, mutation-selection balance, natural selection, neutral mutation, outbreeding, random mating, vulnerable

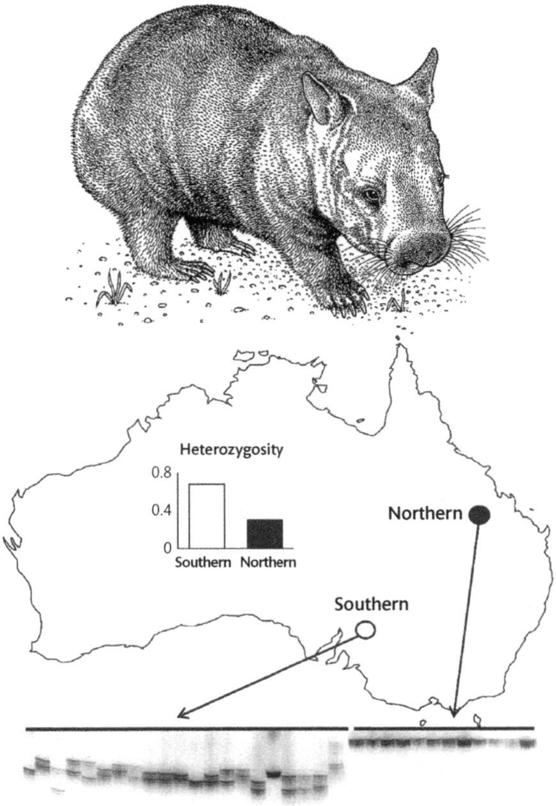

Low microsatellite genetic diversity in the critically endangered northern hairy-nosed wombat (*Lasiorhinus krefftii*), compared to its nearest relative, the southern hairy-nosed wombat (*Lasiorhinus latifrons*) in Australia (Taylor et al. 1994; Beheregaray et al. 2000). The autoradiograph below the map shows microsatellite patterns at a locus for the two species.

Genetic Management of Fragmented Animal and Plant Populations. R. Frankham, J. D. Ballou, K. Ralls, M. D. B. Eldridge, M. R. Dudash, C. B. Fenster, R. C. Lacy & P. Sunnucks. Oxford University Press (2017) © R. Frankham, J. D. Ballou, K. Ralls, M. D. B. Eldridge, M. R. Dudash, C. B. Fenster, R. C. Lacy & P. Sunnucks 2017. DOI 10.1093/oso/9780198783398.001.0001

Background

This book uses theory and observations from conservation and evolutionary genetics, and applies them to the genetic management of fragmented animal and plant populations

Genetic management of fragmented populations is a component of conservation genetics, an applied discipline drawing upon evolutionary, population, molecular, and quantitative genetics and genomics, and applying them to conservation issues with particular emphasis on small populations (Fig. 2.1). These genetic issues are also influenced by demography, ecology, disease biology, etc., and may, in turn, influence them.

This chapter is designed to provide a brief introduction to the critical concepts of conservation and evolutionary genetics that we apply in the remainder of the book.

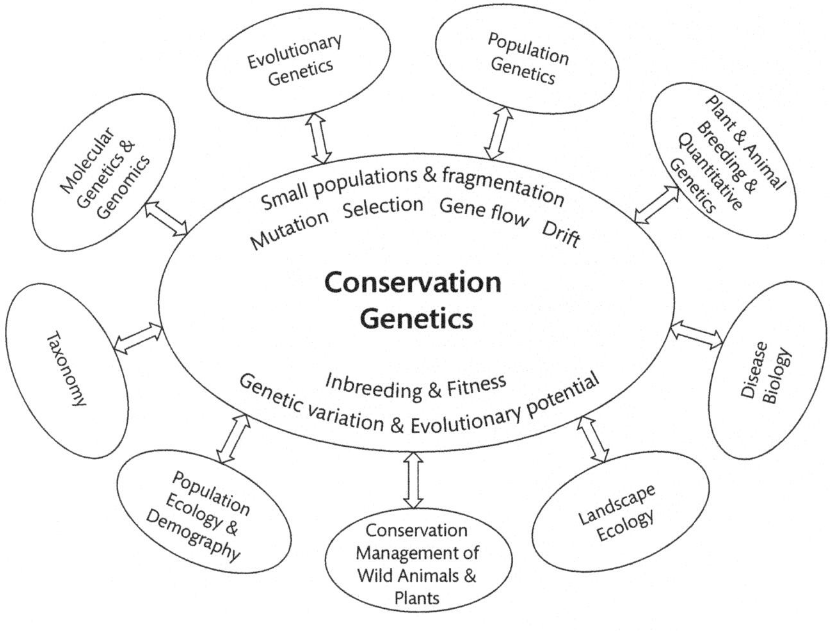

Fig. 2.1 Structure of conservation genetics and its relationship to other disciplines.

What factors control the evolution of small populations?

Populations evolve through the impacts of mutation, selection, chance (genetic drift), and gene flow (migration)

Mutation, selection, chance, and gene flow have the following roles in the evolution of populations:

- mutation is the ultimate source of genetic variation for evolution
- selection is the only force causing adaptive evolutionary change (Chapter 4)
- chance effects (genetic drift) deplete genetic diversity by random sampling of gametes in finite populations (Chapter 4)
- gene flow (also referred to as migration) moves genetic variants from one population fragment to another (we defer further consideration of this until Chapter 5).

These forces operate in the context of the species' reproductive system and mode of inheritance. We concentrate in Chapters 2–7 on the outbreeding diploids that are usually the chief concern of conservation managers, and consider the implications of alternative reproductive systems and modes of inheritance in Chapter 8.

Next, we elaborate on the operation of the first three variables listed above.

Mutation

Mutations are the ultimate source of genetic diversity for evolution

We focus on the effects of mutations on reproductive fitness and their frequency of occurrence.

Mutations are predominantly harmful, few have beneficial effects, and some are neutral

Mutations have effects that are either (Remold & Lenski 2001; Vale et al. 2012):

- harmful
- advantageous
- conditional on the environment
- conditional on other alleles at the same locus (dominant, recessive, or heterozygote advantage)
- conditional on allelic variation at other loci (genetic background or epistasis) or
- neutral (no selective advantage or disadvantage).

In a population that is adapted to its environment, most new mutations are harmful, few are advantageous (Fig. 2.2), some are conditional, and the proportion that are neutral depends on the character being considered (see Fig. 2.3). Harmful mutations are pivotal to the effects of inbreeding on reproductive fitness (Chapter 3). Advantageous mutations and some conditional ones are utilized in adaptive evolution (Chapter 4). Geneticists use neutral mutations to characterize genetic diversity, genetic population structure, and rates of gene flow (Chapters 4, 5, and 10), and to build phylogenetic trees used in taxonomy (Chapter 9).

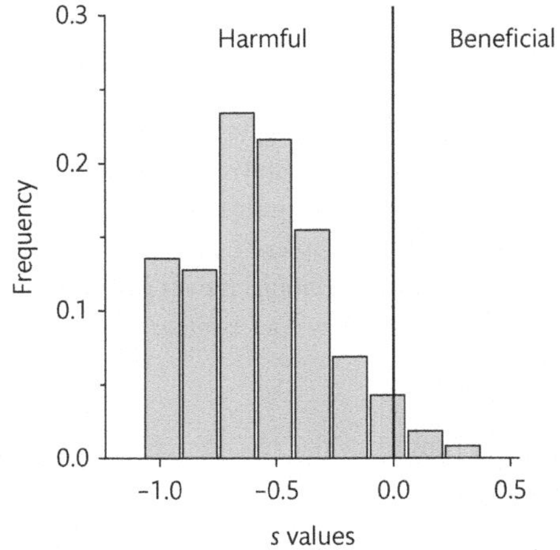

Fig. 2.2 Most new mutations are harmful in the environment to which the population is adapted. Distribution of mutational effects in bacteriophage (after Vale et al. 2012). The scale at the bottom is the selection coefficient (*s*) where values of -1.0 are lethal, < 0 harmful, 0 neutral, and > 0 beneficial.

Frequencies of mutations

Mutations occur at low rates at the many nucleotides that comprise the genome

New mutations arise at individual loci at low rates that vary for different loci and characters. For example, mutation rates/locus/generations are ~ 1×10^{-5} for morphological mutations in mice (*Mus musculus*), maize (*Zea mays*), and *Drosophila* (Frankham et al. 2010). Rates for DNA nucleotides in the nucleus are ~ 1.2×10^{-8} in humans. Mutation rates are 9–25 times higher for mitochondrial DNA (mtDNA) nucleotides than for nuclear ones in mammals, but lower than nuclear rates in plants (Lynch et al. 2006; Charlesworth & Charlesworth 2010). Mutations can arise at all nucleotides in the genome, including protein coding regions, regulatory regions, non-translated regions, and non-transcribed regions.

There are ~ 25,000 coding loci in mammals, 14,000–20,000 in invertebrates, and ~ 26,000 to ~ 45,000 in plants, in addition to many regulatory sequences that control the behavior of coding loci (Frankham et al. 2010; Mattick et al. 2010). Given the huge size of most genomes (e.g. humans have 3.2×10^9 base pairs), most gametes contain a new mutation somewhere. For example, human gametes, on average, carry ~ 35 new base mutations per generation somewhere in their genome, and new human zygotes contain ~ 70 (Kong et al. 2012).

Selection

Selection is the only force that produces adaptive evolutionary change

Selection increases the frequencies of genetic variants associated with greater reproductive fitness, leading to improved evolutionary adaption to the environment that a population inhabits. Selection may be:

- directional
- balancing or
- diversifying.

Directional selection decreases the frequency of harmful alleles (Chapter 3), and increases the frequency of favorable ones (adaptive evolution: Chapter 4). Balancing selection actively maintains genetic diversity in populations when heterozygotes are advantageous or rare alleles are favored. For example, the small population of the San Nicolas Island fox (*Urocyon littoralis dickeyi*) on the California Channel Island has no genetic diversity for microsatellites (~ neutral), but considerable diversity at the major histocompatibility complex (MHC), presumably due to balancing selection associated with disease resistance (Funk et al. 2016).

Diversifying selection involves alleles that are advantageous in some habitats, but harmful in others (Kondrashov & Houle 1994). It results in adaptively differentiated populations in diverse environments (home site advantage or genotype × environment interaction). Some conditional alleles are beneficial at some times and harmful at others, and occasionally result in balancing selection.

The strength of selection on individual loci is measured by the selection coefficient (s). This is the reduction in fitness of a homozygous genotype, compared to that of the fittest genotype. For example, if genotypes AA and Aa produce 10 fertile offspring that survive to reproductive age, while the aa genotype produces 9 (a relative fitness of $9/10 = 0.9$), then the selection coefficient against aa is $s = 0.10$.

To understand the evolutionary behavior of different characters, we need to specify the selective forces acting on them, as addressed in the next section.

Selection intensities vary among characters

The selective forces on alleles vary from very weak or absent, to strong, depending on the character

Selective values for different characters range from negligible or weakly selected, to strongly selected, as illustrated in Fig. 2.3. Most mutations in untranscribed DNA are believed to be neutral, or nearly so, while those in untranslated DNA were previously believed to be neutral, but this is now controversial (Mattick et al. 2010). Mutations resulting in amino acid substitutions in proteins are generally harmful in the environment to which the population is adapted, and are removed by natural selection (see Kimura 1983; Gillespie 1991; Hey 1999). For those changes in amino acid sequence that persist as protein polymorphisms, some may be neutral, and some subject to balancing selection, but the selective forces are usually weak (Brookfield & Sharp 1994; Kreitman & Akashi 1995).

Selection is stronger on groups of loci found in clusters, such as mtDNA, chloroplast DNA (cpDNA), the vertebrate major histocompatibility complex (MHC) (important in fighting diseases), and self-incompatibility loci (SI) in plants (genetically based incompatibility that prevents self-fertilization). Selection is also strong on inherited morphological (visual) polymorphisms (e.g. banded versus non-banded snail shells, as illustrated in the margin), and reproductive fitness (Endler 1986; Mousseau et al. 2000; Hereford et al. 2004).

Visual polymorphism in *Cepea nemoralis* snails (Europe)

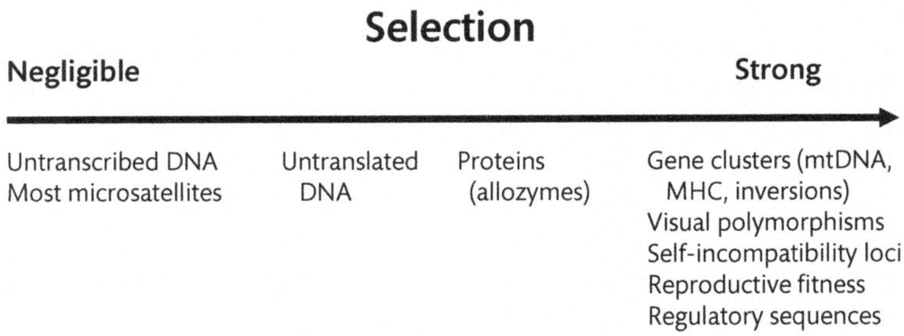

Fig. 2.3 Intensities of selection on different characters (after Frankham et al. 2010, Fig. 9.4).

Chance (genetic drift)

Chance effects arise from random sampling of gametes during reproduction in finite populations, and are greater in smaller than in larger populations

Chance effects have only minor impacts on the evolution of large populations. However, conservation genetics is concerned predominantly with small populations where chance effects are much more important.

Chance sampling in finite populations (referred to as genetic drift or drift) results in effects that are **random in direction** and dispersive, in contrast to the directional effects of mutation, migration, and selection. Chance results in five related genetic consequences (Falconer & Mackay 1996; Frankham et al. 2010):

1. Random genetic drift: random changes in allele frequencies across generations within populations (e.g. asterisked population in Fig. 2.4)
2. Genetic differentiation among isolated population fragments. As drift in different isolated sub-populations is random, they differentiate over generations (Fig. 2.4). This differentiation is not adaptive
3. Increased genetic uniformity within isolated population fragments. Increased homozygosity is connected with inbreeding and its harmful effects on fitness (see "Why is inbreeding important in conservation?")
4. Reduced heterozygosity across the combined population fragments. As alleles drift and are lost, heterozygosity across sub-populations is reduced on average
5. Selection is less effective in small than large populations because chance plays a greater role in driving changes in the small populations.

Effects 1, 2, 3, and 5 are evident in experiments on red flour beetle (*Tribolium*) populations (Fig. 2.4). Allele frequencies in individual populations fluctuated (drift) more from generation to generation in the small ($N = 10$) than large ($N = 100$) populations, the small populations diversified more rapidly than the large ones, and lose genetic diversity faster. Further, selection favored the wild-type allele in the large populations, but was much less effective in the small populations, including one replicate where the favored wild-type allele was lost.

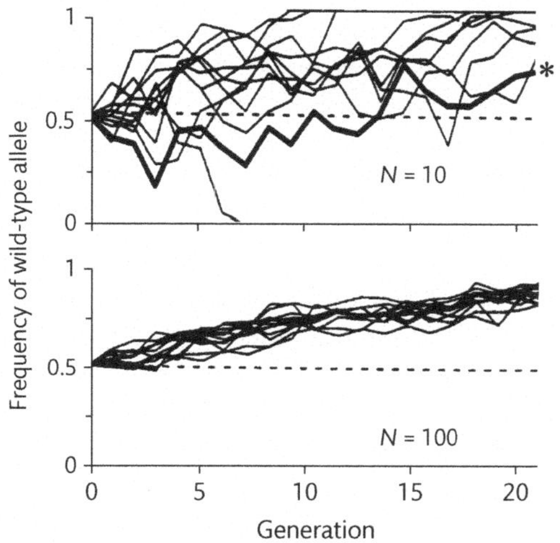

Fig. 2.4 Random genetic drift of the wild-type allele at a body color locus in *Tribolium* flour beetles, along with selection favoring the wild-type allele. All 24 populations began with frequencies of 0.5 of the wild-type and black alleles, and were maintained as isolated populations by random sampling of either 10 or 100 parents in each generation (after Rich et al. 1979; Frankham et al. 2010, Fig. 8.2).

Predicting the effects of genetic drift

The impacts of genetic drift can be predicted using binomial sampling theory

There is a rich body of theory to predict loss of genetic diversity due to drift that we introduce in this chapter and expand upon in Chapters 3–5 (Crow & Kimura 1970). The predictions of this theory have been supported with computer simulations (see later in this chapter) and empirical evidence.

We begin with the simple example of the progeny from self-fertilization in a diploid individual with genotype A_1A_2. The progeny genotypes are A_1A_1, A_1A_2, A_2A_1, and A_2A_2, each with a probability of ¼, assuming Mendelian segregation. Thus, if only one offspring is produced, there is a probability of ½ that one or other allele will be lost (i.e. that the individual is either A_1A_1 or A_2A_2). With N offspring ($2N$ alleles), the probability that all sampled alleles are A_1 (or all A_2) is $(½)^{2N}$. Consequently, it is less likely that an allele is lost if N is large. If zygote genotypes are equally likely to survive, then the mean frequency of A_1 (and A_2) in progeny across many replicates is ½, i.e. there is no change in the mean.

Variation among allele frequencies across replicate populations depends on the initial allele frequency and the parent population size

The distribution of allele frequencies in samples of N individuals from a population with two alleles A_1 and A_2 at initial frequencies of p and q is given by the terms of the binomial expansion $(p + q)^{2N}$ (see List of Symbols located after the Acknowledgments). Thus the probability that a population has only A_1 alleles is p^{2N}, while the chance that it has only A_2 alleles is q^{2N} (these both correspond to fixation, with the rarer allele more likely to be lost). The five possible outcomes and their predicted frequencies for the simple case of populations with two parents per generation ($N = 2$) are shown in Example 2.1.

Example 2.1 Expected distribution of allele frequencies in populations of size *N* = 2

If we take many samples of two individuals (= 4 gametes) from the same population where alleles A_1 and A_2 have frequencies of p and q (where $p + q = 1$), respectively, the expected distribution of allele frequencies for A_1 in the next generation (p') is given by the terms of the binomial expansion $(p + q)^4$, where the power of 4 is the number of gametes sampled. The expected outcomes (allele frequencies in the next generation) and terms of this expansion are given in the following table.

Outcomes	p'	Frequency (f)	Example ($p = 0.7$: $q = 0.3$)	
$4\ A_1: 0\ A_2$	1.00	p^4	0.7^4	= 0.2401
$3\ A_1: 1\ A_2$	0.75	$4p^3q$	$4 \times 0.7^3 \times 0.3$	= 0.4116
$2\ A_1: 2\ A_2$	0.50	$6p^2q^2$	$6 \times 0.7^2 \times 0.3^2$	= 0.2646
$1\ A_1: 3\ A_2$	0.25	$4pq^3$	$4 \times 0.7 \times 0.3^3$	= 0.0756
$0\ A_1: 4\ A_2$	0.00	q^4	0.3^4	= 0.0081
Totals		1		= 1.0000

Thus, the five possible outcomes have frequencies of A_1 of 1, 0.75, 0.5, 0.25, and 0. For example, with initial base population frequencies of $p = 0.7$ and $q = 0.3$, the proportions of each outcome are as shown in the last column.

The mean frequency across all outcomes is the same as in the original population as:

$$\begin{aligned}\text{Mean } p' &= \Sigma p_i' f_i \\ &= [(1 \times 0.2401) + (0.75 \times 0.4116) + (0.5 \times 0.2646) + (0.25 \times 0.0756) + 0] \\ &= 0.7\end{aligned}$$

The average heterozygosity is reduced by drift, as the first and last outcomes lack variation and heterozygosity is reduced from $2pq$ in the parents to $1.5pq$ in the progeny, a proportionate reduction of ¼ (and from 0.42 to 0.315 in the numerical example). In general, the reduction in heterozygosity due to a single generation of sampling is $1/(2N)$ for different sized populations.

The variance in allele frequencies among replicate populations (σ_p^2 for A_1) after one generation is given by the binomial sampling variance (adjusted for diploids carrying two copies of each locus):

$$\sigma_p^2 = \frac{pq}{2N} \quad 2.1$$

Thus, the variance depends on the allele frequencies and the population size. Variances are greater in smaller than larger populations (Fig. 2.5) and are greatest when both alleles have frequencies of 0.5. When there are two alleles, the variances for the A_1 and A_2 alleles are equal ($\sigma_p^2 = \sigma_q^2$).

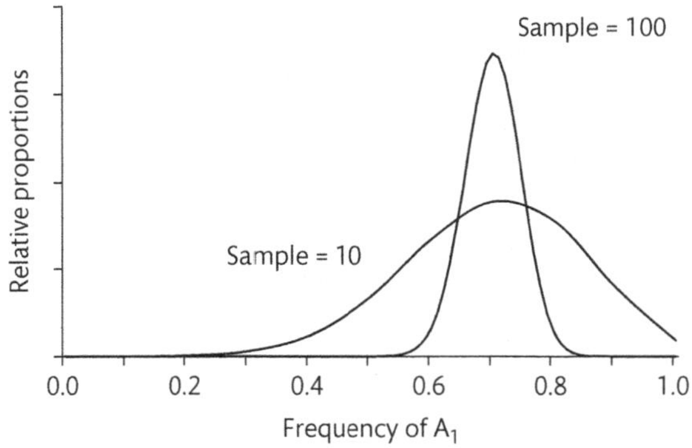

Fig. 2.5 Expected distribution of allele frequencies when many replicate samples of sizes 10 and 100 are taken from a population with an allele frequency of 0.7.

Empirical studies confirm that population bottlenecks result in loss of alleles (especially rare ones), reduced genetic diversity, and random drift in allele frequencies

In experimental populations of *Drosophila* fruit flies, single pair bottlenecks led to the loss of microsatellite (see Appendix 1) alleles (particularly rare ones), changes in allele frequencies, and elevated variation in allele frequencies among replicates, as predicted (England et al. 2003). Heterozygosity was reduced from 0.61 in the base population to 0.44 in the bottlenecked populations (close to the predicted 25% decline), and the mean number of alleles declined from 12 to 3.75. Population bottlenecks have also been observed to result in reduced genetic diversity in threatened species, including Ethiopian wolves (*Canis simensis*), Mauritius kestrels (*Falco punctatus*), and Florida torreya trees (*Torreya taxifolia*) (Spielman et al. 2004a).

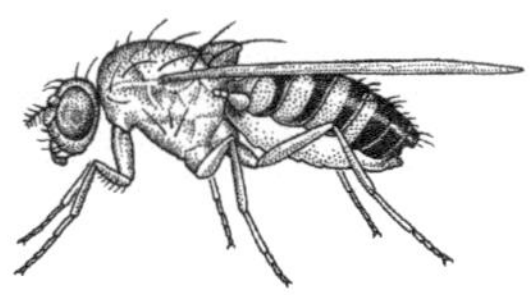

Drosophila fruit fly

Having discussed the individual effects of mutation, migration, selection, and chance on the genetic composition of populations, we now illustrate the combined effects of mutation and selection.

Mutation-selection balance and the mutation load

The opposing forces of mutation adding harmful alleles and selection removing them results in a load of rare harmful alleles in populations

As mutation adds harmful alleles to populations and selection removes them, equilibria result in low frequencies (typically < 1%) of harmful alleles. Table 2.1 presents the equilibrium frequencies for harmful mutations in outbreeding diploids due to the balance between mutation and selection for loci with different degrees of dominance. As expected from theory, most mutations observed in populations are recessive or partially recessive, and observed frequencies are generally low, in accord with predicted equilibria (Crow & Kimura 1970; Frankham et al. 2010).

Table 2.1 Predicted equilibrium frequencies of harmful mutations in populations ($\hat{q}$) as a result of mutation-selection balance for loci with different degrees of dominance. Mutation rates are *u*, selection coefficients s, and dominance *h*. The examples assumes $u = 10^{-6}$, $s = 0.01$, and $h = 0.025$.

Dominance	Fitness of genotypes			Equilibrium	Example
	AA	Aa	aa	$\hat{q}$	
Recessive	1	1	$1 - s$	$\sqrt{\frac{u}{s}}$	3×10^{-3}
Partially dominant	1	$1 - hs$	$1 - s$	$\frac{u}{hs}$	4×10^{-4}
Dominant	1	$1 - s$	$1 - s$	$\frac{u}{s}$	10^{-5}

Next we will consider what genetic markers are used in conservation genetics to measure and describe genetic diversity.

What genetic markers are used in conservation genetics?

Conservation genetics utilizes a range of genetic markers and characters, including DNA, microsatellites, allozymes, morphological traits, and chromosomal variants

For those not familiar with molecular genetic markers, including short variable number DNA sequence repeats (microsatellites), single nucleotide polymorphisms (SNPs), variation in DNA sequences (nuclear, mitochondrial, and chloroplast), and protein variants detected by electrophoresis (allozymes), details are given in Appendix 1.

How do we measure and describe genetic diversity?

We describe genetic diversity at individual loci in a population using the proportion of heterozygotes and the number of alleles, and average these over multiple loci

Multiple individual loci

Following genotyping of individuals from a population for a given locus, we summarize the variation based on the proportion of individuals that are heterozygous (having more than one genetic variant [allele] at a locus) and the number of alleles at the locus, as illustrated in Box 2.1 for the Ethiopian wolf. Mean genetic diversity is characterized using observed heterozygosity (H_o), expected heterozygosity (H_e), and allelic diversity (A). Observed heterozygosity is simply the number of heterozygotes as a proportion of the total individuals typed. Expected heterozygosity is the proportion of heterozygotes expected from the allele frequencies under random mating, based on the Hardy–Weinberg equilibrium (Box 2.2). The latter measure is preferentially used to indicate genetic diversity as it has more desirable statistical properties.

Box 2.1 Measuring and describing genetic diversity in the critically endangered Ethiopian wolf and in non-endangered wolves and dogs

(after Gottelli et al. 1994; Anonymous 2003)

The Ethiopian wolf has a total population of < 400 individuals in only six isolated areas of Ethiopia. Genetic diversity for nine microsatellite loci in two populations is presented in the following table, one on the Sanetti Plateau, where few domestic dogs were present, and the other in Web Valley, where dogs were abundant.

Taxon	A	H_o	H_e	n
Ethiopian wolf				
Sanetti	2.0	0.150	0.138	16
Web	2.8	0.313	0.271	23
Domestic dogs	6.4	0.516	0.679	35
Gray wolf	4.5		0.620	18
Coyote	5.9		0.675	17

Three points are of note for conservation. First, the Ethiopian wolf populations have lower genetic diversity than the taxonomically related, non-endangered gray wolves, coyotes, and domestic dogs. Second, the relatively "pure" Sanetti population has less genetic diversity than the Web Valley population that co-exists and is known to breed with domestic dogs. Third, mating in the Ethiopian wolves is approximately random within populations, as observed and expected heterozygosities were similar in both populations.

Ethiopian wolf

Box 2.2 Hardy-Weinberg equilibrium genotype frequencies in large random mating populations

(Hardy 1908; Weinberg 1908)

Genotypes at a diploid locus attain Hardy–Weinberg equilibrium frequencies within one generation in a large random mating population, if there is no mutation, migration, or selection. For example, if there are two alleles, A_1 and A_2, at frequencies of p and q (where $p + q = 1$), the gametes, the zygotes, and their frequencies are given in the table below:

Allele and frequency		Ova	
		$p\ A_1$	$q\ A_2$
Sperm	$p\ A_1$	$p^2\ A_1A_1$	$pq\ A_1A_2$
	$q\ A_2$	$pq\ A_1A_2$	$q^2\ A_2A_2$

Thus, the zygotic frequencies for A_1A_1, A_1A_2, and A_2A_2 genotypes are p^2, $2pq$, and q^2, respectively. For example, if the two alleles have frequencies of 0.7 and 0.3, respectively, A_1A_1 has an expected frequency of $0.7^2 = 0.49$, A_1A_2 $2 \times 0.7 \times 0.3 = 0.42$, and A_2A_2 $0.3^2 = 0.09$ (and they sum to 1.0, as they must).

DNA sequence diversity is described using measures related to heterozygosity and allelic diversity. For example, nucleotide diversity (π) is heterozygosity at the nucleotide level, and haplotype diversity (ht) is heterozygosity for a segment of DNA (Charlesworth & Charlesworth 2010).

We are now in a position to consider the consequences of the joint operation of all the forces on the genetic diversity of populations, referred to as standing genetic diversity. This predicts the ability of populations to evolve (evolutionary potential) and the extent of harmful impacts of inbreeding on fitness.

How much standing genetic diversity do species and populations contain?

Large populations of sexually outbreeding species typically contain high levels of genetic diversity, while small populations typically have reduced levels

In general, large populations of outbreeding species contain large stores of genetic diversity for DNA sequences, microsatellites, allozymes, harmful alleles, and quantitative genetic variation (Lewontin 1974; Frankham et al. 2010; Leffler et al. 2012; Romiguier et al. 2014). Conversely, small populations have lower levels of all forms of genetic diversity (Frankham et al. 2010; Frankham 2012). Further, meta-analyses revealed that levels of genetic diversity are positively correlated with population sizes across species and for populations within species (Frankham 1996, 2012).

Threatened species (critically endangered, endangered, and vulnerable species) typically have lower genetic diversity than non-threatened ones

For example, threatened species have small or declining populations, and they have on average 25–35% less microsatellite genetic diversity than non-threatened species (Spielman et al. 2004a; Evans & Sheldon 2008; Flight 2010) (Table 2.2).

Table 2.2 Microsatellite genetic diversity (average number of alleles per locus [*A*] and heterozygosity [*H*]) for a range of globally threatened taxa compared to one or more closely related, but non-threatened taxa (after Frankham et al. 2010).

Threatened species			Non-threatened species			Reference
	A	*H*		*A*	*H*	
Mammals						
African wild dog (*Lycaon pictus*)	3.5	0.56	Gray wolf (*Canus lupus*)	8.0	0.77	1, 2
Cheetah (*Acinonyx jubatus*)	3.4	0.39	Puma (*Puma concolor*)	4.9	0.61	1
Chimpanzee (*Pan troglodytes*)		0.64	Human (*Homo sapiens*)		0.78	1
Ethiopian wolf (*Canis simensis*)	2.4	0.21	Coyote (*Canis latrans*)	5.9	0.68	1
Giant panda (*Ailuropoda melanoleuca*)	3.7	0.44	Brown bear (*Ursus arctos*)	6.8	0.66	3, 4
Mexican wolf (*Canis lupus baileyi*)	2.7	0.42	Gray wolf (*Canis lupus*)	4.5	0.62	1
Long-footed potoroo (*Potorous longipes*)	3.7	0.56	Allied rock-wallaby (*Petrogale assimilis*)	12.0	0.86	1
N. hairy-nosed wombat (*Lasiorhinus krefftii*)	2.1	0.32	S. hairy-nosed wombat (*Lasiorhinus latifrons*)	5.9	0.71	1
Birds						
Mariana crow (*Corvus kubaryi*)	1.8	0.16	American crow (*Corvus brachyrhynchos*)	6.0	0.68	1

Cheetah (Africa)

Threatened species	A	H	Non-threatened species	A	H	Reference
Mauritius kestrel (*Falco punctatus*)	1.4	0.10	European kestrel (*Falco tinnunculus*)	5.5	0.68	1
Seychelles kestrel (*Falco araea*)	1.3	0.12	Greater kestrel (*Falco rupicoloides*)	4.5	0.59	1
Reptiles						
Komodo dragon (*Varanus komodoensis*)	4.0	0.31	Ridge-tailed monitor (*Varanus acanthurus*)	30.1	0.93	1, 5
Plants						
Ziyuan fir (*Abies ziyuanensis*)	2.0	0.34	Norway spruce (*Picea abies*)	25.0	0.79	6
Honduran mahogany (*Swietenia macrophylla*)	9.7	0.55	Royal mahogany (*Carapa guianensis*)	9.3	0.67	1

References: 1. Frankham (2000); 2 Vilà et al. (2003); 3. Lu et al. (2001); 4. Waits et al. (2000); 5. Fitch et al. (2005); 6. Tang et al. (2008).

Komodo dragon (Indonesia)

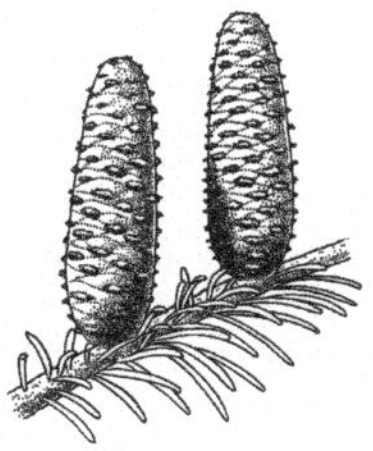

Ziyuan fir (China)

The load of harmful alleles

Outbreeding populations contain a load of rare harmful alleles that reflect a balance between mutation and selection

As predicted by theory, large populations of outbreeding species have a large store of harmful alleles that can be made homozygous by inbreeding (Charlesworth & Willis 2009: Chapter 3). While harmful alleles are typically rare at individual loci, they are found at many loci across the genomes of animal and plant species. For example, over 6,000 single locus gene genetic disorders are known in humans, many due to rare harmful recessive alleles, e.g. phenylketonuria and cystic fibrosis (Amberger et al. 2015). Further, single locus chlorophyll deficiencies are found in many plant species due to mutations across multiple loci (Willis 1992).

Adaptive genetic variation

Outbreeding species typically have adaptive genetic variation in the form of beneficial alleles, conditional alleles, and those subject to balancing selection

Adaptive genetic variation is found for most fitness characters in most species. For example, it has been found for fecundity/seed output in house mice, rats (*Rattus* sp.), invertebrates, and maize, for fitness on altered diets/soils in humans (*Homo sapiens*), invertebrates, and plants, for resistance to disease in humans, European rabbits (*Oryctolagus cuniculus*), and plants, for beak shape and size in Galapagos finches (*Geospiza* sp.), for tolerance of ocean acidity in marine algae, for predator avoidance in fish, and for tolerance of grazing and droughts in plants (Lewontin 1974; Frankham et al. 2010; Herron et al. 2014; Lohbeck et al. 2014). Adaptive genetic variation is most readily observed when the environment is changed, as many previously harmful conditional alleles become beneficial (Montgomery et al. 2010).

Since population size is a critical issue in conservation genetics, we next specify how we define and measure the (genetically) effective population size, and contrast it with the census population size.

How do we measure population size?

Census population size

Census population size is the number of individuals in a population

The census population size (*N*) is the number of individuals in a population, determined by counting or estimation from mark–recapture, line-transects, etc. It is variously defined to include all adult and juvenile individuals, all adults, or all breeding pairs (and this needs to be specified when a size is reported).

Genetically effective population size

Genetic impacts of small population size depend upon the genetically effective population size, rather than the census population size

All evolutionary genetic processes depend upon the genetically effective population size (N_e) (Wright 1969; Charlesworth 2009). N_e is defined by equating the population size of the target populations to that of an idealized population that results in the same genetic consequences (Box 2.3). It is based on the number of fertile adults, but accounts for unequal contributions to the next generation from different breeders, variations in sex ratios of breeders, and fluctuation in population size across different generations.

Box 2.3 Effective population size (N_e) and the idealized population

(Wright 1969; Crow & Kimura 1970; Falconer & Mackay 1996; after Frankham et al. 2010)

The effective size of a population (N_e) is the size of an idealized population (see "The idealized population", below) that would lose genetic diversity, or become inbred, or drift, or coalesce at the same rate as the actual population. For example, if a real population loses genetic diversity at the same rate as an ideal population of 50, then we say the real population has an effective size of 50, even if it contains 500 potentially reproductive adults.

It is crucial that all relevant variables be included in estimates of N_e (termed composite estimates). There are different N_e concepts, including inbreeding, variance, eigenvalue, and coalescence, and these are not necessarily equivalent, especially if population sizes are changing over generations (Crow & Kimura 1970; Charlesworth 2009). Furthermore, there are many methods for estimating N_e, based on different assumptions and reflecting different time frames (see Wang 2005; Charlesworth 2009; Luikart et al. 2010; Tallmon et al. 2010; Gilbert & Whitlock 2015). In particular, single-generation estimates are typically much larger than multigenerational (composite) ones, and the former are not appropriate for the many conservation applications where we deal with multiple generations.

The idealized population

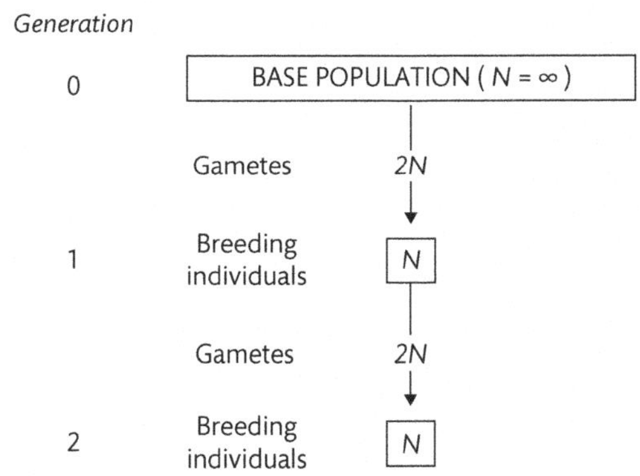

(Frankham et al. 2010, Fig. 8.8)

We begin by assuming a large (essentially infinite) random mating base population, from which we take a sample of size N adults to form the idealized population (figure above). This population (also referred to as the Wright–Fisher population) is maintained as a random mating, closed population in succeeding generations. Alleles may be lost by chance and allele frequencies may fluctuate due to genetic drift (Wright 1969; Charlesworth 2009). The simplifying conditions applied to the idealized population are:

- number of breeding individuals is constant in all generations
- generations are distinct and do not overlap
- no gene flow with other populations
- all adults are potential breeders
- all individuals are hermaphrodites (possess both female and male sex organs)
- union of gametes is random, including the possibility of uncommon self-fertilization
- no selection at any stage of the life cycle
- no mutation
- within the population, breeding individuals contribute gametes equally to a pool from which zygotes are formed, but the number of offspring per adult varies randomly.

Effective population sizes in natural populations

The effective size of wild unmanaged populations is usually much less than the number of breeding adults

Real populations deviate in structure from the assumptions of the idealized population by having variable numbers in successive generations, high variation in family sizes, unequal sex ratios, non-random mating, and overlapping generations. Consequently, their effective population sizes are usually less than census sizes. For example, the current human population totals ~ 7 billion people, of which ~ 4 billion are potentially breeding adults (United States Census Bureau 2014), but the long-term effective size for humans is only ~ 10,000 (Charlesworth 2009).

For most species, N_e is unknown, so inferences are often based upon census sizes (N) and N_e/N ratios for those species where it is known, especially closely related taxa. What are typical values for this ratio?

N_e/N ratios

Long-term N_e/N ratios are typically much less than unity, averaging 0.1–0.2

In unmanaged wild populations, the N_e/N ratio with all relevant variables included averages 0.11–0.14, based on three meta-analyses (Frankham 1995a; Palstra & Ruzzante 2008; Palstra & Fraser 2012). N_e/N ratios vary widely with life history, especially age at sexual maturity and adult lifespan (Waples et al. 2013). In particular, species with high fecundity have lower ratios in the order of 10^{-3} to 10^{-6}, based on data from fish, oysters, shrimp and seaweed (Frankham 2012).

Consequently, adverse genetic effects in finite populations (loss of genetic diversity and inbreeding) typically occur sooner and faster than expected from census population sizes.

Why is inbreeding important in conservation?

Inbreeding adversely impacts fitness of individuals, populations, and species, and increases extinction risk

Inbreeding is one of the two major genetic concerns in relation to population persistence, along with loss of genetic diversity. Offspring that are inbred typically have reduced survival and reproduction compared to non-inbred ones (termed inbreeding depression), especially in naturally outbreeding species (Ralls & Ballou 1983; Crnokrak & Roff 1999; Keller & Waller 2002; Chapter 3).

Having introduced inbreeding, it is now time to define what it is and how we measure it.

What is inbreeding?

Inbreeding is the production of offspring by individuals related by descent

Inbreeding encompasses the production of offspring by related individuals, whether they are closely related (e.g. self-fertilization, brother–sister, and parent–offspring matings), or more distantly related (e.g. first cousins, second cousins, etc.) (Box 2.4). The degree of inbreeding depends on how related the parents are through having ancestors in common.

How do we measure inbreeding?

The inbreeding coefficient *F* is the probability that two alleles at a gene locus are identical by descent

The extent of inbreeding is measured using Wright's inbreeding coefficient (*F*: Wright 1969). It is defined as the probability that two alleles at a locus in an individual are identical by descent from an allele in an ancestor of both parents (Malécot 1969). This can be assessed directly from pedigrees, either via direct probability calculations (Example 2.2) or by using computer software (e.g. PMx: Ballou et al. 2011). However, for most wild populations, pedigrees are lacking and inbreeding must be estimated from molecular marker data (see Chapters 3, 11, and 13; Wang 2016).

Example 2.2 Computing inbreeding coefficients for individuals resulting from self-fertilization and full-sib mating

(Frankham et al. 2010, Fig. 12.2)

To compute the inbreeding coefficient of individual X in the examples of self-fertilization and full-sib mating, we attribute distinct genotypes to the common ancestors (A_1A_2 and A_3A_4) and compute the probability that two alleles in X are identical-by-descent assuming Mendelian segregation. The lines indicate potential transmission pathways of the alleles from the offspring's common ancestor(s).

Self-fertilization

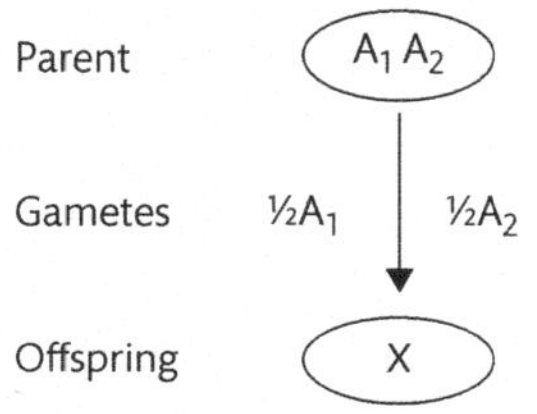

$$F_X = \Pr(x = A_1A_1) + \Pr(x = A_2A_2)$$
$$= \tfrac{1}{4} + \tfrac{1}{4} = \tfrac{1}{2}$$

Full-sib mating

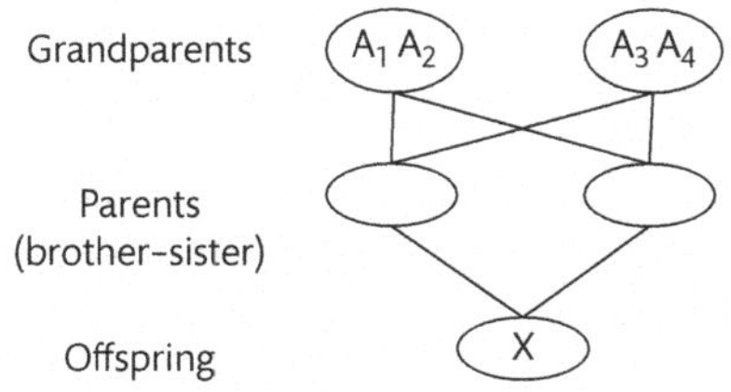

$$F_X = \Pr(x = A_1A_1 \text{ or } A_2A_2 \text{ or } A_3A_3 \text{ or } A_4A_4)$$
$$= \tfrac{1}{16} + \tfrac{1}{16} + \tfrac{1}{16} + \tfrac{1}{16} = \tfrac{1}{4}$$

(Frankham et al. 2010, Fig. 12.2)

Thus, the zygotic inbreeding coefficient of an individual resulting from self-fertilization is ½, and that for an individual resulting from full-sib mating is ¼.

As inbreeding increases, homozygosity for harmful alleles increases, reducing reproductive fitness, a major issue in conservation genetics. Box 2.4 illustrates the pedigree for a highly inbred giraffe (*Giraffa camelopardalis*) and provides more information on inbreeding, as a prelude to detailed treatment of its consequences in Chapter 3.

Nigerian giraffe

Box 2.4 Inbreeding and its consequences

(after Frankham et al. 2010)

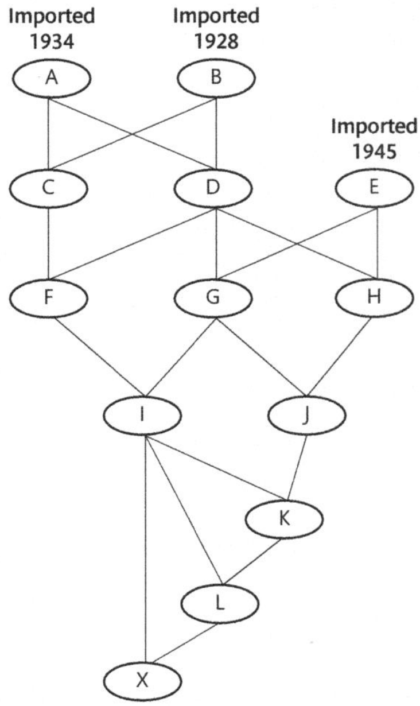

(Frankham et al. 2010, Box 2.1)

The Nigerian giraffe X born in Paris Zoo in 1992 shown at the bottom of the above pedigree resulted from repeated matings between relatives and thus was highly inbred (Bingaman Lackey 1999).

The inbreeding coefficient (*F*)

The inbreeding coefficient of an individual refers to how closely related its parents are. Levels of inbreeding in offspring for different kinds of relationships among parents (over one generation) are shown in the table below:

Parents	Offspring *F* (parents not inbred)
Self-fertilization (or selfing)	0.5
Brother–sister, mother–son, or father–daughter	0.25
Half-brother–half-sister (half sibs)	0.125
First cousins	0.0625
Unrelated	0

Repeated inbreeding can produce even higher inbreeding coefficients: an *F* of 0.999 is reached after 10 generations of continuous self-fertilization, while an *F* of 0.986 occurs after 20 generations of brother–sister mating. The final Nigerian giraffe X in the pedigree above had an inbreeding coefficient of 0.52 and died shortly after birth.

Levels of inbreeding can be determined from pedigrees, or inferred from heterozygosities for genetic markers (Chapters 3, 11, and 13).

Inbreeding in random-mating populations

Inbreeding is unavoidable in small closed random-mating populations

While inbreeding due to self-fertilization and brother–sister mating is obvious, less obvious inbreeding is unavoidable in small genetically isolated random-mating populations, albeit occurring at a slower rate. Even if a population is founded by completely unrelated individuals, in time all individuals will become related. Over a single generation in a closed random-mating population of size N_e, the increase in inbreeding is $1/(2N_e)$ (the same as the loss of heterozygosity). Thus, in a population of $N_e = 10$, inbreeding increases by $1/(2 \times 10) = 5\%$ in the first generation. Inbreeding increases progressively across generations, as illustrated in Fig. 2.6 for different sized populations: note that inbreeding increases faster in smaller than larger populations.

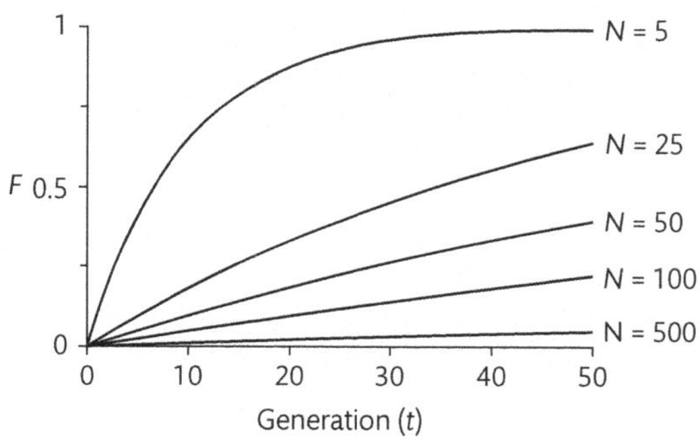

Fig. 2.6 Increase in inbreeding coefficient (*F*) with generations (*t*) in closed populations of different sizes, as predicted by eqn 3.2 in Chapter 3 (Frankham et al. 2010, Fig. 12.4).

Why do we use computer simulations in conservation genetics?

Computer simulation is used to investigate problems in conservation genetics that are difficult to solve mathematically, such as the impact of selection or mutation in small populations, or the complex interactions of demographic, ecological, and genetic factors determining population persistence

Due to the stochastic (random) nature of genetic drift, the impacts of selection on allele frequencies in small populations are difficult to model algebraically. Consequently, computer simulations are often used to study the impacts of chance and selection (Frankham et al. 2010; Hoban et al. 2012; Hoban 2014). For example, the *Tribolium* flour beetle experiment described in Fig. 2.4 has been simulated in Fig. 2.7. Note the similarity of the results with the experimental data.

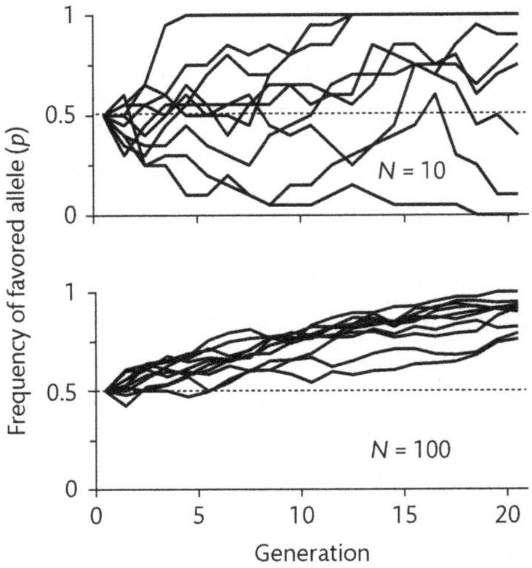

Fig. 2.7 Computer simulation illustrating the operation of selection in replicate populations with sizes of $N = 10$ and $N = 100$ (Frankham et al. 2010, Fig. 8.10). Selection is for a locus where heterozygotes and the better homozygote have advantages of 5% and 10%, respectively, over the poorer homozygote in reproductive fitness.

Computer simulations are used in several different ways in conservation and evolutionary genetics, as follows:

- check the validity of results from mathematical models
- provide numerical solutions for expressions produced by stochastic mathematical models
- evaluate the validity of approximate mathematical solutions to problems
- provide predictions against which empirical results can be compared
- investigate problems that are too complex to solve with mathematical models.

In the latter context, computer simulation provides links between simple tractable mathematical models (with many simplifying assumptions) and experiments with real living organisms in all their complexity (see Hoban et al. 2012 for a review of software). For example, Lacy (1987) used computer simulations to evaluate the likely effects of drift, selection, migration, and population subdivision on small populations of endangered species. Notably, Ballou & Lacy (1995) used computer simulation to evaluate the effects on retention of genetic diversity and inbreeding of alternative genetic management schemes proposed for endangered species. Their work led to the recommended strategy (minimizing mean kinship) for genetic management of captive populations that we apply to managing fragmented wild populations in Chapter 13.

We now proceed to discuss details of genetic problems in small populations, first, inbreeding depression (Chapter 3), then loss of genetic diversity and reduced ability to evolve (Chapter 4), and finally the genetic impacts of population fragmentation and isolation (Chapter 5).

Summary

1. Genetic management of fragmented populations involves the application of evolutionary genetic theory and knowledge to alleviate problems due to inbreeding and loss of genetic diversity in small populations.
2. Populations evolve through the effects of mutation adding genetic diversity, gene flow (migration) spreading it among populations, selection favoring particular variants over others, and chance (genetic drift) leading to random changes within and among populations, and to loss of genetic diversity.
3. Outbreeding, sexually reproducing populations of large size typically contain substantial genetic diversity, including neutral alleles, harmful alleles in mutation-selection balance, and advantageous alleles, some with effects conditional on the environment and some subject to balancing selection.
4. Small populations, including threatened species, typically contain less genetic diversity than large populations and non-threatened species.
5. Genetic impacts of small population size (inbreeding, loss of genetic diversity, and population differentiation) are determined by the effective population size, rather than the census number of individuals.
6. The effective population size is the size of an idealized population that results in the same inbreeding, loss of neutral genetic diversity, or fluctuations in allele frequencies as that of the target population.
7. Inbreeding is the production of offspring from matings between relatives.
8. Small closed random-mating populations become inbred and lose genetic diversity at rates proportion to $1/(2N_e)$.

FURTHER READING

Allendorf et al. (2013) *Conservation and the Genetics of Populations*: A comprehensive textbook on conservation genetics, encompassing the topics in Section I of this book.

Frankham et al. (2004) *A Primer of Conservation Genetics*: A simple introduction to the topics in Section I of this book.

Frankham et al. (2010) *Introduction to Conservation Genetics*: A comprehensive treatment of the background to the topics in this book.

Freeman & Herron (2013) *Evolutionary Analysis*: Wide-ranging textbook that includes coverage of evolutionary genetics.

Hoban et al. (2012) A review of computer simulation software for evolutionary genetics.

Thompson (2013) *Relentless Evolution*: Excellent review of evolution, documenting its pervasive nature and its frequent rapidity.

SOFTWARE

NeEstimator v2: reimplementation of free software for the estimation of effective population size (N_e) from genetic data, using a variety of methods (Do et al. 2014). http://molecularfisherieslaboratory.com.au/neestimator-software

PMx: a program for computing inbreeding coefficients and other measures from pedigrees (Ballou et al. 2011; Lacy et al. 2012). www.vortex10.org/PMx.aspx

Inbreeding reduces reproductive fitness

CHAPTER 3

Inbreeding reduces survival and reproduction (inbreeding depression), and thereby increases extinction risk. Impacts are generally greater in naturally outbreeding than inbreeding species, in stressful than benign environments, for fitness than peripheral traits, and for total fitness compared to its individual components. Inbreeding depression is due to increased homozygosity for harmful alleles and at loci exhibiting heterozygote advantage. Inbreeding depression is near universal in sexually reproducing organisms that are diploid or have higher ploidies.

TERMS

Additive, catastrophes, demographic stochasticity, directional selection, disruptive selection, dominance, environmental stochasticity, extinction vortex, heterosis, heterozygote advantage, lethal equivalents, mutational meltdown, overdominance, partial dominance, polyploid, purging, reproductive fitness, stabilizing selection

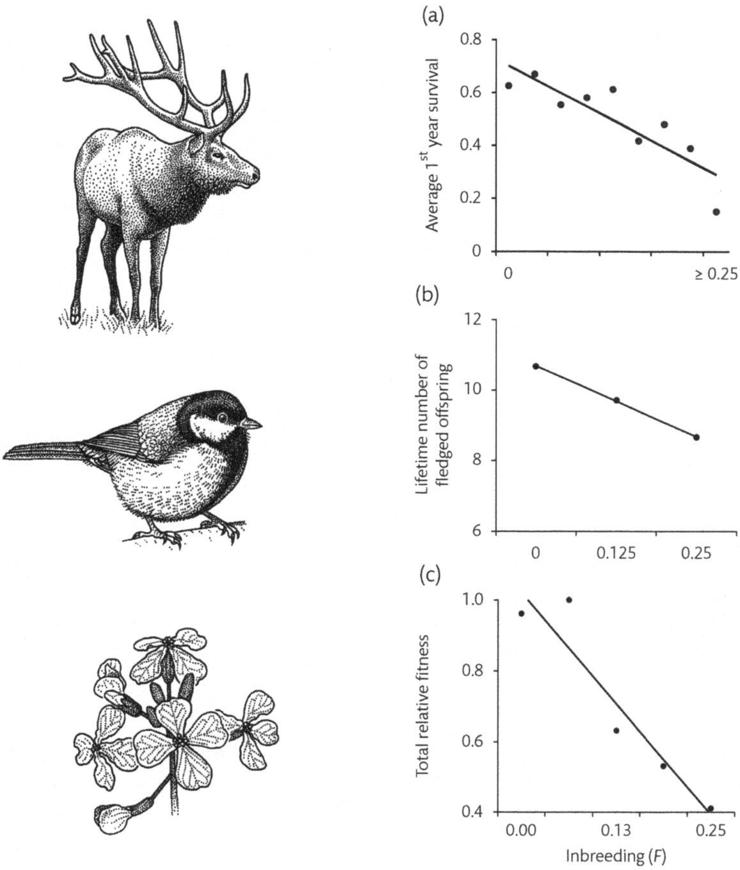

Inbreeding depression for fitness in the wild for (a) red deer (*Cervus elaphus*) (Scotland), (b) great tits (*Parus major*) (UK), and (c) wild radish (*Raphanus sativus*) (USA) (Nason & Ellstrand 1995; Szulkin et al. 2007; Walling et al. 2011). In all three, fitness declines progressively with increasing inbreeding.

Genetic Management of Fragmented Animal and Plant Populations. R. Frankham, J. D. Ballou, K. Ralls, M. D. B. Eldridge, M. R. Dudash, C. B. Fenster, R. C. Lacy & P. Sunnucks. Oxford University Press (2017) © R. Frankham, J. D. Ballou, K. Ralls, M. D. B. Eldridge, M. R. Dudash, C. B. Fenster, R. C. Lacy & P. Sunnucks 2017. DOI 10.1093/oso/9780198783398.001.0001

> "that any evil directly follows from the closest interbreeding has been denied by many persons, but rarely by any practical breeder, and never, as far as I know, by one who has largely bred animals which propagate their kind quickly..." (Darwin 1868).

Why is inbreeding an important issue in conservation?

Inbreeding depresses means of reproductive fitness traits (termed inbreeding depression) and increases extinction risks

Inbreeding has long been known to reduce reproductive fitness in naturally outbreeding species, and to a lesser extent in naturally inbreeding species. For example, Darwin (1876) found that self-fertilization reduced seed production by an average of 41%, compared to cross-fertilization, based on studies of 23 species of plants. Subsequently, inbreeding has also been shown to reduce fitness in laboratory and domesticated animals and plants, and humans (Charlesworth & Charlesworth 1987; Bittles & Neel 1994; Falconer & Mackay 1996; Lynch & Walsh 1998).

While there was initially skepticism about whether inbreeding had harmful effects in wild species, Ralls & Ballou (1983) showed that inbreeding reduced juvenile survival in 41 of 44 wild mammal species in captivity compared to that in outbred individuals (Fig. 3.1). For example, mortality in the pygmy hippopotamus (*Choeropsis liberiensis*) was 55% in inbred offspring versus 25% in outbred ones.

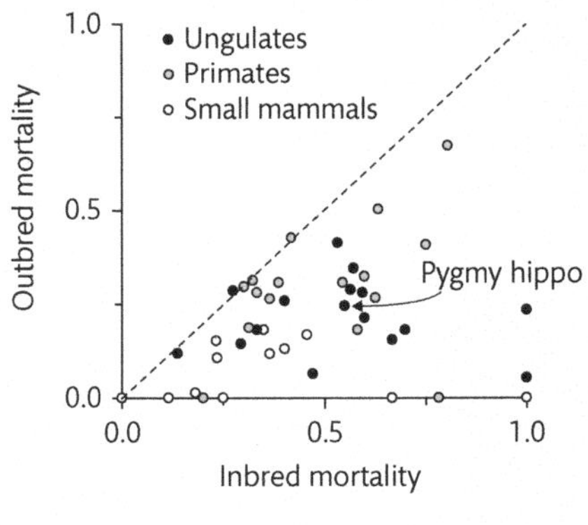

Fig. 3.1 Inbreeding depression for juvenile survival in 44 captive mammal populations (Ralls & Ballou 1983). Juvenile mortality in outbred individuals is plotted against that in inbred individuals from the same populations; inbreeding is harmful below the dotted line (Frankham et al. 2010, Fig. 12.1).

Inbreeding depression also occurs in wild species in natural habitats, and its impacts are typically more severe than found in captivity. For example, inbred individuals showed inbreeding depression in 90% of 157 valid data sets from a broad array of animal and plant taxa in natural environments, and the average impact was seven-fold greater than in captivity (Crnokrak & Roff 1999; Frankham et al. 2010).

The harmful impacts of inbreeding elevate extinction risks, as illustrated in Box 3.1. We will elaborate on this later in the chapter.

Box 3.1 Inbreeding increased extinction risk in butterfly populations in Finland

(Saccheri et al. 1998; Nieminen et al. 2001; Hanski 2011)

Levels of molecular heterozygosity were determined in 42 Glanville fritillary butterfly populations in Finland in 1995, and their fate recorded the following year: 35 populations persisted and seven went extinct by autumn 1996 (Saccheri et al. 1998). Extinction rates were higher for populations with lower heterozygosity (more inbred) (graph, below), even after accounting for the effects of demographic and environmental variables known to affect extinction risks (population size, time trend in population size, and area). Inbreeding explained 26% of the variation in extinction rate.

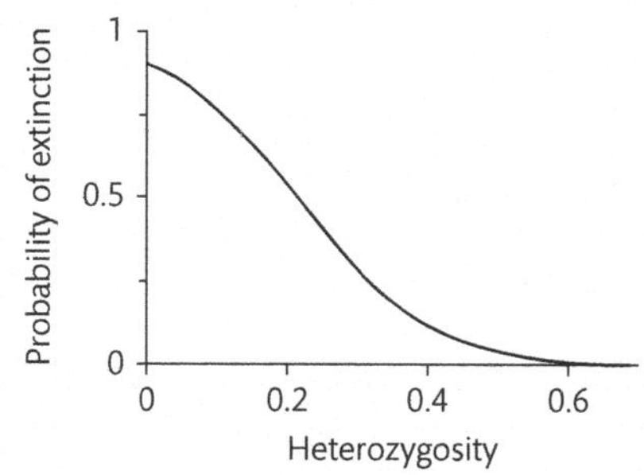

(Frankham et al. 2010, p. 24)

The causal link between inbreeding and extinction risks was confirmed by placing butterflies that were outbred, or inbred by full-sib mating, in the field. All six inbred populations went extinct within one generation, while 4 of 6 outbred populations persisted (Nieminen et al. 2001).

Glanville fritillary butterfly (Europe)

Why is inbreeding harmful?

Inbreeding increases homozygosity for harmful recessive alleles and for alleles at loci exhibiting heterozygote advantage, thereby reducing reproductive fitness

Inbreeding depression arises because all naturally outbreeding populations contain a load of harmful recessive alleles (Chapter 2). Consequently, inbred individuals have an elevated probability of being homozygous for harmful recessive alleles, or alleles at loci showing heterozygote advantage. For example, the tiger (*Panthera tigris*) father in Fig. 3.2 is heterozygous for a recessive white mutation, and progeny of a mating with his daughter have a 1/8 chance of being homozygous for it. Genome sequencing has revealed that threatened bird species (inbred) exhibit elevated homozygosity and higher levels of harmful alleles than non-inbred bird species, as do inbred gorilla (*Gorilla gorilla*) and cheetah populations (Li et al. 2014; Dobrynin et al. 2015; Xue et al. 2015).

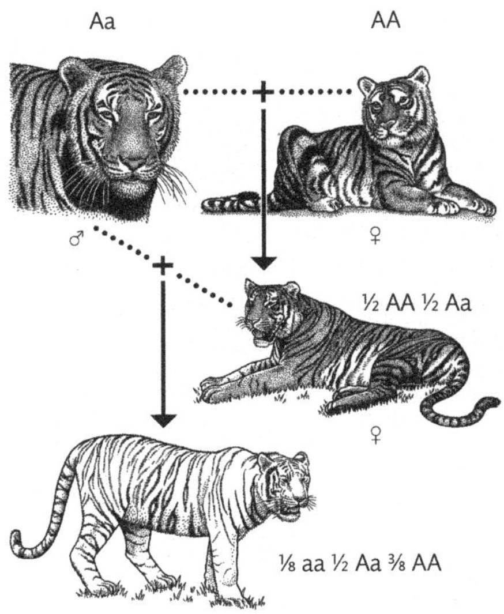

Fig. 3.2 Risk of homozygosity for a recessive white mutation (a) carried by the tiger father (genotype Aa) in the progeny of a father-daughter mating. In the final generation, aa offspring will be white and the AA and Aa genotypes phenotypically normal (after Frankham et al. 2010, p. 260).

What factors affect the magnitude of inbreeding depression?

Inbreeding depression depends upon directional dominance of beneficial alleles, the increase in the inbreeding coefficient, and the consequent increased homozygosity for harmful alleles

To identify the variables affecting the magnitude of inbreeding depression, we consider a simple model of two equivalent populations, one with random mating and the other with an inbreeding coefficient of F. Both have a single locus with two alleles A_1 and A_2, at frequencies of p and q, and no selection (Fig. 3.3 and Table 3.1). To determine the impact of these genotypes on the phenotype, we attribute arbitrary genotypic values of a, d, and $-a$, respectively, to represent the average fitnesses of the three genotypes. Thus, loci exhibiting intermediate phenotypes in heterozygotes (additive) have $d = 0$, those with dominance $d = a$, and those with heterozygote advantage $d > a$, as illustrated in Fig. 3.3.

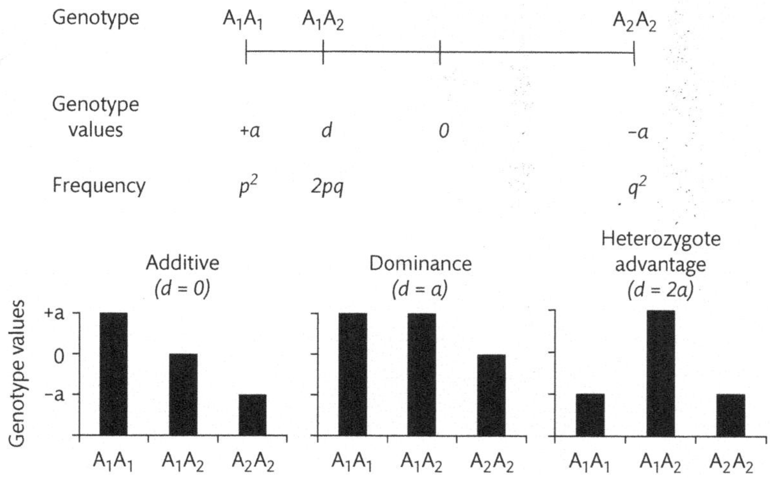

Fig. 3.3 Arbitrary genotype values assigned to the three genotypes at a locus, and examples of loci exhibiting additive, dominant, and heterozygote advantage effects (Frankham et al. 2010, Figs 5.2 and 5.3, after Falconer & Mackay 1996).

Table 3.1 Impact of inbreeding versus random mating on the mean fitness of a population: single locus model (Falconer & Mackay 1996).

Genotype	Fitness value	Genotype frequencies		Genotype frequency × value	
		Random mating	Inbred	Random mating	Inbred
A_1A_1	a	p^2	$p^2 + Fpq$	p^2a	$p^2a+Fpqa$
A_1A_2	d	$2pq$	$2pq(1 - F)$	$2pqd$	$2pqd(1 - F)$
A_2A_2	$-a$	q^2	$q^2 + Fpq$	$-q^2a$	$-q^2a - Fpqa$
			Means Outbred = $a(p - q) + 2pqd$ Inbred = $a(p - q) + 2pqd - 2dpqF$ $ID^a = 2dpqF$		

[a] *ID* = inbreeding depression

The mean fitness of the inbred population is reduced by $2pqdF$ for this locus, compared to the random mating populations (Table 3.1). However, the effects of inbreeding have to be summed across all loci (Σ) in the genome. Thus, inbreeding depression (ID) for total fitness is:

$$ID = \Sigma 2pqdF \qquad 3.1$$

Consequently, the magnitude of inbreeding depression depends upon the:

- increase in the inbreeding coefficient (F)
- decrease in heterozygosity ($2Fpq$) for harmful alleles
- directional dominance deviation (d) of beneficial alleles
- number of loci segregating for harmful alleles or exhibiting heterozygote advantage (reflected in Σ).

Since inbreeding depression is a function of change in heterozygosity, it does not occur in haploids (no heterozygotes) or asexuals (no change in heterozygosity) (Chapter 8).

This mathematical modeling leads us to predict that inbreeding depression will have the following characteristics:

- be ubiquitous in sexually reproducing species that are diploid or have higher ploidies
- increase with the change in inbreeding (ΔF)
- be greater for fitness than peripheral traits
- occur for all components of fitness
- be greater for total fitness than its components
- be worse in more stressful environments
- vary across lineages, families, and species
- be reduced by natural selection under some circumstances (purging)
- occur in small isolated random mating populations.

In what follows, we present empirical evidence relating to these predictions.

Ubiquity of inbreeding depression

Inbreeding depression is essentially ubiquitous in naturally outbreeding populations and species that are diploid or have higher ploidies

Almost all studies in outbreeding species report harmful effects of inbreeding on reproductive fitness. In fact, Lacy (1997) concluded that "no species of mammal has been shown to be unaffected by inbreeding," given that studies reporting no statistical evidence of inbreeding depression had sample sizes or levels of inbreeding that were too low to have adequate statistical power to detect the expected fitness depression. Angeloni et al. (2011) reached a similar conclusion for plants.

Thus, the assumption for an unstudied outbreeding species must be that it will suffer reduced fitness if inbred, or it will be suffering inbreeding depression if it has previously been inbred, or both.

Inbreeding depression affects mammals, birds, reptiles, amphibians, fish, invertebrates, and plants

Inbreeding depression affects all major eukaryotic taxa (Crnokrak & Roff 1999). Species exhibiting inbreeding depression for fitness in the wild include many mammals (e.g. African lions, Florida panthers, golden lion tamarins, gray wolves, Mexican wolves, white-footed mice [*Peromyscus leucopus*], common shrew [*Sorex araneus*], and Soay sheep [*Ovis aries*]); birds (e.g. greater prairie chicken, Mexican jay [*Aphelocoma wollweberi*], song sparrow [*Melospiza melodia*], red-cockaded woodpecker [*Picoides borealis*], and great reed warbler [*Acrocephalus arundinaceus*]); fish (e.g. Atlantic salmon [*Salmo salar*], desert topminnow, and rainbow trout [*Oncorhynchus mykiss*]); reptiles (e.g. Swedish adder); invertebrates (e.g. snails and insects); and many species of plants (see Fenster & Dudash 1994; Frankham 1995b; Keller & Waller 2002; Liberg et al. 2005; Hedrick & Fredrickson 2008; Johnson et al. 2010; Angeloni et al. 2011).

Naturally inbreeding species and small isolated populations are often already highly homozygous and inbred. Consequently, they will typically be suffering inbreeding depression, and will exhibit less inbreeding depression if further inbred (due to small ΔF and little increase in homozygosity) (Chapter 8).

Inbreeding depression increases with ΔF

Fitness typically continues to decline as inbreeding increases

The simple model in Table 3.1 predicts a linear relationship between mean fitness and inbreeding coefficient, and a similar linear relationship is expected if the effects of different loci combine in an additive fashion. However, some other models for combining the effects of different loci yield non-linear relationships (Crow & Kimura 1970; Charlesworth & Charlesworth 1987).

Most empirical studies exhibit an approximately linear relationship between quantitative trait mean and *F* (Lynch & Walsh 1998). For example, grain yield (a fitness trait) and height (a peripheral trait) in maize show essentially linear declines with inbreeding (Fig. 3.4), and the chapter frontispiece illustrates three further examples. However, a minority of cases show non-linear (but declining) relationships between mean and *F*.

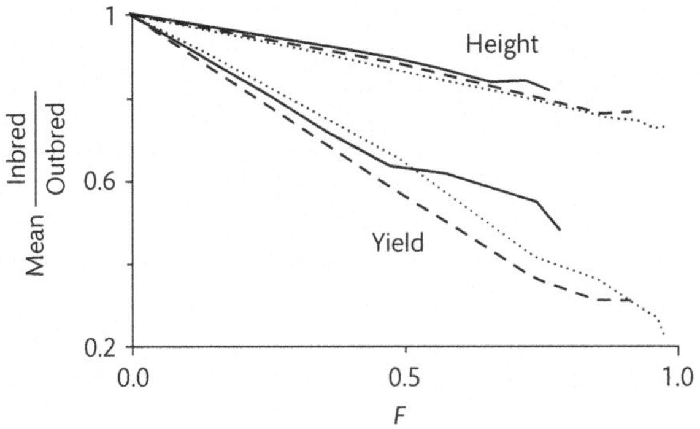

Fig. 3.4 Approximately linear declines of height and grain yield in maize (Frankham et al. 2010, Fig. 13.4, after Falconer & Mackay 1996). The dotted and dashed lines refer to populations inbred by consecutive generations of selfing (two studies) and the solid line to those inbred by continuing full-sib mating.

Inbreeding depression occurs in small closed random mating populations

Inbreeding due to small population size over multiple generations results in inbreeding depression

For most naturally outbreeding species, inbreeding occurs slowly in small closed populations. The cumulative inbreeding over t generations (F_t) in finite random mating, populations of size N_e is (Falconer & Mackay 1996):

$$F_t = 1 - (1 - \frac{1}{2N_e})^t \quad 3.2$$

For example, an inbreeding coefficient of 0.25 is reached after 57 generations in a population with an N_e of 100.

This inbreeding results in inbreeding depression, but it typically takes much longer to accumulate than with selfing or full-sib mating. Box 3.2 details inbreeding depression for litter size and proportion of abnormal offspring in a small Swedish adder population. Similarly, slow inbreeding has reduced population fitness in small populations of black-footed rock-wallabies, greater prairie chickens, topminnow fish, fruit flies, house flies (*Musca domestica*), butterflies, and plants (Frankham 1995b; Heschel & Paige 1995; Madsen et al. 1996; Fischer & Matthies 1998; Westemeier et al. 1998; Bryant et al. 1999; Eldridge et al. 1999; Mattila et al. 2012).

Swedish adder

Box 3.2 Inbreeding depression in a small isolated population of adders in Sweden

(Madsen et al. 1996, 1999, 2004)

In Sweden, a small isolated population of adders (< 40 individuals), separated from the main distribution of the snake for at least a century, has low molecular genetic diversity, and so is inbred, relative to the main population.

The small population exhibited inbreeding depression for litter size and proportion of abnormal offspring, compared to the larger northern population. Different environmental conditions were ruled out as an explanation for abnormal offspring results, as the progeny of an introduced male from the large population, when mated to females from the small population, produced progeny with a greatly reduced frequency of abnormalities.

Inbreeding depression is greater for fitness than for peripheral traits

A locus will contribute to inbreeding depression only if $d > 0$. Thus, harmful alleles must be partially or completely recessive (and favorable alleles dominant, or partially so), or show heterozygote advantage (see Fig. 3.3), to contribute to inbreeding depression. Overall, there must be a consistency in the direction of the dominance across loci (harmful alleles recessive and beneficial ones dominant) for there to be substantial inbreeding depression. Such directional dominance is a feature of fitness traits measured in the environment to which the population is adapted, as fitness is typically subject to directional selection (Hoekstra et al. 2001). Conversely, peripheral traits are typically subject to selection favoring phenotypic intermediates (stabilizing selection), or selection that changes direction across time or space (disruptive selection: Schluter 1988; Kingsolver & Pfennig 2007), and will typically show little consistent dominance across loci and, thus, limited inbreeding depression.

Empirical results show that inbreeding depression is typically less for traits more peripherally related to fitness (DeRose & Roff 1999). For example, it is less for height than grain yield in maize (Fig. 3.4), and abdominal bristle number in *Drosophila* exhibits none, yet fitness traits show strong inbreeding depression (Mackay 1986).

Inbreeding depression occurs for all aspects of reproductive fitness

Inbreeding has been observed to have harmful consequences for all aspects of reproductive fitness in animals and plants (e.g. sperm production, sperm quality, mating ability, female fecundity, juvenile survival, mothering ability, age at sexual maturity, predator avoidance, adult survival and longevity in animals, and related components in plants: Frankham et al. 2010; Losdat et al. 2014; Møller & Nielsen 2015). For example, inbreeding depresses male gametic performance across diverse systems and traits in animals and plants, based on a meta-analysis across 37 species and 183 study traits (Fig. 3.5).

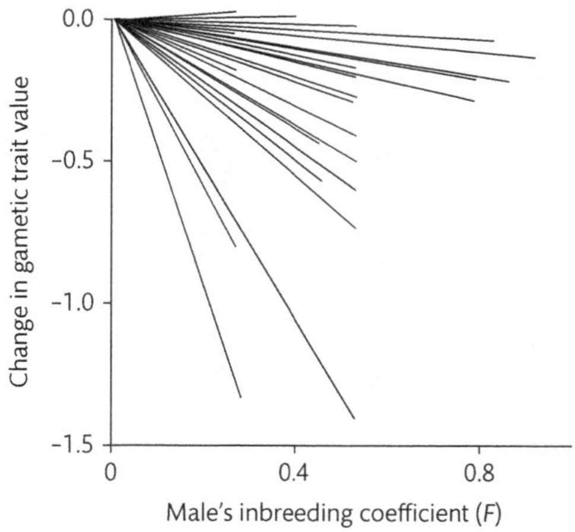

Fig. 3.5 Inbreeding depresses male gametic performance across diverse systems and traits as indicated by the plots of mean sperm or pollen quality, quantity, and competitiveness against the inbreeding coefficient *F* (Losdat et al. 2014). Each line represents a separate study.

Not all studies on all species report inbreeding depression for all characters studied, but virtually all show it for most reproductive fitness characters (Darwin 1876; Crnokrak & Roff 1999).

Inbreeding depression is greater for total fitness than its components

Since more loci are expected to affect total fitness than its components, total fitness is expected to exhibit more inbreeding depression than its components. Greater inbreeding depression for overall fitness than for its components has been observed in many studies, as for example in plants, old-field mice (*Peromyscus polionotus*), house mice, chickens (*Gallus gallus domesticus*), turkeys (*Meleagris gallopavo*), Japanese quail (*Coturnix japonica*), chukar partridges (*Alectoris chukar*), great tits, song sparrows, and takahe (*Porphyrio hochstetteri*) (Beilharz 1982; Abplanalp 1990; Dudash 1990; Lacy et al. 1996; Keller 1998; Meagher et al. 2000; Szulkin et al. 2007; Grueber et al. 2010).

This means that inbreeding depression will accumulate in successive stages across the life cycle, as observed in animals such as the great tit (Fig. 3.6), takahe, and the red deer, and many plant species (Husband & Schemske 1996; Grueber et al. 2010; Huisman et al. 2016). **Thus, information on total fitness is required to capture the full impacts of inbreeding.**

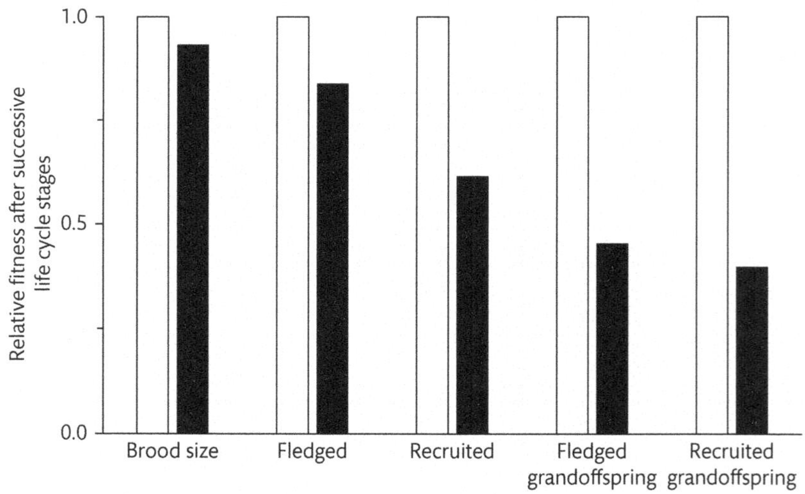

Fig. 3.6 Cumulative inbreeding depression across the life cycle in the great tit (Szulkin et al. 2007). Black bars represent the relative fitness of inbred offspring (F = 0.25), compared to outbred offspring (white bars, F = 0) for successive life history traits.

Inbreeding depression is greater in more stressful environments

The effects of homozygosity for harmful alleles are typically greater in more stressful environments (Kondrashov & Houle 1994). Consequently, inbreeding effects on fitness are expected to be more harmful in stressful than benign environments, and this is generally observed. Inbreeding depression was on average 69% greater in stressful than in benign environments, and increased linearly with the stressfulness of environments across an array of taxa, based on meta-analyses (Armbruster & Reed 2005; Fox & Reed 2011; Enders & Nunney 2012; Reed et al. 2012). For example, inbreeding depression in the rose pink plant (*Sabatia angularis*) was 29% greater for total fitness in the field than in the greenhouse (Fig. 3.7a) and greater in bad than good environments in the great tit (Fig. 3.7b).

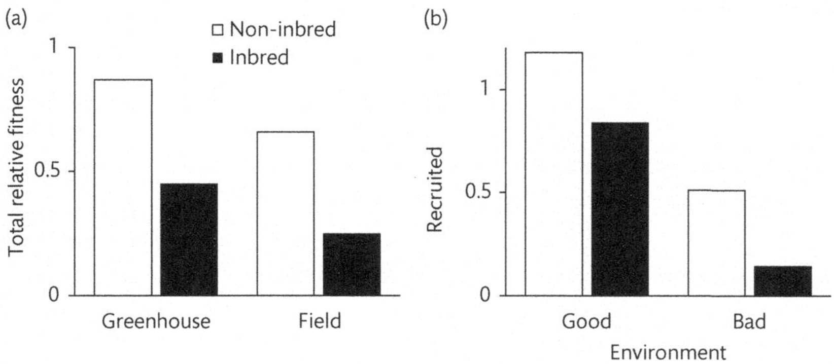

Fig. 3.7 Inbreeding depression is greater in stressful than benign conditions in (a) the rose pink plant grown in the glasshouse versus field conditions (after Dudash 1990), and (b) great tit birds in good and bad environments (Szulkin & Sheldon 2007).

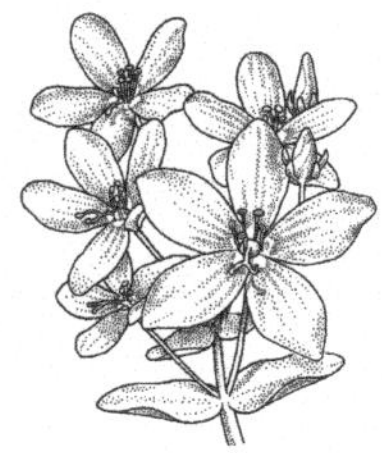

Rose pink plant (USA)

Great tit (UK)

Harmful impacts of inbreeding vary widely

The harmful effects of inbreeding on fitness vary among individuals, families, populations, and species

Since inbreeding depression depends on the frequency of homozygotes for harmful alleles, it is expected to exhibit large stochastic effects. Empirical studies reveal variation in inbreeding depression among lineages within species for plants, old-field mice, dairy cattle (*Bos taurus*), *Drosophila* fruit flies, and flour beetles (Dudash et al. 1997; Reed et al. 2002; Armbruster & Reed 2005). Inbreeding depression varied among three sub-species of *Peromyscus polionotus* old-field mice, and among three replicates within each sub-species (Fig. 3.8). Further, Kärkkäinen et al. (1996) reported geographic variation in inbreeding depression in the Scots pine (*Pinus sylvestris*).

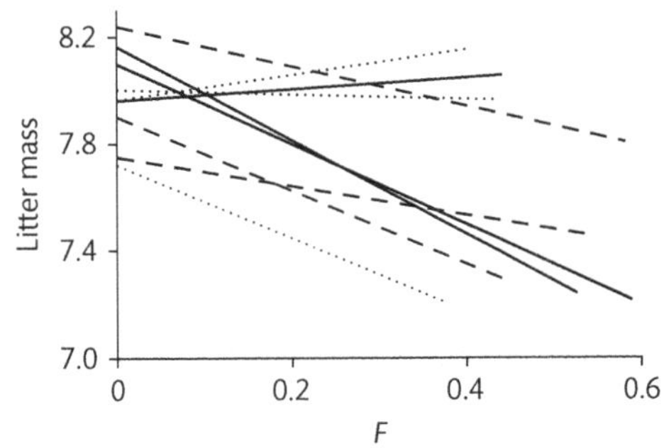

Fig. 3.8 Differences in inbreeding depression among three sub-species of old-field mice (solid, dashed, and dotted lines) and of three samples within each sub-species for total mass of progeny weaned per pair (Frankham et al. 2010, Fig. 13.3, after Lacy et al. 1996). Regression lines relating fitness and inbreeding coefficient for each replicate of each sub-species are plotted.

Species and populations also vary in the components of fitness that are affected by inbreeding. For example, captive populations of Mexican and red wolf (*Canis lupus rufus*) did not exhibit inbreeding depression for juvenile survival, but exhibited it for adult survival (Kalinowski et al. 1999; Wilcken 2002).

Natural selection can reduce inbreeding depression

Inbreeding depression may be reduced (purged) by selection against harmful alleles, especially under inbreeding or in small populations, but it is rarely eliminated

Selection can reduce the frequency of harmful recessive alleles (purging), but it cannot reduce inbreeding depression due to heterozygote advantage (Dudash & Carr 1998). As homozygosity for harmful recessive alleles makes a much greater contribution to inbreeding depression than do alleles involved in heterozygote advantage (Charlesworth & Charlesworth 1999; Swanson-Wagner et al. 2006; Charlesworth & Willis 2009; but see Cheptou & Donohue 2011), many harmful alleles are potentially subject to purging.

The factors affecting the efficiency of purging of harmful alleles are (Glémin 2003; Frankham et al. 2014a):

- genetic basis of inbreeding depression (dominance > heterozygote advantage) (Charlesworth & Charlesworth 1987)
- mating system (inbreeding > random mating) (Lande & Schemske 1985)
- effects of prior inbreeding (prior > none)
- population size (large N_e > small)
- strength of natural selection (strong > weak).

Partial purging of the genetic load of harmful alleles has been documented in many species, including plants, mice, gorillas, birds, and fruit flies (Dudash & Carr 1998; Byers

& Waller 1999; Crnokrak & Barrett 2002; Boakes et al. 2007; Leberg & Firmin 2008; Li et al. 2014; Xue et al. 2015). The effects of prior inbreeding are not always consistent, but this is not surprising given the number of variables affecting the efficiency of purging and the diversity of regimes that have been used (Glémin 2003); in three old-field mouse populations, effects ranged from reduced inbreeding depression, through no effect, to enhanced inbreeding depression (Fig. 3.8).

Even if purging reduces inbreeding depression for a population in one environment, it often does not lead to any benefits in other environments (Bijlsma et al. 1999).

A prior history of small population size is often associated with reduced subsequent inbreeding depression, but does not usually remove it completely

Experimental evidence indicates that purging effects are modest and that small partially inbred populations usually continue to exhibit inbreeding depression when inbred further, even when they have low genetic diversity (Ballou 1997; Dudash & Carr 1998; Boakes et al. 2007; Angeloni et al. 2011; Mattila et al. 2012).

The persistence of inbreeding depression in small populations, even those which habitually inbreed, is due to fixation at loci exhibiting heterozygote advantage and fixation of new harmful mutations (Chapter 2).

Maternal and zygotic inbreeding both affect fitness

Inbreeding depression for fitness depends on both an individual's inbreeding coefficient (zygotic) and its mother's inbreeding coefficient (maternal)

The fitness of offspring depends on their zygotic genotype and maternal provisioning during embryonic development, as well as subsequent parental care and feeding in some animals. As inbreeding reduces maternal fitness, maternal provisioning and care are reduced, and offspring fitness suffers. Maternal inbreeding effects on progeny fitness have been documented in many species of vertebrates, invertebrates, and plants and are expected to varying degrees in all animal and plant species (Frankham 2015; Huisman et al. 2016; Chapter 6). Consequently, the impacts of inbreeding on fitness from mating a pair of siblings from a non-inbred mother must be carried through to the grand-offspring to encompass the full inbreeding depression.

How large are the impacts of inbreeding on total fitness?

The impacts of inbreeding on total fitness are typically large and extremely harmful

While inbreeding depression effects on individual fitness components may be modest, effects on total fitness in the wild are typically substantial (Table 3.2). The bird studies are likely underestimates due to extra-pair paternities (15% for the great tit) and half the estimates do not include the impacts of maternal inbreeding.

Table 3.2 Inbreeding depression (δ) for total fitness in wild species of animals and plants due to a 25% increase in the inbreeding coefficient expressed as % reduction in mean of inbred progeny compared to outbred progeny in similar environments. Lethal equivalents (L.E.), described later in the chapter are also reported.

Common name	Genus and species	δ %	L.E.	References
Red deer	*Cervus elaphus*	99[a]	18.7	Huisman et al. (2016)
Collared flycatcher	*Ficedula albicollis*	94[b]	7.5	Kruuk et al. (2002)
Great tit	*Parus major*	55	3.2[a]	Szulkin et al. (2007)
Song sparrow	*Melospiza melodia*	79	6.2[a]	Keller (1998)
Takahe	*Porphyrio hochstetteri*	88	8.0	Grueber et al. (2010)
Deerhorn clarkia	*Clarkia pulchella*	100[a,b]	39.2[a]	Newman & Pilson (1997)
Rose pink plant	*Sabatia angularis*	38[a,b]	1.9[a]	Dudash (1990)
Wild radish	*Raphanus sativus*	58[b]	3.5[a]	Nason & Ellstrand (1995)

[a] Our calculation from the original data, using methods detailed later.
[b] Maternal inbreeding contribution not included.

Does inbreeding increase extinction risks?

Inbreeding increases the risk of extinction in outbreeding diploid and polyploid species

Evidence from captive and wild populations, and computer modeling all show that inbreeding increases extinction risk in outbreeding diploid and polyploid populations and species, as detailed in what follows (Frankham 2005).

Inbreeding causes extinction in captive populations

Deliberately inbred populations of laboratory and domestic animals and plants show greatly elevated extinction rates

Sustained inbreeding has been shown to increase extinction rates in captive populations of many species, including *Drosophila*, house flies, mice, Japanese quail (*Coturnix coturnix japonica*), maize, Italian ryegrass (*Lolium multiflorum*), and *Mimulus guttatus* plants (Frankel & Soulé 1981; Frankham 1995b; Dudash et al. 1997; Bijlsma et al. 1999, 2000; Reed & Bryant 2000; Reed et al. 2002, 2003a, 2003b; reviewed by Frankham 2005). For example, inbreeding clearly increased the risk of extinction in captive populations of *Drosophila* and mice (Fig. 3.9a). In Italian ryegrass, 19 of 20 populations were extinct after four generations of self-fertilization (Polans & Allard 1989), while all replicates of the least inbred treatment survived. Even slow inbreeding due to finite population size (N_e = 10 or 20) causes extinctions, but at a lower rate than with full-sib mating (Fig. 3.9b).

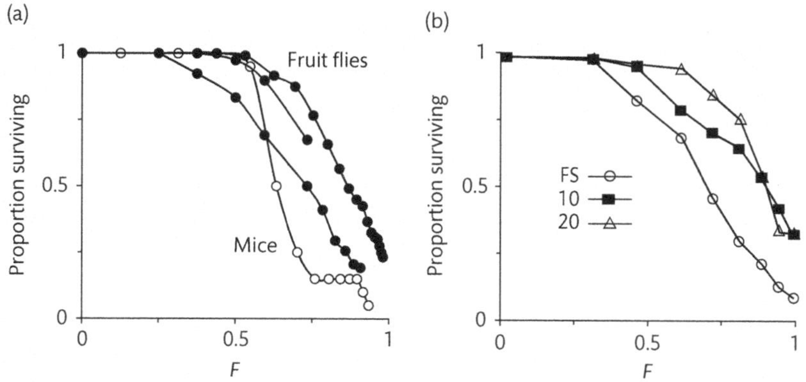

Fig. 3.9 Proportion of populations going extinct rises with inbreeding (*F*). (a) Populations of mice and two species of *Drosophila* fruit flies (one with two populations) were inbred using continuous brother-sister mating (Frankham et al. 2010, Fig. 2.1, after Frankham 1995c). Other causes of extinction can be ruled out in these populations. (b) Populations of *Drosophila* maintained using full-sib mating (FS) or genetically effective sizes of 10 and 20 (Frankham et al. 2010, Fig. 13.5, after Reed et al. 2003a).

Inbreeding interacts with other threats: the extinction vortex

The effects of inbreeding depression and loss of genetic diversity frequently interact with demographic, environmental, and catastrophic factors in a feedback process termed the "extinction vortex"

Small populations typically have elevated extinction risks from the combined impacts of diverse deterministic and stochastic variables. However, genetic and demographic factors are expected to interact in a negative feedback that accelerates the rate of decline in an "**extinction vortex**" (Gilpin & Soulé 1986; Fig. 3.10). If populations become small for any reason (human impacts, demographic or environmental stochasticity, or catastrophes), they become more inbred and less demographically stable, further reducing population size and increasing inbreeding further. Such extinction vortices have been observed in several populations of vertebrates and plants (Matthies et al. 2004; Fagan & Holmes 2006; Blomqvist et al. 2010). For example, strong genetic effects were identified in the rapidly declining southern dunlin (*Calidris alpina schinzii*), an endangered shorebird (Blomqvist et al. 2010).

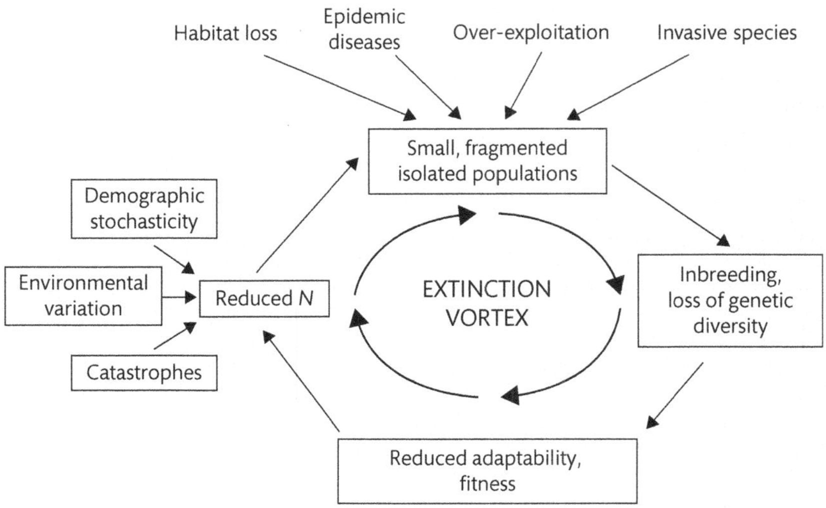

Fig. 3.10 The extinction vortex describes the possible interactions between human impacts, inbreeding, loss of genetic diversity, and demographic instability in a downward spiral towards extinction (Frankham et al. 2010, Fig. 2.2).

Notably, the complicated interactions between genetic, demographic, and environmental factors can make it extremely difficult to identify the immediate cause(s) for any particular extinction event (Lande 1988; Frankham 2005).

Inbreeding elevates extinction risks in wild populations

Despite multiple deterministic and stochastic threats experienced by wild populations, two lines of evidence demonstrate that inbreeding elevates the extinction risk of wild populations in natural settings:

- direct evidence shows that inbreeding and loss of genetic variation contribute to the extinction of populations in nature, and
- computer projections with input parameters from real species show that inbreeding will increase extinction risks for wild populations.

Studies in wild conditions have demonstrated the involvement of inbreeding in extinctions of animal and plant populations

Inbreeding was a significant predictor of extinction risk for butterfly populations in Finland after the effects of all other ecological and demographic variables had been removed (Box 3.1). Similarly, experimental populations of the deerhorn clarkia plant with higher inbreeding ($F = 0.08$) exhibited 69% extinction rates over three generations in the wild, while populations with lower inbreeding ($F = 0.04$) showed only a 25% extinction rate (Newman & Pilson 1997). Thus, very small differences in inbreeding translated into large differences in population extinction rates. Further, a study in the shore campion plant (*Silene littorea*) also documented adverse impacts of inbreeding on extinction rates in wild populations in Spain (Vilas et al. 2006).

Computer projections show that inbreeding elevates extinction risk for most outbreeding populations in the wild

Computer projections incorporating factual life history information are often used to assess the combined impact of all deterministic and stochastic factors on the probability of population extinctions (population viability analysis: Appendix 3; Lacy & Pollak 2014).

Almost all computer projections, using a range of outbreeding birds, mammals, reptiles, amphibians, and plants, yielded substantial increases in extinction risk when the effects of inbreeding were included, as compared to runs where they were excluded (Oostermeijer 2000; Brook et al. 2002; O'Grady et al. 2006; Schiegg et al. 2006). Results for four different species are shown in Fig. 3.11. Inbreeding depression was projected to reduce median times to extinction for vertebrates in the wild by an average of 37% for populations with carrying capacities of 100, 500, and 2,000 (O'Grady et al. 2006).

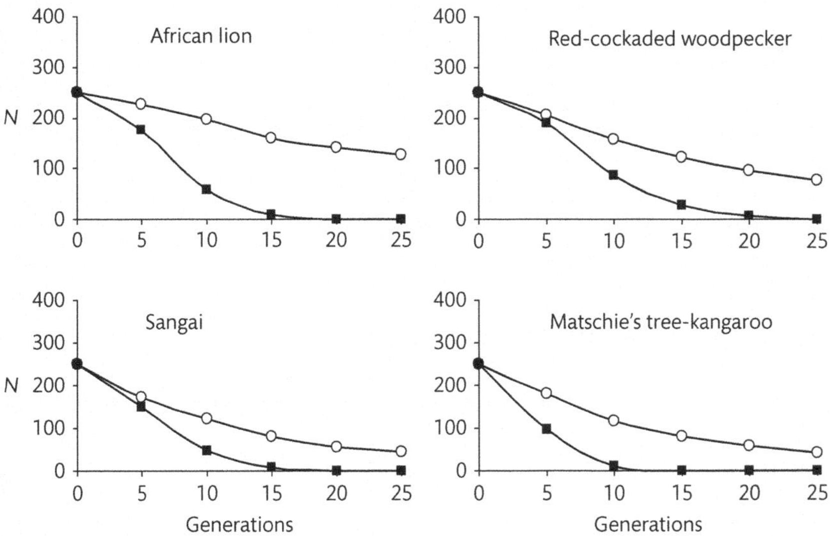

Fig. 3.11 Inbreeding is projected to substantially increase extinction risk in wild populations. Computer projections of population sizes for threatened wild populations of African lion, red-cockaded woodpecker, Matschie's tree-kangaroo (*Dendrolagus matschiei*), and Sangai (*Rucervus eldii eldii*) when the harmful effects of inbreeding are included (■) or excluded (○) in addition to all other stochastic and deterministic threats (Frankham et al. 2010, Fig. 2.4, after O'Grady et al. 2006).

How do we detect inbreeding depression?

Comparisons of inbred and outbred individuals (or populations) maintained under the same environmental conditions are required to detect inbreeding depression

As survival and reproduction are strongly influenced by environmental conditions, tests for inbreeding depression require comparisons of the fitness of inbred and non-inbred individuals under the same environmental conditions (Falconer & Mackay 1996). For example, the effects of inbreeding on captive mammals were studied by comparing juvenile survival of inbred and outbred offspring matched for zoo, enclosure in zoo, year of birth, and density of population (Ballou & Ralls 1982). The effects of inbreeding for white-footed mice in the wild were documented by simultaneously releasing inbred and outbred offspring of the same age into the wild, and following their subsequent survival and weights (Jiménez et al. 1994). Alternatively, characteristics of inbred and outbred individuals from different times can be compared to the same control population under similar environmental conditions (Vrijenhoek 1994; Chapter 6).

The existence of inbreeding depression can also be inferred if the fitness of crossed progeny is greater (**heterosis** or **genetic rescue**) than for contemporary inbred offspring in the same environment (Falconer & Mackay 1996; Chapter 6).

If pedigrees are not available to determine *F*, genetic markers, such as multiple microsatellite loci or genomic sequence data on individuals, can be used to infer the degree of inbreeding of individuals, and inbreds and outbreds compared in the same environment (Hoffman et al. 2014; Wang 2014a, 2016; Kardos et al. 2015, 2016; see also Chapters 11 and 13). Such methods have been applied to detect inbreeding depression in harbor seals (*Phoca vitulina*), Soay sheep, red deer, black grouse (*Lyrurus tetrix*), Attwater's prairie chickens (*Tympanuchus cupido attwateri*), and eelgrass (*Zostera marina*) (Coltman et al. 1999; Slate et al. 2000; Höglund et al. 2002; Hämmerli & Reusch 2003; Hammerly et al. 2013; Hoffman et al. 2014; Huisman et al. 2016; see also Chapter 13).

How do we quantify inbreeding depression?

Inbreeding depression is usually quantified as the proportionate decline in mean phenotype due to a change in inbreeding coefficient

Delta (δ)

A general measure of inbreeding depression (δ) is the proportionate decline in mean of a phenotypic trait (e.g. fitness) due to inbreeding, as follows:

$$\delta = 1 - \frac{\textit{inbred fitness}}{\textit{outbred fitness}} \qquad 3.3$$

This is simply the measure *ID* defined in Table 3.1, divided by the mean of the outbred population. Example 3.1 illustrates the use of eqn 3.3 to estimate inbreeding depression in rose pink plants. This formula does not specify the level of inbreeding, but in plants (where it is widely used) it typically involves comparing selfed ($F = 0.5$) and outcrossed progeny ($F = 0$). The compilation of estimates of inbreeding depression due to sib-mating ($F = 0.25$) in Table 3.2 is presented in this form.

Example 3.1 Inbreeding depression in rose pink plants

Total fitness in the field of propagules resulting from selfing and outcrossing were 0.25 and 0.66, respectively (Dudash 1990). Thus, the inbreeding depression (δ) for total fitness in this species is:

$$\delta = 1 - \frac{inbred\,fitness}{outbred\,fitness} = 1 - \frac{0.25}{0.66} = 0.62$$

Thus, δ is 62% due to a difference of 0.5 in the inbreeding coefficient between the inbred and outbred offspring.

Lethal equivalents

Lethal equivalents are used for quantifying the extent of inbreeding depression for survival, especially in animals

A lethal equivalent refers to a group of detrimental alleles that would cause death if homozygous, e.g. one lethal allele, two alleles that cause a 50% probability of death, etc. (Morton et al. 1956). The probability of surviving (S) can be expressed as a function of inbreeding F (Morton et al. 1956):

$$S = e^{-(C+BF)} \qquad 3.4$$

where e^{-C} is survival in an outbred population, e^{-BF} the decrement in survival due to inbreeding, F the inbreeding coefficient, and B the rate at which survival declines with a change in inbreeding (lethal equivalents). B measures the additional genetic damage that would be expressed in a complete homozygote ($F = 1$) and is the number of lethal equivalents per gamete, while $2B$ is the number per individual.

Equation 3.4 can be converted into a linear form by taking natural logarithms (ln) of both sides, yielding:

$$\ln S = -\ C - BF \qquad 3.5$$

Lethal equivalents are estimated from the slope of the regression of natural logarithm of survival on level of inbreeding, using data on survival rates of individuals with different levels of F. For the golden lion tamarin (Table 3.3 and Fig. 3.12) the slope of the regression of ln (7-day survival) on F is $B = -1.717F$. Thus, the population contains 1.72 haploid and 3.43 diploid lethal equivalents.

Table 3.3 Data on survival to 7 days for offspring with different levels of inbreeding (*F*) in the golden lion tamarin in natural habitats in Brazil (Dietz et al. 2000).

F	Survived	Died	Total
0	411 (94.7%)	23 (5.3%)	434
0.017	11 (84.6%)	2 (15.4%)	13
0.063	5 (71.4%)	2 (28.6%)	7
0.125	8 (88.9%)	1 (11.1%)	9
0.243	11 (61.1%)	7 (38.9%)	18

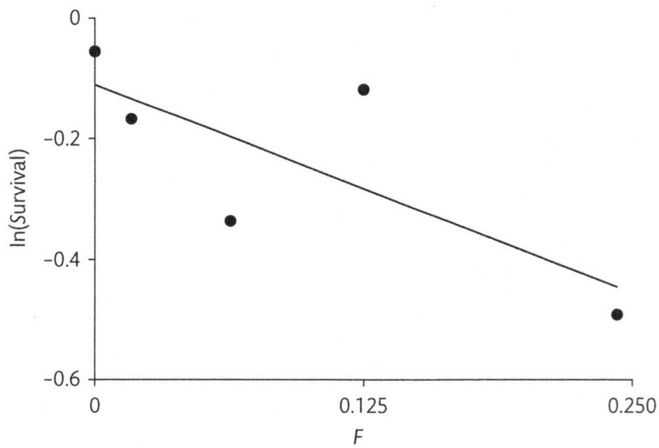

Fig. 3.12 Relationships between survival to 7 days and inbreeding coefficient in golden lion tamarins in natural habitats in Brazil (Dietz et al. 2000). The natural logarithm of survival (ln*S*) is plotted against the inbreeding coefficient (*F*) along with the best fitting linear regression line: $\ln S = -0.057 - 1.717\,F$ ($P = 0.01$, $r^2 = 0.87$).

Values of zero survival create problems in this approach as we cannot take the logarithm of 0. This can be overcome by binning such observations with those for individuals with non-zero survival and similar inbreeding. Alternatively, Kalinowski & Hedrick (1998) devised a maximum likelihood method of estimation that avoids the problem.

Lethal equivalents can also be measured for other fitness components by expressing the means for inbred progeny as a proportion of that for outbred progeny, and thus estimating lethal equivalents using the methods just described (O'Grady et al. 2006).

A point estimate of lethal equivalents can also be obtained from data on the fitness of inbreds (W_I) and outbreds (W_O), and the inbreeding coefficient of the inbreds, using a variation of eqn 3.4, as follows:

$$\frac{W_I}{W_O} = e^{-FB} \qquad 3.6$$

By taking natural logarithms and rearranging, we obtain an estimate of lethal equivalents, as:

$$B = \frac{-\ln(W_I/W_O)}{F} \qquad 3.7$$

Use of this approach is illustrated in Example 3.2.

Example 3.2 Point estimate of lethal equivalents

In the song sparrow, Keller (1998) found that the total fitness impact of a full-sib mating ($F = 0.25$) was a 79% reduction in fledged offspring from eggs, compared to that for non-inbred eggs. Thus:

$$\frac{W_I}{W_O} = 1 - 0.79 = 0.21$$

and

$$B = \frac{-\ln(W_I/W_O)}{F} = \frac{-\ln(0.21)}{0.25} = 6.2$$

Thus, the song sparrow population contains 6.2 haploid lethal equivalents.

How many lethal equivalents do outbreeding species exhibit in the wild?

A meta-analysis found that wild populations of outbreeding species averaged ~ 6 haploid lethal equivalents for total fitness (O'Grady et al. 2006). The known estimates for individual species (Table 3.2) range from 1.9 to 39.2, with a median of 6.85. Sampling variation explains much of the wide range, but there are also likely true differences among species (Grueber et al. 2010).

What is mutational meltdown?

Mutational meltdown describes the accumulation of new harmful mutations in small populations, and their subsequent adverse effects on fitness and population persistence

Mutational meltdown is a phenomenon closely related to inbreeding depression in that both involve exposing harmful recessive alleles. They differ in that inbreeding depression mainly increases homozygosity for pre-existing harmful mutations, while mutational meltdown focuses on accumulation and fixation of new harmful mutations (Lynch et al. 1995a).

While harmful alleles are kept at low frequencies in large populations due to the balance between mutation and natural selection, in small populations genetic drift becomes the primary determinant of whether alleles become more (or less) common each generation. Thus, mildly harmful alleles become effectively neutral, and some increase in frequency and reduce reproductive fitness (Lande 1995; Lynch et al. 1995b). Over many generations, sufficient harmful alleles may accumulate to cause negative population growth and a decline to extinction, especially in asexual species (Chapter 8).

Mutational meltdown appears to be minor compared to other genetic threats in outbreeding sexually reproducing species

The contribution of mutational accumulation to extinction risk in small populations is controversial, especially in naturally outbreeding species (Charlesworth et al. 1993; Lande 1995; Lynch et al. 1995b). Its impact depends critically on the distribution of mutational effects, being important when there are many mutations of predominantly small harmful effect, but only minor when a substantial proportion of mutations have large effects, such that they are removed by natural selection even in small populations (Garcia-Dorado 2003). Further, fixation of beneficial mutations reduces fitness loss (Whitlock 2000). Empirical tests indicate that mutation accumulation is of minor importance in outbreeding sexual species within normal time spans of conservation concern, compared to inbreeding depression (Gilligan et al. 1997).

What are the relationships between harmful mutation, drift, inbreeding, fixation, and fitness decline?

The harmful alleles that cause inbreeding depression are continually added to populations by mutation, removed by selection and drift, or made homozygous by inbreeding or drift

The processes we have described throughout this chapter change over time (Fig. 3.13). Reduction in reproductive fitness in small populations is due to chance fixation, either of harmful alleles (most important) or homozyosity at loci exhibiting heterozygote advantage (less important).

The current fitness of large outbreeding populations will typically be high in their usual niche, but they likely carry a high load of individually rare harmful mutations and would suffer a large fitness reduction if inbred (inbreeding load). Conversely, a population that has been small for many generations is likely to have reduced current fitness due to drift and fixation of harmful alleles (past inbreeding depression or fixed genetic load), but will probably have a low load of unfixed harmful alleles and show only modest inbreeding depression from inbreeding in the current generation. For example, a small population of Glanville fritillary butterflies on Pikku Tytärsaari Island exhibits low fitness from past inbreeding, but exhibited little further inbreeding depression when deliberately inbred. Conversely, the large population on Åland has high current fitness, but exhibited substantial inbreeding depression when deliberately inbred (Saccheri et al. 1998; Nieminen et al. 2001; Hanski 2011; Mattila et al. 2012). Related findings have been reported in the German gentian (*Gentianella germanica*) and highlands scrub hypericum (*Hypericum cumulicola*) plants (Paland & Schmid 2003; Oakley & Winn 2012).

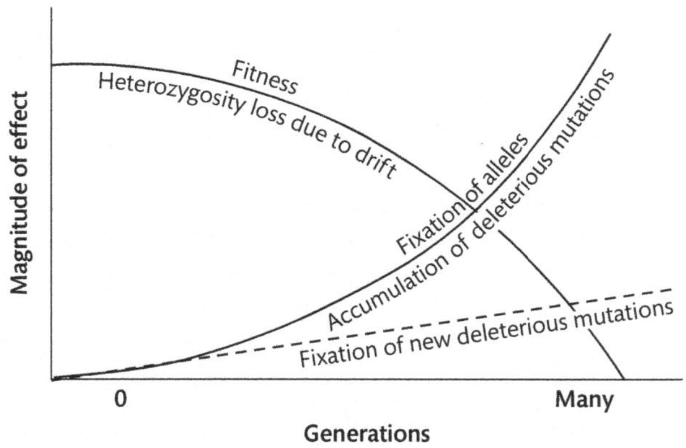

Fig. 3.13 The occurrence of harmful mutations, followed by drift, fixation, inbreeding depression, and fitness decline are continuing processes.

Different terms are used to describe prior inbreeding depression

Reduced fitness due to prior fixation of harmful alleles is variously referred to as (prior) inbreeding depression, or fixed genetic load due to drift, and sometimes these are considered a different process to current inbreeding depression (e.g. Paland & Schmid 2003). However, as both are due to fixation of harmful alleles, or alleles at loci showing heterozygote advantage, we refer to both as inbreeding depression.

Can we reverse inbreeding depression and mutational accumulation?

Inbreeding depression and mutational accumulation can usually be reversed by augmenting gene flow between the target population and another isolated population within the species

Since most harmful alleles involved in inbreeding depression and mutational accumulations are recessive, or exhibit heterozygote advantage, crossing to unrelated, isolated populations within species can usually reverse the harmful effects, as described in Chapter 6.

Summary

1. Inbreeding reduces reproductive fitness and increases the risk of extinction in essentially all well-studied populations of naturally outbreeding sexually reproducing diploid and polyploid species.
2. Inbreeding depression is due to increased homozygosity for harmful recessive alleles and at loci exhibiting heterozygote advantage.
3. Inbreeding depression is typically greater (a) in stressful than in benign conditions, (b) for total fitness than for individual fitness components, and (c) in naturally outbreeding than inbreeding species.
4. Natural selection may remove (purge) the alleles that cause inbreeding depression, especially following inbreeding or population bottlenecks, but it has limited effects in small populations and usually does not completely eliminate inbreeding depression.
5. Some new mildly harmful alleles increase in frequency by chance in small populations and reduce fitness (mutational meltdown), but their impacts are probably minor in sexual populations over the usual time frames of conservation concern.

FURTHER READING

Charlesworth & Willis (2009) Review on the genetics of inbreeding depression.
Frankham (2005) Review on the role of genetics factors in extinctions.
Keller & Waller (2002) Review on inbreeding effects in wild populations.
Li et al. (2015) Genome sequencing study reporting that threatened bird species had lower heterozygosity, and more harmful alleles than non-threatened bird species.
O'Grady et al. (2006) Realistic computer projections showed that inbreeding depression substantially reduces median times to extinction for most threatened outbreeding species.

SOFTWARE

COANCESTRY: a program for simulating, estimating, and analyzing relatedness and inbreeding coefficients (Wang 2011). http://www.zsl.org/science/software/coancestry
INBREEDR: an R package for the analysis of inbreeding based on genetic markers (Stoffel et al. 2016). https://CRAN.R-project.org/package=inbreedR
PMx: a program for demographic and genetic management, including computing inbreeding coefficients from pedigrees (Ballou et al. 2011; Lacy et al. 2012). http://www.vortex10.org/PMx.aspx

Loss of genetic diversity reduces ability to adapt

CHAPTER 4

Species face ubiquitous environmental change and must adapt or they will go extinct. Genetic diversity is the raw material required for evolutionary adaptation. However, loss of genetic diversity is unavoidable in small isolated populations, diminishing their capacity to evolve in response to environmental changes, and thereby increasing extinction risk.

TERMS

Adaptive evolution, additive genetic variation, cline, common garden, ecotype, epigenetic, evolutionary potential, heritability, maximum likelihood, mutational variance, partial dominance, phenotypic plasticity, quantitative character, quantitative genetic variation, quantitative trait loci, relative fitness, selection coefficient, selection differential, self-incompatibility

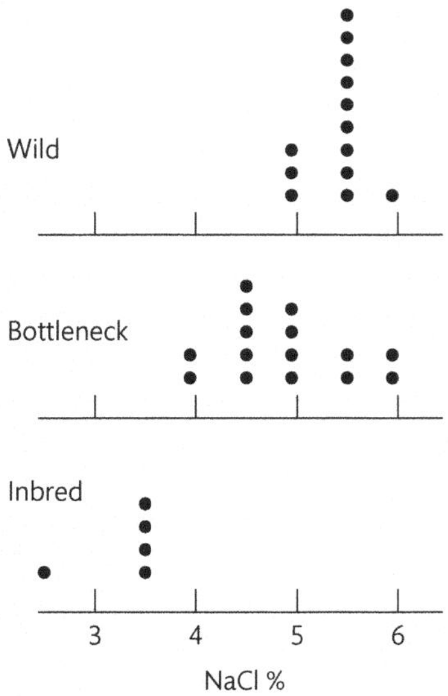

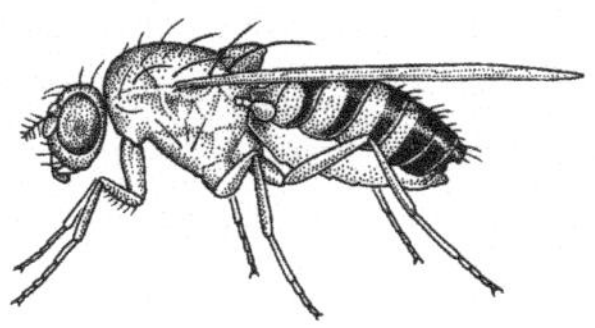

Population size bottlenecks reduce genetic diversity and evolutionary potential in *Drosophila* fruit flies. The black circles represent the concentrations of NaCl at extinction for each replicate population from stocks that were (a) the wild outbred, (b) subjected to a single-pair bottleneck for one generation (bottleneck), or (c) inbred by full-sib mating for 35 generations (inbred). All replicates were expanded to the same population size, and subjected to increasing concentrations of NaCl until extinct (Frankham et al. 2010, Fig. 8.6, after Frankham et al. 1999).

Genetic Management of Fragmented Animal and Plant Populations. R. Frankham, J. D. Ballou, K. Ralls, M. D. B. Eldridge, M. R. Dudash, C. B. Fenster, R. C. Lacy & P. Sunnucks. Oxford University Press (2017) © R. Frankham, J. D. Ballou, K. Ralls, M. D. B. Eldridge, M. R. Dudash, C. B. Fenster, R. C. Lacy & P. Sunnucks 2017. DOI 10.1093/oso/9780198783398.001.0001

Why should we be concerned about conserving the ability of species to adapt?

Environmental change is a ubiquitous feature of the conditions experienced by species. Consequently, populations and species need to evolve to avoid extinction, especially in the long term

Species have to cope with many changes over time in physical and biotic environments (pests, parasites, diseases, and competitors), and especially changes wrought by human activities. Global climate change is now occurring as a consequence of the burning of fossil fuels (UNEP 2007; IPCC 2014). Many species are now experiencing increased temperatures to which they are poorly adapted, as indicated by coral reef bleaching, mass die-offs of mangroves, changes in plant communities (affecting food supply for herbivores), altered hibernation in mammals, reduced population growth rates in animals and plants, etc. (Lane et al. 2012; Cahill et al. 2013; Selwood et al. 2015; Fairfax 2016; Chapter 14). On the geological time scale, there are major climatic shifts between ice ages and warm periods.

Disease organisms evolve new strains and spread to new locations, new diseases arise, and pathogens switch hosts providing selective forces for hosts to adapt (Garrett 1994). Further, adaptations in competitors, pests, and parasites mean that species must continually evolve to avoid falling behind competing organisms (the "Red Queen" hypothesis; Van Valen 1973).

Adaptive evolutionary changes may allow populations to cope with conditions that no individual could previously tolerate

Adaptation may take the form of immediate morphological, physiological, or behavioral modifications, and/or **evolutionary (genetic) adaptation** where natural selection alters the genetic composition of populations over multiple generations. Physiological adaptations by individuals (phenotypic plasticity) include immune responses against diseases, induction of enzymes to utilize altered diets, modification in hemoglobin levels with altitude, etc. (Chapter 14). Behavioral adaptations may include altered food preferences, predator avoidance behavior, etc. However, there is a limit to non-genetic adaptation, as it is typically confined to a single generation. Epigenetic changes due to DNA methylation, histone modification, and micro-RNAs probably fall into a similar category to non-genetic adaptation, but they are transmitted over generations to some extent, especially in plants (Becker et al. 2011; Donohue 2014). If environmental changes are greater than individuals can tolerate, then the species goes extinct.

Conversely, adaptive evolutionary change through natural selection continuing over generations may allow a population to prosper under conditions more extreme than any individual could originally tolerate (Freeman & Herron 2013; Thompson 2013).

In the remainder of this chapter we provide evidence for the ubiquity of adaptive evolutionary change, evaluate its rapidity and magnitude, consider the factors controlling the evolution of populations, and examine the impacts of loss of genetic diversity in small populations on their ability to evolve in response to altered environments.

How common is evolutionary adaptation?

Genetic adaptation of species to their abiotic and biotic environmental conditions is **ubiquitous** in animal and plant species with genetic diversity

Adaptive evolutionary changes have allowed species to inhabit almost every imaginable niche on Earth: altitudes from 6,500 m on Mount Everest to deep ocean trenches, arctic saline pools at −23°C to boiling thermal springs and deep sea vents, and from oceans and freshwater to deserts. Further, plants have adapted to grow in almost every soil on the planet, even growing without soil in some cases (Dobzhansky et al. 1977; Thompson 2013).

Adaptive evolutionary changes have been documented in animal morphology, behavior, color, prey size, body size, life history attributes, disease resistance, predator avoidance, tolerance to pollutants, biocide resistance, etc. (Thompson 1998; Mousseau et al. 2000; Hoekstra 2006). Adaptive evolutionary changes in plants include those to soil conditions, water stress, flooding, light regimes, exposure to wind, resistance to disease, grazing, air pollution, and herbicides (Rockwood 2015; Briggs & Walters 2016). Populations adapted to different ecological conditions are so common that they have their own term (ecotypes: Turesson 1922).

If we are to understand the potential of evolutionary adaptation to allow persistence in the face of environmental change, we need to understand how rapidly adaptation can occur and how large the changes are.

How rapidly does adaptation occur?

Some adaptive evolutionary changes have occurred rapidly

While Darwin (1859) concluded that evolutionary change was typically slow, rapid evolutionary changes have been documented in many invertebrates, vertebrates, and plants, especially in response to human-caused environmental change (Hendry & Kinnison 1999; Kinnison & Hendry 2001; Reznick & Ghalambor 2001). For example, rapid adaptive evolution in response to environmental changes has occurred in beak and body dimensions in Darwin's medium ground finch (*Geospiza fortis*), migration patterns and rates in birds and plants, life history strategies in Trinidadian guppies (*Poecilia reticulata*), and flowering time in many species of plants (Berthold et al. 1992; Grant & Grant 1995; Cody & Overton 1996; Reznick et al. 1997; Franks et al. 2007).

Many rapid evolutionary changes have been observed in response to human caused environmental changes

Rapid adaptive evolutionary changes have been recorded in many human-impacted species (Palumbi 2001). European rabbits in Australia evolved resistance to the myxoma virus within a few generations after the virus was introduced as a control measure (Box 4.1). Further, length of mouthparts has altered in soapberry bugs (*Leptocoris tagalicus* and *Jadera haematoloma*) in response to utilization of non-native plants (Carroll et al. 2005), while several species of plants evolved heavy metals tolerance within a few generations during colonization of polluted mine wastes (Bradshaw 1984).

Box 4.1 Rapid adaptive evolutionary changes in rabbits in Australia following the introduction of myxoma virus as a control agent

(Fenner & Ratcliffe 1965; Nowak 2004; after Frankham et al. 2010)

European wild rabbits rapidly reached plague proportions when introduced into Australia in the 19th century, and a variety of control measures failed to halt their spread.

Myxoma virus was released into Australia as a biocontrol agent in 1950. Initially, mortality of infected rabbits was over 99%, but genetic resistance of rabbits to the virus evolved rapidly. Mortality to a standard virus strain declined from an initial value of ~ 90% to 25% in 1958 (the 6th disease outbreak), as shown in the figure below.

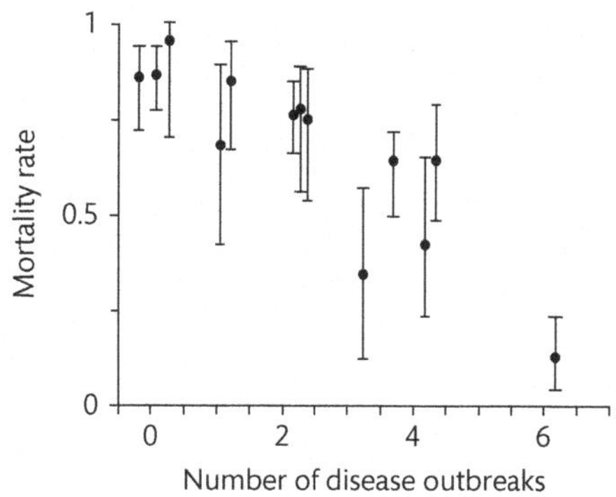

(Frankham et al. 2010, p. 119, after Fenner & Ratcliffe 1965)

Recently, calicivirus escaped from rabbit control trials on Wardang Island, South Australia into the Australian mainland rabbit population and increased resistance to this virus has also evolved (Elsworth et al. 2012).

Many species have rapidly evolved resistance to biocontrol agents (insecticides, pesticides, antibiotics, etc.) (Georghiou 1986). For example, hundreds of insect species have evolved resistance to insecticides (McKenzie 1996), while rats (*Rattus norvegicus* and *R. rattus*) and mice have evolved resistance to warfarin and other anti-coagulant rodenticides (Pelz et al. 2005). Further, almost 200 species of plants have evolved resistance to herbicides (Heap 2007; Briggs & Walters 2016).

Adaptive evolutionary changes as a result of global climate change have already been observed in invertebrates, vertebrates, and plants (see Chapter 14).

How large are adaptive evolutionary changes?

Many cases of large evolutionary changes have been documented in the fossil record and in contemporary studies of adaptation in response to environmental change

Evolution may result in changes that are many-fold in magnitude, as revealed by the fossil record, contemporary data (previous section), and directional selection studies (Futuyma 1998; Frankham et al. 2010). In the fossil record, the height of the molar tooth in the horse lineage increased more than 10-fold (~ 4.7 mm to 52.5 mm) over about 40 million years (Manly 1985; Futuyma 1998). Brain size increased more than three-fold in the human lineage (~ 400 cm^3 to 1,400 cm^3) in 3 million years (Futuyma 1998), while the Haast eagle of New Zealand (*Harpagornis moorei*) increased 10-fold in size in about one million years (Bunce et al. 2005).

Rates and magnitudes of change wrought by artificial selection are typically far greater than usual in evolutionary history (Futuyma 1998).

Evolutionary changes due to artificial selection

Directional selection over many generations in large populations may produce extremely large genetic changes

Large changes in phenotype over time have been observed in many populations subjected to directional selection, especially in laboratory and domestic species (Fig. 4.1). For example, flying speed in *Drosophila* increased 85-fold over 100 generations of directional selection (Fig. 4.1b), while oil percentage in maize seed increased ~ four-fold over 80 generations (Fig. 4.1a).

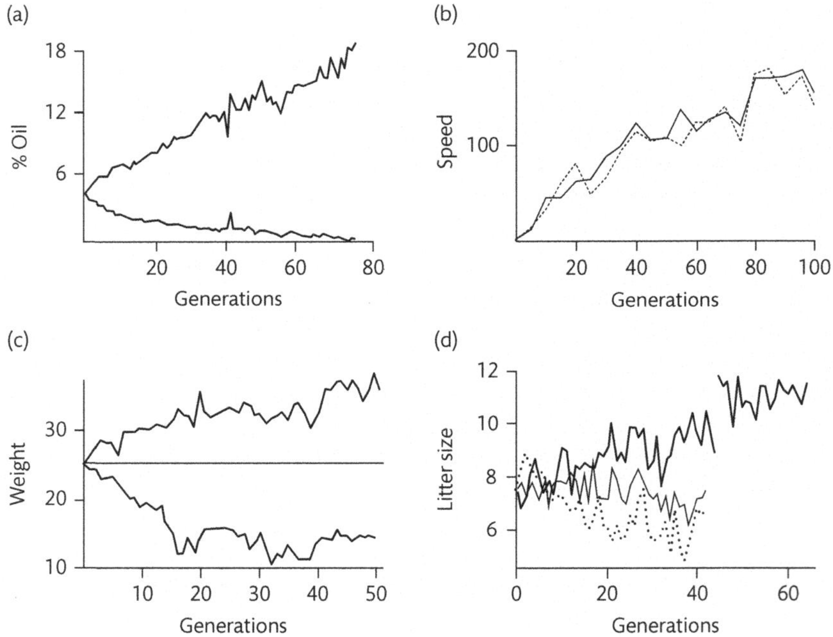

Fig. 4.1 Long-term response to directional selection. (a) Selection for high and low oil-content in maize seeds (after Dudley 1977). (b) Selection for increased flying speed in replicate populations of *Drosophila* fruit flies (Weber 1996). (c) Selection for high and low six-week body weight in mice (after Roberts 1966). (d) Selection for litter size in mice (after Falconer 1977); the bold solid line shows the impacts of selection for increased litter size, the dotted line selection for decreased litter size, while the fine solid line is an unselected control population (Frankham et al. 2010, Fig. 6.7).

To predict the ability of populations to adapt to environmental change, we need to understand the variables determining the rate and magnitude of change over time scales of a few generations.

What determines the ability to undergo adaptive evolution?

To evolve, species need genetic diversity, reproductive excess, and a selective force

Naturally outbreeding species with large populations normally possess genetic diversity among individuals (Chapter 2), allowing evolutionary adaptation through natural selection (Hoffmann & Sgrò 2011). For example, over 200 species of moth have evolved black body colors (industrial melanism) in response to industrial pollution (Box 4.2). The evolution of industrial melanism illustrates several of the issues discussed in this chapter.

Box 4.2 Evolution of industrial melanism in moths

(Haldane 1924; Kettlewell 1973; Majerus 1998; Cook et al. 2005; de Roode 2007; van't Hof et al. 2011, 2016; after Frankham et al. 2010)

(From Plate 8.2 from The Evolution of Melanism by Kettlewell (1973). By permission of Oxford University Press)

The best studied example of industrial melanism is the peppered moth (*Biston betularia*) in Great Britain. Prior to the industrial revolution, the peppered moth was well camouflaged as it rested on speckled lichen-covered trees (right-hand photograph above). However, when sulfur pollution killed most lichen and soot darkened trees, the speckled moth (typical) became clearly visible (left-hand photograph). The previously rare dark variants (melanics) were better camouflaged on the blackened trees and suffered less predation, leading to higher frequencies of the melanic allele in the industrial areas, than in relatively unpolluted areas.

The melanic form of the peppered moth was first recorded in 1848, but by 1900 they represented about 99% of all moths in the polluted Midlands of England. The melanic form remained at low frequencies in unpolluted areas as indicated by the frequencies on the map (insularia is a less dark form than the melanics). There were gradations in frequencies of melanics from polluted to unpolluted areas (clines) due to differential selection across the gradient and migration of moths between habitats.

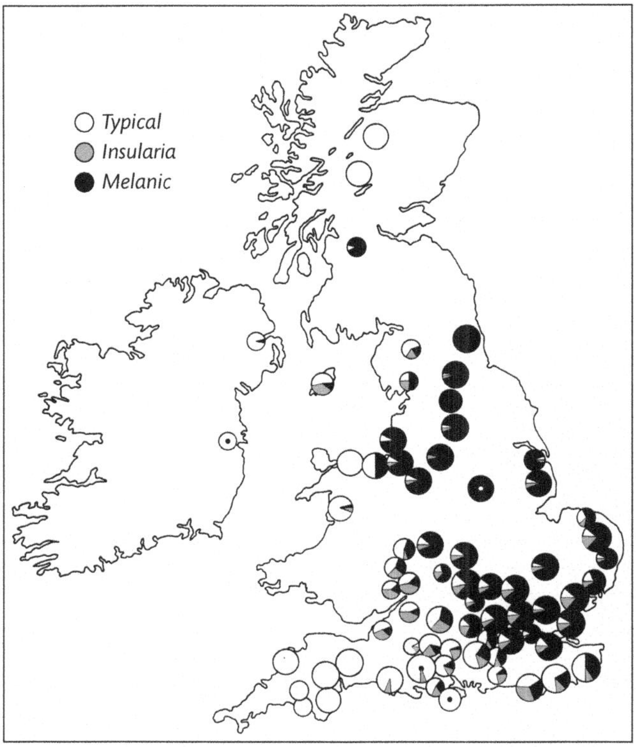

(Frankham et al. 2010, p. 125, after Kettlewell 1973. Reprinted by permission from Macmillan Publishers Ltd: Kettlewell, H. B. D. (1958). A survey of the frequencies of *Biston betularia* (L.) (Lep.) and its melanic forms in Great Britain. Heredity, 12(1), 51–72., copyright 1958.)

The melanic form is due to a dominant allele at a single locus. Recent DNA sequencing studies have shown that melanic peppered moths all arose from a single mutation around 1819 that has been subjected to recent strong selection (van't Hof et al. 2011, 2016). Based on modeling of the rise in frequency of melanics, Haldane (1924) estimated that they had a selective advantage of 32%, and empirical studies by Kettlewell (1956) and Majerus (de Roode 2007) revealed a large selective advantage, as shown in the table. Kettlewell captured melanic and typical moths, marked and released them, and recaptured them shortly afterwards.

Recapture rates were approximately twice as high for melanics as for typicals in the polluted area, yielding a selection coefficient of ~ 50%. Selection in the unpolluted area was of a similar magnitude, but in the opposite direction. Selection was somewhat higher than predicted by Haldane, but this was attributed to the use of a very heavily polluted area for the experiment.

	Polluted area		**Unpolluted**	
	Melanic	**Typical**	**Melanic**	**Typical**
Numbers marked and released	447	137	473	496
Released moths recaptured	27.5%	13.1%	6.34%	12.5%

The frequencies of melanic moths have subsequently declined following pollution control legislation that led to return of lichens (Grant & Wiseman 2002; Cook et al. 2005).

Most adaptive evolution is for fitness, a quantitative character (and this typically involves alleles at many quantitative trait loci [QTL]), rather than a single locus as was the case for industrial melanism.

Adaptive evolution for fitness

The extent of adaptive evolution is determined by selection, genetic diversity for fitness, mutation rates, effective population size, and generations

Cumulative genetic adaptation (ΣGA) over multiple generations (t) is predicted by the following equation:

$$\sum GA_{\mathrm{t}} = Sd \times h^2 \sum_{i=1}^{t} (1 - \frac{1}{2N_e})^{i-1} \qquad 4.1$$

where Sd = selection differential (difference in traits mean between parents and non-parents), and h^2 = heritability, the proportion of phenotypic variance for the character that is due to additive genetic effects (related to genetic diversity for the character: Falconer & Mackay 1996). The summation term on the right-hand side reflects erosion of genetic diversity each generation due to genetic drift (see eqn 4.3). As this standing variation is exhausted, the rate of response slows and reaches a plateau (Gilligan et al. 2003). The total response to selection in the long term, derived from genetic variation in the initial population, is predicted to be ~ $2N_e\ S\ h^2$, and to take ~ 1.4 N_e generations to get halfway to the selection limit for additive loci and up to $2N_e$ generations if recessive alleles are being favored by selection (Robertson 1960; Falconer & Mackay 1996).

For evolutionary change due to new mutations, the asymptotic rate of genetic adaptation per generation after mutation and drift reach equilibrium is predicted to be (Hill 1982):

$$GA_{\text{mutation}} \approx \frac{2N_e S d V_m}{V_P} \quad 4.2$$

where V_m is the increase per generation in additive genetic variation due to mutation (mutational variance) and V_P is the phenotypic variance. This provides an approximate prediction of response (Frankham 1983; Mackay et al. 1994). Importantly, both prediction equations show that evolutionary potential is higher in populations with large than small N_e.

In experimental studies in the laboratory and field, selection response increases with population size, as predicted by these equations (Fig. 4.2). Merilä et al. (2001) claimed that such predictions do not apply in the wild (but many of the characters they applied it to are size measures known or suspected to be subject to stabilizing selection). Conversely, Hoffmann et al. (2016) reported positive relationships between heritability and evolveability across more than 100 traits in farmed cattle. Further, Grant & Grant (1995) found a close correspondence between observed and predicted selection response for beak characters in Galápagos finches in the wild (*Geospiza fortis*). Notably, studies where populations have moved to new environments (our context) do report rapid selection response in the wild (Losos et al. 1997; Reznick et al. 1997; Huey et al. 2000).

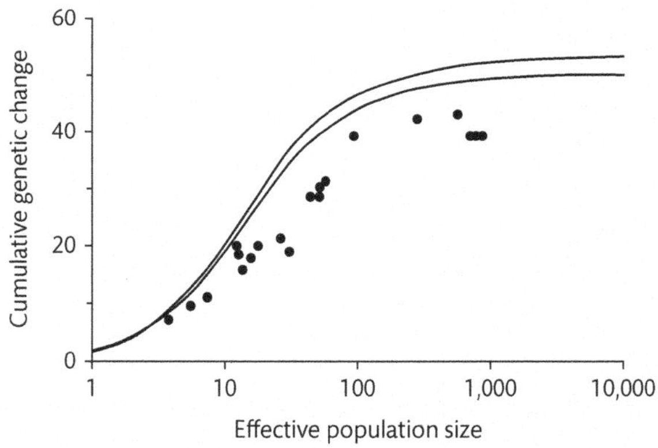

Fig. 4.2 Long-term genetic change due to directional selection increases with effective population size. Cumulative genetic change (mean at generation 50, divided by that in the first generation) for selection plotted against N_e for a diversity of characters in mice, *Drosophila*, red flour beetles, and maize. The asymptotic curves represent 50 generation predictions based on eqn 4.1 (lower curve) and with mutation added (higher curve) (Frankham et al. 2010, Fig. 15.1, after Weber 2004).

We now deal in more detail with each of the components of adaptive evolution, namely loss of genetic diversity over generations, heritability, and selection.

Loss of genetic diversity in small populations

Loss of genetic diversity is unavoidable in small closed populations

All closed finite populations lose genetic diversity due to chance sampling of alleles during reproduction, with the rate being faster in smaller than larger populations (Chapter 2). Over a single generation in a random mating population of size N_e, the loss of heterozygosity is $1/(2N_e)$. For example, a closed population of size $N_e = 50$ loses 1% of its genetic diversity in a single generation. Over several generations, loss of genetic diversity for neutral loci is described by the following equation:

$$\frac{H_t}{H_0} = \left(1 - \frac{1}{2N_e}\right)^t \qquad 4.3$$

where H_t is heterozygosity (Hardy–Weinberg expected heterozygosity) at generation t, H_0 initial heterozygosity, and N_e the genetically effective population size. This equation predicts an exponential decay of genetic diversity with generations, as the right-hand part in the equation is approximately $e^{-t/2N_e}$ (Fig. 4.3).

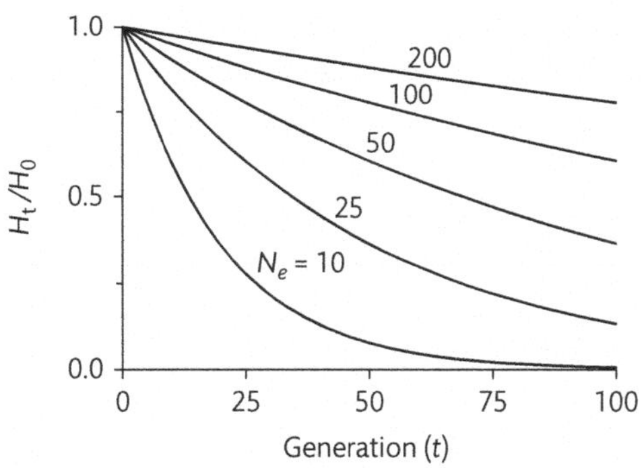

Fig. 4.3 Predicted decline in heterozygosity over generations in different sized populations (Frankham et al. 2010, Fig. 11.1, after Foose 1986).

Observed declines in microsatellites and allozymes in experimental populations of *Drosophila* over 50 generations were close to expectations (Fig. 4.4), while quantitative genetic variation declined at a slower rate (Montgomery et al. 2000, 2010; Gilligan et al. 2005). Further, positive correlations between genetic diversity and population size (as predicted by this theory) have been observed in studies across a variety of taxa for many measures of genetic diversity, both near-neutral and ones subject to selection (e.g. MHC variation) (Frankham 2012).

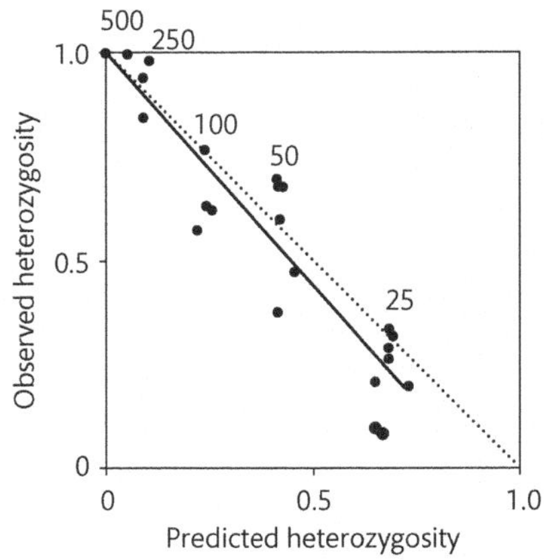

Fig. 4.4 Microsatellite genetic diversity is lost at approximately the rate predicted by neutral theory. Comparisons of observed and predicted proportions of initial microsatellite heterozygosities retained after 48 generations in experimental *Drosophila* populations of different effective sizes (numbers above the dots) plotted against predicted heterozygosity (Frankham et al. 2010, Fig. 11.2, after Montgomery et al. 2010). The solid line indicates the regression line of best fit and the dotted line equality of observations and predictions.

The ability to evolve depends on genetic diversity for fitness, a quantitative character. This genetic diversity is characterized by the heritability.

Heritability

Genetic diversity affects the ability to evolve via the heritability, the proportion of phenotypic variation (V_P) for the trait (fitness in our case) determined by additive genetic effects (V_A), namely:

$$h^2 = \frac{V_A}{V_P} \qquad 4.4$$

The heritability has a range of 0–1, because it is a proportion. It is a reflection of genetic diversity as V_A is a function of heterozygosity for loci affecting the quantitative traits, as follows (Falconer & Mackay 1996):

$$V_A = \sum 2pq[a + d(q - p)]^2 \qquad 4.5$$

where a reflects the average effect of alleles and d their dominance, as defined in Chapter 3.

Heritabilities are lower in more variable environments, other things being equal. For example, Geber & Griffen (2003) found heritabilities in plants averaged 0.42, 0.22, and 0.12 in greenhouse, outdoor common garden, and field, respectively—the pattern expected if environmental variability is greater in more natural environments.

We discuss means for estimating heritabilities later in this chapter.

How large are heritabilities?

Heritabilities are typically lower for fitness than peripheral traits, but are similar across major animal taxa

Heritability estimates are similar among birds, mammals, fish, and reptiles (Postma 2014), and between plants and animals in outbreeding species (Geber & Griffen 2003). Empirical studies of heritabilites in wild environments consistently report lower heritabilities for fitness (life history) traits than for traits peripherally related to fitness (e.g. morphology, behavior, and physiology) (Postma 2014; Mittell et al. 2015; Wood et al. 2016). For example, heritability estimates for life history and morphology averaged 0.15 and 0.32, respectively, using the most reliable methods (the animal model, described later) (Wood et al. 2016). Notably, studies that estimated heritabilities for total fitness (lifetime reproductive success) in mammals and birds reported values close to zero (see Gustafsson 1986; Kruuk et al. 2000; McCleery et al. 2004; Brommer et al. 2007; Foerster et al. 2007; McFarlane et al. 2014).

There are several hypotheses as to why heritability estimates for lifetime fitness are so low, namely:

- exhaustion of genetic diversity for fitness due to selection
- high environmental variance
- conflict between the sexes in effects of loci
- trade-offs in fitness effects of alleles on different fitness components
- high non-additive genetic variation (dominance and interactions between loci [epistasis]).

All of these factors are likely involved.

How can evolution occur if there is no heritability for fitness?

Heritabilities cannot be extrapolated from one environment to another

When a population moves from the environment to which it is adapted, the effects of conditional alleles will often change (genotype $\times$ environment interactions), with some previously harmful alleles becoming beneficial, some previously neutral alleles now having effects, etc. In general, we expect this to increase V_A, but a meta-analysis indicates little change for fitness characters (Charmentier & Garant 2005). However, the effect on heritability also depends on V_P in the two environments, and this will often be higher in the wild, thus reducing the heritability (Simons & Roff 1994).

Despite presumed low heritabilities of fitness in the environments to which they were adapted, many populations have shown rapid responses to selection in novel environments, often for fitness (Briscoe et al. 1992; Frankham & Loebel 1992; Losos et al. 1997; Reznick et al. 1997; Frankham et al. 1999, 2010; Huey et al. 2000; England et al. 2003; Reed et al. 2003b).

Contributions to adaptation from pre-existing genetic diversity and new mutations

Selection response in the short term operates primarily on pre-existing genetic diversity, rather than on new mutations, as mutation is a rare event

Adaptive evolutionary change may utilize either pre-existing genetic diversity, or that arising due to mutation. However, most mutations are harmful when populations are well adapted to their current environment, such that the frequency of beneficial mutations is often low (Chapter 2). Consequently, theory and empirical results indicate that new mutations will generally make little contribution to evolutionary adaptation for about the first 20 generations, but become increasingly important subsequently as mutations accumulate and standing variation is depleted (Robertson 1978; Hill 1982; Bradshaw 1991; Hartley et al. 2006). However, where environmental change is large, a higher proportion of mutations are expected to be beneficial, such that mutational contributions in large populations may occur sooner than they do with no environmental change (Remold & Lenski 2001; Tenaillon 2014; Stearns & Fenster 2016).

Inability to adapt in some populations

Some populations are unable to adapt to new environments, either due to lack of relevant genetic diversity, weak selection, rapid environmental changes, small N_e, or combinations of these

The inability to adapt to altered environments is strongly associated with small isolated populations where loss of genetic diversity is high and selection efficiency low, because drift effects overwhelm selection. For example, small plant populations are much less likely to show adaptive differentiation (home site advantage) than large ones (Leimu & Fischer 2008). Adaptation is more likely with greater habitat differences, an indicator of the effects of selection intensity (Hereford 2009a). However, lack of ability to adapt to thermal stress was found in two rainforest species of *Drosophila*, despite normal (high) levels of microsatellite variation and ability to evolve in response to other stresses (Hoffmann et al. 2003; Kellerman et al. 2006; however, see van Heerwaarden & Sgrò 2014). In these cases, natural selection appears to have exhausted genetic diversity for thermal stress, but not for other traits. Etterson & Shaw (2001) concluded that partridge pea (*Chamaecrista fasciculata*) populations had limited capacity to evolve in response to climate change, but this was due to unfavorable genetic correlations among different fitness components, rather than lack of genetic diversity.

Reduced evolutionary potential in endangered species

Threatened species typically have substantially compromised ability to evolve in response to environmental change, due to low effective population sizes (N_e), low genetic diversity, and reduced reproductive rates

Many threatened species will likely have lowered heritabilities for fitness characters, as threatened species have on average 25–35% lower genetic diversity than taxonomically related non-threatened taxa (Frankham 2012). This discussion would be most appropriately based on heritabilities in threatened and non-threatened species, but there are insufficient data to compare them, so we rely on information from multiple near neutral molecular markers (Harrisson et al. 2014).

While a meta-analysis by Reed & Frankham (2001) reported a non-significant relationship between heterozygosity and heritability for life history traits, this may be due to the small number of markers and the high sampling variances in most of those studies. Conversely, heritabilities have been shown to decline in small populations and to be related to selection response and molecular genetic diversity, based on comparison involving isolated populations within species under controlled environmental conditions (Frankham et al. 1968; Briscoe et al. 1992; Frankham et al. 1999; Reed et al. 2003b; Gilligan et al. 2005). Many threatened species probably have reduced reproductive fitness due to inbreeding depression, compared to taxonomically related, non-threatened species (Spielman et al. 2004a). This will often reduce the selection differential in threatened species, because with fewer progeny produced per generation, the intensity of selection is reduced (Box 4.3).

We provide a hypothetical example of the potential impacts of the above factors on evolutionary potential for the small Isle Royale population of gray wolves and predict that they have ~ 44% poorer ability to evolve over five generations than the large mainland population (Box 4.3). We are not aware of any comparative empirical data on the ability of threatened and non-threatened populations or species to evolve.

Box 4.3 Potential impact of small population size on evolutionary potential in Isle Royale gray wolves

(after Frankham et al. 2010)

We illustrate the hypothetical impact of restricted population size on evolutionary potential, using the Isle Royale gray wolf (Wayne et al. 1991; Adams et al. 2011). This population on Isle Royale in Lake Superior was founded by one female and two males (Adams et al. 2011). Consequently, heterozygosity of the island population will initially be ~ $(1 - 1 / [2N_e]) = (1 - 1 / [2 \times 2.67]) = 81\%$ of that on the mainland. Since heritability is approximately proportional to heterozygosity (Falconer & Mackay 1996), this bottleneck at foundation reduces evolutionary potential by ~ 19% through its impact on h^2.

The selection differential (*Sd*) will be reduced in the Isle Royale population due to inbreeding depression. With an inbreeding coefficient of 0.25, the reduction in lifetime production of offspring reaching sexual maturity is likely to be about 50% (Frankel & Soulé 1981). Consequently, we assume that females in the mainland population have 30 pups over a 5-year reproductive lifespan and the island population females only 15 pups. If the population size is stable, then one pair of pups from each pair contributes to the next generation. This represents 6.7% of the mainland population breeding and 13.3% of the island population, giving *Sd* values of $1.9\sigma_P$ and $1.6\sigma_P$, respectively, where σ_P is the phenotypic standard deviation for litter size (using tables in Becker 1984; Falconer & Mackay 1996). Thus, the island population would have a 16% lower selection differential than the mainland population (provided the environmental variances are the same in the two populations).

For the purposes of this exercise, we assume that the Isle Royale wolf population has been closed since foundation (an assumption recently shown to be untrue: Hedrick et al. 2014). If the mainland wolf population consists of 5,000 breeding individuals per generation and the Isle Royale population 25, then, with a conservative N_e/N ratio of 0.2, the effective sizes are 1,000 and 5. Consequently, the loss of genetic diversity for the mainland population will be very small [1/(2,000) per generation], and, after 20 generations, only 1% of the initial heterozygosity will be lost. However, with N_e ~ 5 on the island, ~ 10% of the existing genetic variation will be lost each generation. From eqn 4.1, the genetic adaptation in the Isle Royale population over the first five generations in units of heritability times the phenotypic standard deviation, will be given by the terms with generations (t) of 1, 2, 3, 4, and 5 (given the short duration, we can ignore contributions from mutations):

$$GA_1 \approx 1.6 \times 0.81 \times h^2 \times \sigma_P \approx 1.30\, h^2 \sigma_P$$

$$GA_2 \approx 1.6 \times 0.81 \times h^2 (1 - \frac{1}{2 \times 5}) \times \sigma_P \approx 1.17\, h^2 \sigma_P$$

Similarly GA_3, GA_4, and GA_5 will be ~ 1.05 $h^2\sigma_P$, 0.95 $h^2\sigma_P$, and 0.85 $h^2\sigma_P$, respectively. Thus, the cumulative genetic adaptation over the five generations will be ~ 5.32 $h^2\sigma_P$ in the Isle Royale gray wolf, while that for the large mainland population, calculated as above, will be ~ 9.5 $h^2\sigma_P$. Consequently, genetic adaptation in the Isle Royale population will be only ~ 5.32/9.5 = 56% that of the mainland population.

The arguments above emphasize the importance of expanding the population sizes of endangered species to minimize inbreeding and loss of genetic variation to improve their ability to evolve in response to environmental changes.

In the preceding discussion, we emphasized the impact of loss of genetic diversity on the ability to evolve. However, for some loci there are direct effects of loss of genetic diversity on fitness, including for the major histocompatibility complex (next), self-incompatibility loci in plants, and the sex locus in Hymenoptera (Chapter 8).

Reduced genetic diversity decreases resistance to diseases, pests, and parasites

Populations with reduced genetic diversity are expected to suffer more seriously from pests, parasites, and infectious diseases than those with high genetic diversity

Loss of genetic diversity for disease, pest, and parasite resistance severely diminishes the capacity of populations to cope with these threats (Hughes & Boomsa 2004; Spielman et al. 2004b; Whiteman et al. 2006; Frankham et al. 2010). Following sequential assaults by different pathogens, populations with high genetic diversity are more likely to persist than populations with low genetic diversity (Fig. 4.5; Penn et al. 2002).

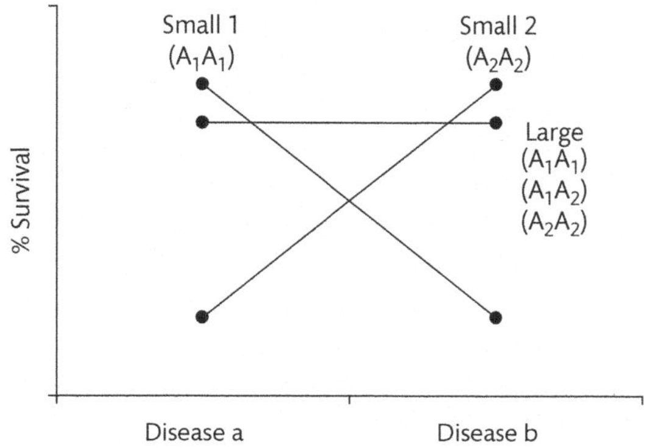

Fig. 4.5 Hypothetical example of the relationship between genetic diversity and disease resistance (Frankham et al. 2010, Fig. 2.5). Two small inbred populations, each homozygous for different alleles are resistant to one pathogen, but susceptible to the other. Conversely, a larger (or crossed) population containing both alleles can resist both pathogens and has better overall survival than the inbreds in the presence of both pathogens. Partial dominance of resistance is assumed.

The MHC in vertebrates is a large cluster of loci involved in recognizing pathogen antigen molecules and regulating immune responses. The more heterozygous an individual is, the more pathogens it can respond against. Genetic diversity at these loci is amongst the highest known for any vertebrate loci, and is maintained by selection favoring heterozygotes, by selection favoring rarer alleles, and perhaps by selection to avoid inbreeding (Garrigan & Hedrick 2003; Sommer 2005; Piertney & Oliver 2006; Sutton et al. 2011). For example, higher heterozygosity at three components of the MHC is associated with longer survival following HIV infection in humans (Carrington et al. 1999). Further, higher heterozygosity at the MHC in populations of lowland leopard frogs (*Lithobates yavapaiensis*) is associated with higher survival from chytrid fungus (*Batrachochytrium dendrobatidis*) infections, and one allele is found only in resistant populations (Savage & Zamudio 2011). There are also other loci affecting disease resistance located outside the MHC (Hill 1998). Despite this selection, alleles for disease resistance are lost by genetic drift in small populations, increasing the chance that a pathogen that can kill one individual can kill all individuals (Sommer 2005).

Having dealt with genetic diversity and heritability, we now turn to the various forms of selection on quantitative traits.

Forms of selection on quantitative characters

Natural selection may favor individuals diverging in one direction, both directions, phenotypic intermediates, or different phenotypes in diverse environments

There are three basic forms of selection that operate on quantitative characters—directional, stabilizing, and disruptive (Fig. 4.6). Directional selection favors phenotypes towards one end of the distribution and results in an evolutionary shift in the mean in this direction (provided there is genetic variation). This is similar to the consequences of directional selection on single loci (Chapter 2). Stabilizing selection favors phenotypic intermediates and results in no change in the mean, but usually results in reduced genetic variation for the trait (Falconer & Mackay 1996). This contrasts with balancing selection on single loci, which generally impedes loss of genetic diversity (Chapter 2). Disruptive (or diversifying) selection favors both phenotypic extremes and does not alter the overall mean, but may lead to increased genetic variation for the trait. If the two phenotypic extremes do not mate with each other, genetically differentiated populations with differential adaptation will result (see later and Chapters 5 and 7), and distinct species may evolve over many generations (Chapter 9).

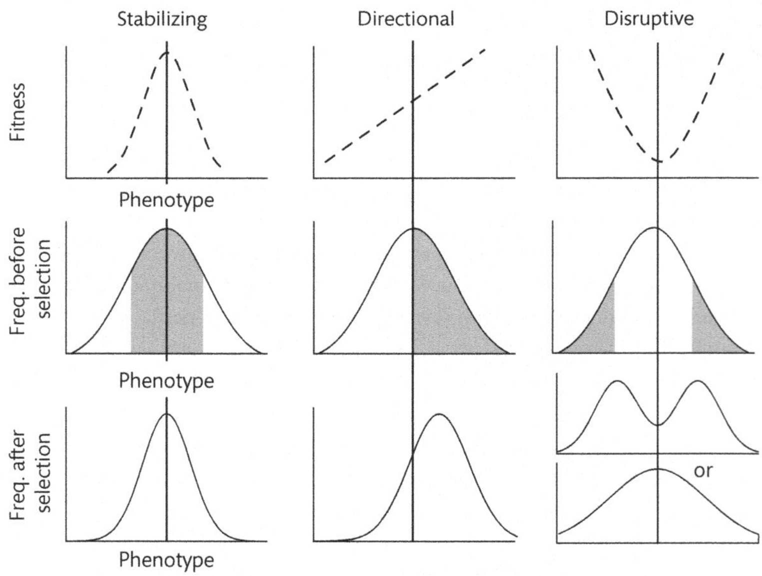

Fig. 4.6 Stabilizing, directional, and disruptive selection operating on quantitative characters (Frankham et al. 2010, Fig. 6.6, after Futuyma 1979). The upper panels indicate the relationships between phenotype and fitness, the middle panels the phenotypic distributions before selection (the shaded areas indicate the portions of the distribution favored by selection), and the bottom panels the distributions after selection in the subsequent generation.

In a constant and relatively uniform environment, reproductive fitness is subject to directional selection (Hoekstra et al. 2001). Conversely, characters more peripheral to reproduction and survival are presumed to experience stabilizing selection and to be showing little evolutionary change (Schluter 1988; Kingsolver & Pfennig 2007).

Different selective forces in different locations with no gene flow between them result in genetically differentiated populations

So far we have dealt with consistent selective forces within a species. However, selection often differs between localities, especially if they have dissimilar environments. When there is no gene flow between localities, such disruptive selection leads to different genetic adaptations, provided the population contains genetic diversity for fitness (Thompson 2013; Chapters 7, 9, 10, and 14). Gene flow between population fragments reduces the extent of adaptive differentiation.

Where there is a selection gradient, as with temperature at different altitudes or latitudes, or day length with latitude, selection and migration may lead to a gradation in genetic adaptation across the landscape (termed a cline: Endler 1977).

Migration–selection balance and clines

Differential selection across the landscape and gene flow often result in clines

Clines can form where there is a balance between selection favoring different alleles in diverse habitats (local adaptation) and migration occurring between the habitats (gene flow). For example, there are clines in leaf waxiness (glaucousness) with elevation in several species of eucalypt trees in Tasmania, Australia, maintained by a balance between migration and selection (Box 4.4). Clines due to migration–selection balance have been found for many characters, including heavy metal tolerance in colonial bent grass plants between mine sites polluted with heavy metals and nearby less polluted pastures, for industrial melanism in peppered moths across gradients from polluted to unpolluted areas (Bishop & Cook 1975; Box 4.1), and for alleles at several allozyme loci (Powers et al. 1991). Clines in morphological characters are relatively common, some, such as increasing body size with increasing latitude, being so pervasive that they are referred to as ecogeographic rules (Futuyma 1998).

Box 4.4 Clines in leaf glaucousness in several species of eucalypt trees in Tasmania, Australia due to differential selection according to altitude, balanced by pollen flow

(Barber 1955; Barber & Jackson 1957; Thomas & Barber 1974; after Frankham et al. 2010)

Some eucalypt trees have leaves with a distinct waxy (glaucous) layer on them. At least eight species of these gum trees in Tasmania have parallel clines of glaucousness, with greater frequencies of waxy leaves at higher, frosty, altitudes. Further, some species show similar clines in different locations.

Eucalypt trees (Australia)

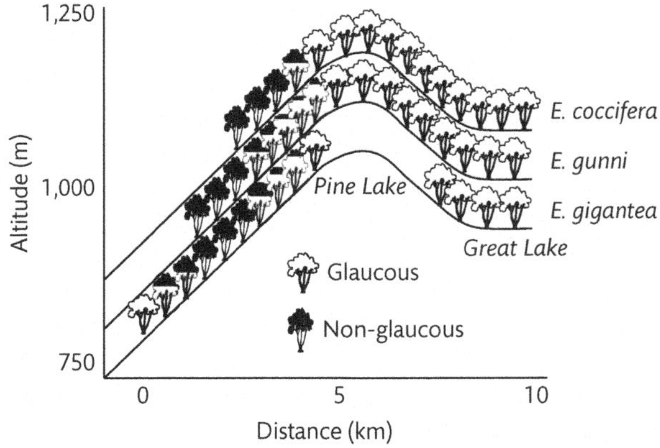

Glaucous leaves have greater survival in heavy frosts. At lower elevations, glaucous plants suffer greater defoliation through insect attacks (next figure). Pollen flow among populations results in mixing of alleles from different elevations, while selection operating between seed and adult stages of the life cycle re-establishes differences among elevations, as shown for the urn gum (*Eucalyptus urnigera*) in the following graph (after Barber 1955).

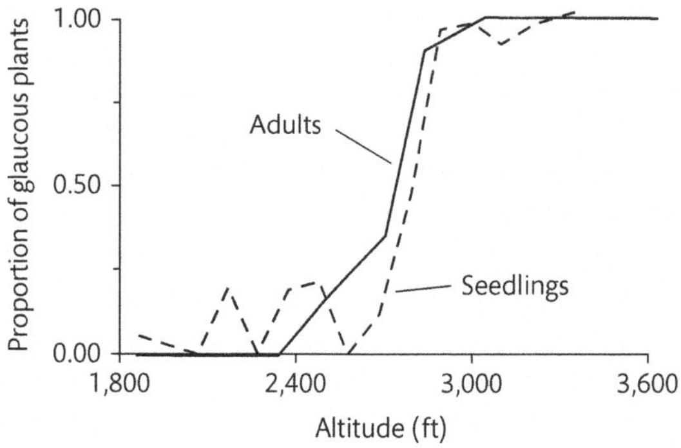

(both figures from Frankham et al. 2010, p. 198)

How can we measure evolutionary potential?

Evolutionary potential can be measured for quantitative traits using the heritability of a particular trait in a particular environment, or indirectly via genomic level measures of overall genetic diversity

Evolutionary potential can be estimated directly as the heritability of the traits of interest in the intended environment. Until recently, this was done using offspring–parent regressions or half-sib correlations for quantitative characters (Falconer & Mackay 1996). These are being replaced by animal model estimates, mixed-model regressions using maximum likelihood procedures on multigenerational relationship data, as these deal better with small samples, unbalanced designs, and complexities generated by environmental heterogeneity (see Kruuk 2004; Kruuk & Hadfield 2007; Wilson et al. 2010; de Villemereuil et al. 2013). However, these approaches are time consuming and only worthwhile with large numbers of offspring (a few hundred at least). Where we wish to determine the heritability for a population in its current environment, the animal model is the recommended method.

This will be of little use where we wish to know the heritability for another environment, as we cannot extrapolate heritabilities across environments. To estimate the heritability of fitness for a forthcoming environment, we need to carry out the assessment in that environment (e.g. higher temperature or increased ocean acidity).

An alternative approach is to measure molecular genetic diversity (SNPs, microsatellites, or allozymes), but this has yielded non-significant correlations with heritabilities for fitness traits, possibly due to the small numbers of marker loci assessed and the high sampling variation on the heritability estimates (Reed & Frankham 2001; Mittell et al. 2015; Wood et al. 2016). Studies using genome-wide genetic diversity among individuals reduce sampling variation for loci (Harrisson et al. 2014). If we cannot estimate heritability in a forthcoming environment of interest, these are the best guide to the heritability in another environment. Much of the variation probably has little functional significance, but it might be representative of variation at loci affecting traits of interest. It may be possible to improve upon this by restricting measures of genetic diversity to functional sequences, but this remains to be determined.

Can we restore the ability to evolve?

Augmenting gene flow between populations can often improve the ability of genetically depleted populations to evolve

Low genetic diversity and compromised ability to evolve in small fragmented populations can typically be reversed by crossing between unrelated populations, as detailed in Chapters 6 and 11.

Summary

1. Environmental change is a ubiquitous feature of the conditions faced by species, so they must evolve, move to avoid threats, or perish.
2. Species require genetic diversity to evolve to cope with environmental change through natural selection (adaptive evolution).

3. The ability of populations to undergo adaptive evolution depends upon the strength of selection, genetic diversity, effective population size, mutation rates, and number of generations.
4. Loss of genetic diversity in small populations reduces their ability to evolve to cope with environmental change, thus increasing their extinction risk.
5. Adaptive evolution in the short to medium term predominantly utilizes pre-existing genetic diversity, but new mutations make increasing contributions in later generations.
6. Evolutionary potential can be estimated from the heritability of fitness in the environment of interest, or by extrapolation from genomic diversity.

FURTHER READING

Anderson et al. (2011) Brief review of the evolutionary genetics of plant adaptation.

de Villemereuil et al. (2013) A comparison of parent–offspring regression and animal model estimates of heritability using simulations.

Falconer & Mackay (1996) *Introduction to Quantitative Genetics*: Provides a very clear treatment of the topics in this chapter.

Mousseau et al. (2000) *Adaptive Genetic Variation in the Wild*: Reviews on natural selection and adaptation in the wild.

Savolainen et al. (2013) Reviews the ecological genomics of local adaptation.

Thompson (2013) *Relentless Evolution*: Excellent recent review of evolution, documenting its pervasive nature and its frequent rapidity.

Wilson et al. (2010) Provides a simple guide to the animal model for estimating heritability in the wild.

SOFTWARE

MCMCglmm: MCMC software for estimating heritability using the animal model (Hadfield 2010). http://cran.r-project.org/web/packages/MCMCglmm/index.html

WOMBAT: REML software for estimating heritability using the animal model (Meyer 2007). http://didgeridoo.une.edu.au/km/wombat.php

CHAPTER 5

Population fragmentation causes inadequate gene flow and increases extinction risk

Most species now have fragmented distributions, often with adverse genetic consequences. The genetic impacts of population fragmentation depend critically upon gene flow among fragments and their effective sizes. Fragmentation with cessation of gene flow is highly harmful in the long term, leading to greater inbreeding, increased loss of genetic diversity, decreased likelihood of evolutionary adaptation, and elevated extinction risk, when compared to a single population of the same total size. The consequences of fragmentation with limited gene flow typically lie between those for a large population with random mating and isolated population fragments with no gene flow.

TERMS

Bayesian, coancestry, dioecious, effectively neutral, effective number of alleles, F_{ST}, F statistics, genetic distance, isolation by distance, matrix, metapopulation, single large or several small (SLOSS), source–sink, Wahlund effect

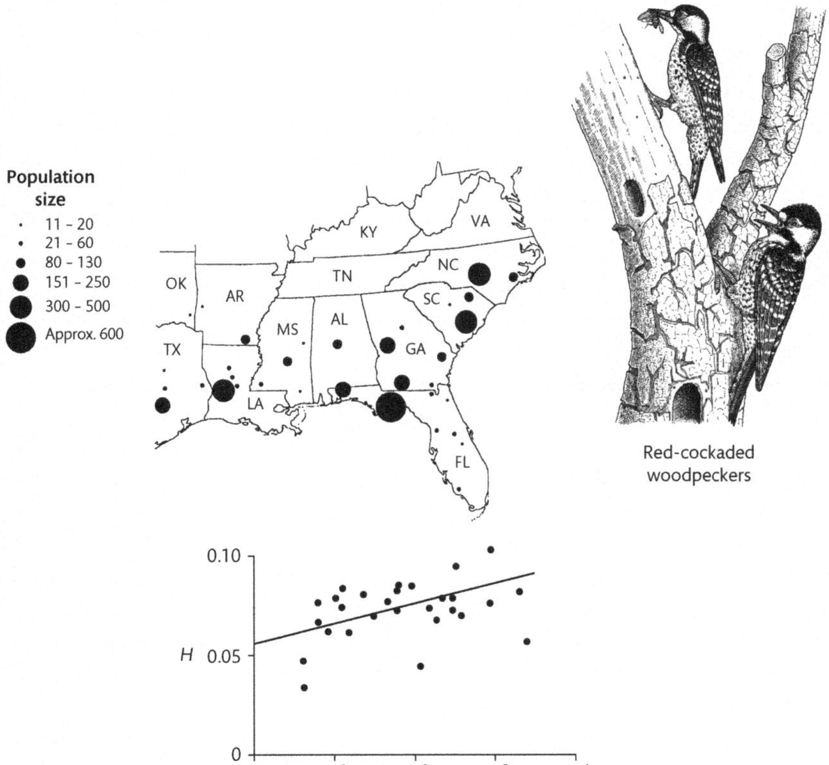

Population fragmentation in the red-cockaded woodpecker in southeastern USA (Kulhavy et al. 1995). The species had an essentially continuous distribution prior to European settlement, but is now highly fragmented (map after James 1995). Genetic diversity (H) is on average less in smaller than larger populations (N), as illustrated in the graph (Frankham et al. 2010, p. 311, after Meffe & Carroll 1997).

Genetic Management of Fragmented Animal and Plant Populations. R. Frankham, J. D. Ballou, K. Ralls, M. D. B. Eldridge, M. R. Dudash, C. B. Fenster, R. C. Lacy & P. Sunnucks. Oxford University Press (2017) © R. Frankham, J. D. Ballou, K. Ralls, M. D. B. Eldridge, M. R. Dudash, C. B. Fenster, R. C. Lacy & P. Sunnucks 2017. DOI 10.1093/oso/9780198783398.001.0001

Why is population fragmentation important in conservation?

Most species have fragmented distributions, many with adverse genetic effects on population persistence

In the previous two chapters, we addressed genetic problems in single small populations. We now turn to the issues facing fragmented populations, as most species consist of multiple population fragments in different habitat patches. For example, Hughes et al. (1997) estimated that the average species had ~ 220 genetically distinct populations (but the numbers will typically be much lower for threatened species). With ~ 8.7 million eukaryotic species (Mora et al. 2011), this implies that there are ~ 1.9 billion genetic populations on the planet (Hughes et al. 1997).

What do we mean by fragmentation? The most obvious characteristic is groups of individuals in physically isolated locations, rather than in a single location. However, the degree of genetic isolation among population fragments depends on the gene flow between them and ranges from totally isolated to effectively a single genetic population. Our concern in this book is genetically distinct population segments within species that are partially to completely isolated genetically, and that have distinct evolutionary fates. The isolating factors between population segments are typically inhospitable habitat, rivers, roads, oceans, and livestock fencing. Conservation fencing also isolates populations, as for large dangerous animals (lions, leopards, elephants, etc.) in South Africa, and to protect threatened marsupials in Australia from introduced predators (see Hayward & Somers 2009).

Human-induced habitat losses and fragmentation are recognized as one of the primary causes of biodiversity loss (UNEP 2007; IUCN 2016). Human impacts on habitat consist of (a) a reduction in total habitat area (and species population size) and (b) generation of separate isolated "island" patches from a larger continuous distribution, usually in a matrix of now-inhospitable terrain. These are evident for the highly speciose, endemic-rich Atlantic forest in São Paulo State, Brazil (Chapter 1 frontispiece), where habitat area for both plants and animals (including the endangered black lion tamarin [*Leontopithecus chrysopygus*]) has been greatly reduced and fragmented (Dietz et al. 1994). Similarly, the habitat of the red-cockaded woodpecker in the eastern coniferous forest of the USA has been reduced in area and fragmented (frontispiece of this chapter), and parallel examples occur throughout the world.

The harmful genetic effects on fitness, evolutionary potential, and extinction risk of reduction in habitat area and species' population sizes have already been addressed in Chapters 3 and 4. This chapter focuses on the genetic effects of population fragmentation per se, the separation of a population into partially or completely isolated fragments for the same total population size. This typically reduces gene flow across the landscape, sometimes to levels with adverse genetic consequences. Our primary focus is on recent ($\leq$ 500 years) anthropogenic habitat fragmentation of previously continuously distributed species, rather than naturally fragmented ones.

Fragmentation reduces gene flow

The adverse genetic effects of fragmentation on extinction risk arise from the associated reductions in gene flow

Fragmentation typically reduces gene flow between population fragments compared to continuous habitat, as for example in European beech trees (*Fagus sylvatica*: Fig. 5.1). Such fragmentation effects are widespread, but the scale over which they are evident depends upon distance between fragments (see later), the dispersal ability of species, and the suitability of the surrounding habitat matrix (Bohonak 1999; Ortego et al. 2015).

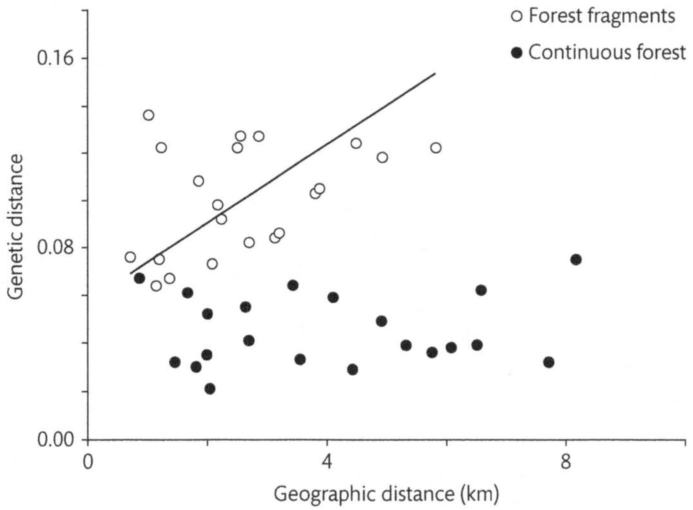

Fig. 5.1 Effects of fragmentation on genetic differentiation among populations of wind-pollinated European beech trees. Genetic differentiation increases with distance in fragmented, but not in continuous forest, because gene flow has been reduced by fragmentation (Jump & Peñuelas 2006).

Gene flow typically declines with distance

Dispersal rates between populations in animals and plants typically decrease with distance (Fig. 5.2). Since gene flow requires dispersal of individuals or gametes, genetic differentiation is often related to geographic distance (isolation by distance). This was predicted theoretically by Wright (1943) and Malécot (1955), and has been observed in many animal and plant species (Sharbel et al. 2000; Crispo & Hendry 2005; Jenkins et al. 2010).

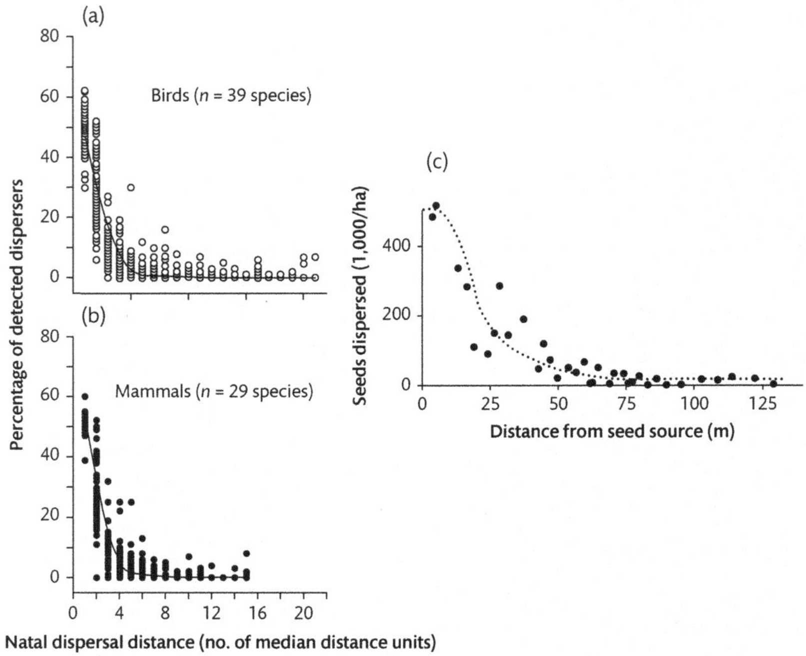

Fig. 5.2 Dispersal rates typically decline with distance in (a) birds, (b) mammals (both after Sutherland et al. 2000), and (c) a Eucalypt tree (from Cremer 1966, reprinted with permission from Taylor & Francis Ltd., from Cremer, K. W. (1966). Dissemination of seed from *Eucalyptus regnans*. *Australian Forestry*, 30(1), 33–37, http://www.informaworld.com) (Frankham et al. 2010, Fig.14.11).

The endangered red-cockaded woodpecker in the eastern USA illustrates many of the genetic features associated with habitat fragmentation for a species with a once continuous distribution, including loss of genetic diversity within population fragments, differentiation among populations, and isolation by distance (Box 5.1).

Box 5.1 Impact of habitat fragmentation on the endangered red-cockaded woodpecker in southeastern USA

(Stangel et al. 1992; Kulhavy et al. 1995; Daniels et al. 2000; Schiegg et al. 2006; after Frankham et al. 2010)

The red-cockaded woodpecker was once common in the near-continuous mature pine forests of southeastern USA, but declined in numbers, primarily due to habitat loss, and now survives in isolated sites (map on chapter frontispiece) with ~ 1% of its original population size. The woodpeckers require old-growth forest, especially for nesting cavities (Schiegg et al. 2006).

As there is little gene flow among isolated sites, populations have lost genetic diversity and diverged genetically (Haig & Avise 1996). Smaller populations show lower genetic diversity than larger ones, as expected for isolated population fragments, but not connected ones (see graph on chapter frontispiece).

Computer simulations indicate that the smallest woodpecker populations are likely to suffer from inbreeding depression in the near future (Daniels et al. 2000; Schiegg et al. 2006).

In this chapter, we consider details of the genetic impacts of restricted gene flow. In the next section, we consider the magnitude of the problem.

How frequently are there problems due to inadequate gene flow?

Many species contain small isolated inbred population fragments with low genetic diversity, due to inadequate gene flow

Frankham et al. (2014a) estimated that 26% of invertebrate, 29% of vertebrate, and 55% of plant species show clear evidence of genetically fragmented populations, as determined from the proportion of species where populations exchange on average less than one effective migrant per generation (Wang 2004a). These are likely to be underestimates for several reasons, including incomplete sampling of species distributions. However, scaling even these to the ~ 8.7 million species on the planet indicates that there are millions of populations with inadequate gene flow that are expected to be suffering genetic problems (Chapter 1).

What are the genetic consequences of completely isolated fragments?

Completely isolated fragments suffer loss of genetic diversity over generations, increased inbreeding, inbreeding depression, reduced ability to evolve, and elevated extinction risk, compared to a single large population of the same total size

The most extreme consequences of fragmentation occur when there is complete genetic isolation (no gene flow) among population fragments. In this case, the effective population size determining inbreeding and loss of genetic diversity is that for each fragment, rather than for the species. Consequently, inbreeding and loss of genetic diversity occur more rapidly than in unfragmented populations. The harmful effects of multiple generations of population fragmentation with total isolation on genetic diversity, fitness, evolutionary potential, and extinction risk have been documented in several *Drosophila* and house fly studies (Borlase et al. 1993; Margan et al. 1998; Bryant et al. 1999; Frankham et al. 1999; Montgomery et al. 2000, 2010; Reed & Bryant 2000; Woodworth et al. 2002; Reed et al. 2003a, 2003b; Frankham et al. 2010, p. 238).

At the species level, fragmented populations may initially retain more overall genetic diversity than a single large population, but in the long term the effects on evolutionary potential are harmful, as many of the small populations become extinct (Fig. 5.3).

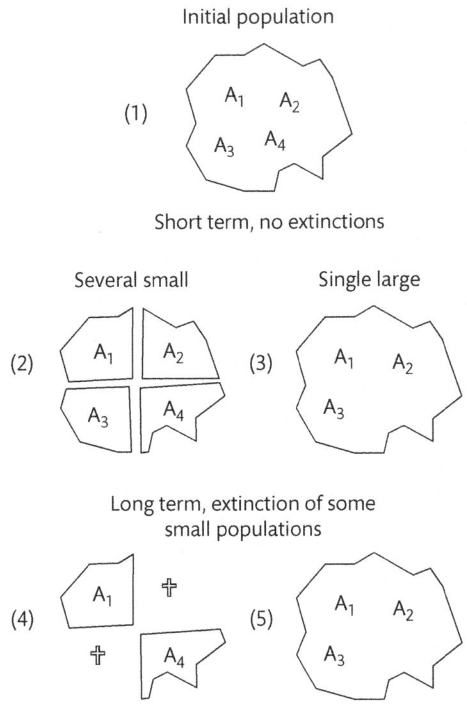

Fig. 5.3 Genetic consequences of a single large population (SL) versus several small (SS) completely isolated population fragments of initially the same total size over different time frames. (1) A_1–A_4 represent four alleles initially present in the population. In the short term, without extinctions, the SS populations (2) are individually expected to become fixed more rapidly, but can retain greater overall genetic diversity than the SL population (3) (Kimura & Crow 1963; Robertson 1964). In the longer term, when some of the small populations have gone extinct, the sum of the SS populations (4) will retain less genetic diversity than the persisting SL population and be more inbred than it (5) (Frankham et al. 2010, Fig. 14.3).

We next introduce theory to quantify these effects.

Theory for completely isolated idealized populations

We evaluate the impacts of an idealized population with multiple isolated fragments of equal sizes (Fig. 5.4) on (a) inbreeding, (b) diversity in allele frequencies among population fragments, and (c) heterozygosity across all population fragments.

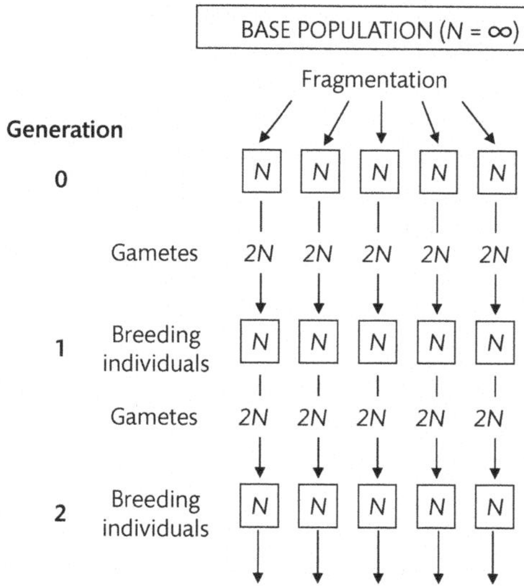

Fig. 5.4 Model of population fragmentation in an idealized population. The original source population is infinite, the isolated population fragments founded from it are of equal size. Each of the individual fragments is itself a random mating idealized population ($N_e = N$). (Frankham et al. 2010, Fig. 14.4).

Impact of population fragmentation on inbreeding

Smaller population sizes in isolated population fragments cause inbreeding levels to increase at a greater rate

The increase in inbreeding coefficients in single isolated population fragments over generations is faster than in a random mating population of the same combined population size. For example, if a population of $N_e = 100$ is split into five equal-sized, totally isolated population fragments of 20, inbreeding increases at a rate of 1/40 (2.5%) per generation in the small fragments versus 1/200 (0.5%) in the single large population. After 10 generations, the inbreeding coefficient is 40% in the small fragments but only 4.9% in the single large population (based on eqn 3.2), with an elevated extinction risk in the small, but not large populations.

Genetic differentiation among fragments

Differentiation in allele frequencies among a group of population fragments is greater for small than for large fragments and increases across generations

When a population is fragmented by random sampling, different fragments will usually have slightly different initial allele frequencies. This diversification is typically measured as variance in allele frequencies (σ_p^2) and can be predicted from the binomial sampling variance (Chapter 2). For a single locus with two alleles, A_1 and A_2, at frequencies of

p and q in the base population, we sample N individuals ($2N$ gene copies) for each fragment. After a single generation, the variance in allele frequency among population fragments is $\sigma_p^2 = pq / 2N$ (note that $N = N_e$ as we are considering idealized populations).

Diversification in allelic frequencies continues, generation after generation, until eventually all populations are fixed (p = 1, or 0). By extending eqn 2.1 over multiple generations, the expected variance in allele frequencies among fragments after t generations is (Falconer & Mackay 1996):

$$\sigma_p^2 = p_0 q_0 [1 - (1 - \frac{1}{2N})^t] \quad 5.1$$

where p_0 and q_0 are the initial allele frequencies and the fragment size (N) is constant over time. From this equation, we predict that variances in allele frequencies will (Charlesworth & Charlesworth 2010):

- increase with generations (t) (Fig. 5.5)
- increase faster in smaller than in larger population fragments (Fig. 5.6)
- reach a maximum of $p_0 q_0$ when all populations have become fixed.

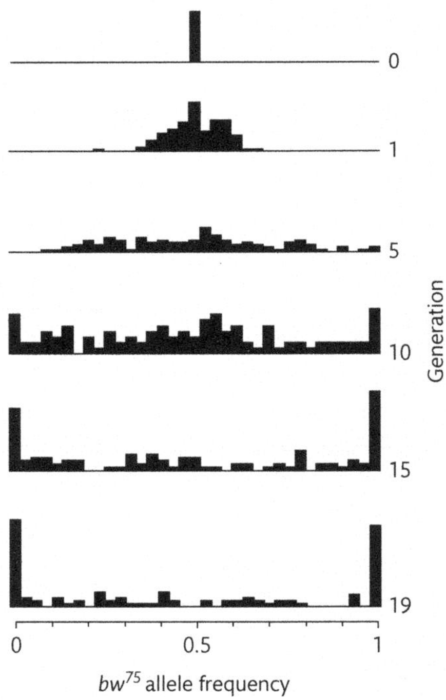

Fig. 5.5 Allele frequencies divergence across generations for totally isolated population fragments of *Drosophila*. The frequency distributions for the bw^{75} allele are shown across 19 generations in 105 replicate populations maintained with 16 parents per generation (but N_e ~ 11.5). All populations began with initial frequencies of 0.5 (Frankham et al. 2010, Fig. 14.6, after Buri 1956).

The distributions of allele frequencies become flatter with increasing generations as allele frequencies diverge (Fig. 5.5), resulting in an essentially uniform distribution after $2N$ generations for an initial frequency of 0.5, and about $4N$ generations for an initial frequency of 0.1 (excluding the fixed populations). Such variation in allele frequencies among population fragments is found in many species, including red-cockaded woodpeckers (Box 5.1), giant pandas (Box 1.1), black-footed rock-wallabies (Eldridge et al. 1999), and many plants (Hamrick & Godt 1996a).

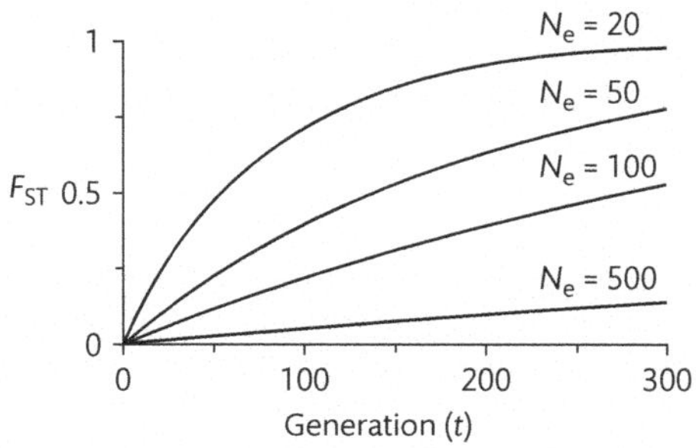

Fig. 5.6 Standardized genetic divergence among populations ($F_{ST} = \sigma_p^2 / p_0 q_0$, as explained later) increases more rapidly over generations in small than in large populations, based on computer simulations (Frankham et al. 2010, Fig. 14.8).

Effects of population fragmentation on genotype frequencies

Genetic drift among population fragments reduces heterozygosity across all fragments to less than the Hardy–Weinberg equilibrium expectation

As the frequencies in population fragments drift apart, the heterozygosity in the entire population declines when compared to Hardy–Weinberg expectations (Table 5.1), referred to as the Wahlund effect (Wahlund 1928). For example, if multiple isolated random mating fragments are each founded with initial frequencies $p = q = 0.5$, heterozygosity is maximal ($2pq = 0.5$) at the outset. However, as frequencies deviate from $p = q = 0.5$ due to drift, heterozygosity is inevitably reduced (e.g. if two fragments drift to $p = 0.3$, $q = 0.7$ and $p = 0.7$, $q = 0.3$, then average heterozygosity is reduced to 0.42). For example, the endangered spreading avens plant (*Geum radiatum*) has a heterozygosity of only 0.052, averaged across five populations in the eastern USA, but an expected heterozygosity of 0.098, due to both population fragmentation and inbreeding within populations (Hamrick & Godt 1996b).

Heterozygosities lower than Hardy–Weinberg expectations are widely used to diagnose populations that are genetically fragmented and/or inbred (see later and Chapter 10).

Population fragments simultaneously become genetically differentiated, lose genetic diversity, and become inbred due to drift

A critical, but little appreciated, expectation in conservation genetics is that population fragments simultaneously become genetically differentiated, lose genetic diversity, and become inbred due to drift (Table 5.1). When the variance in allele frequency rises by a given amount, heterozygosity declines by twice this amount and inbreeding increases correspondingly. For example, 18 isolated populations of mice, each maintained with 8 pairs of parents per generation for 27 generations, had a variance in allele frequencies of 0.069 for five allozyme loci, a reduction in heterozygosity of 0.122 (34%), and an inbreeding coefficient of 0.38 (Falconer & Mackay 1996, pp. 70–71; see also Coleman et al. 2013; Weeks et al. 2016).

Table 5.1 Predicted genotype frequencies before and after fragmentation. A random mating base population is used to found multiple isolated random mating population fragments that go through several generations (as in Fig. 5.4). Identical scenarios are treated as either genetic drift or inbreeding processes (p_0 and q_0 are allele frequencies before fragmentation) (after Falconer & Mackay 1996).

Genotype	Frequency before fragmentation	Frequencies in the total population after fragmentation	
		Genetic drift	Inbreeding
A_1A_1	p_0^2	$p_0^2 + \sigma_p^2$	$p_0^2 + Fp_0q_0$
A_1A_2	$2p_0q_0$	$2p_0q_0 - 2\sigma_p^2$	$2p_0q_0\,(1 - F)$
A_2A_2	q_0^2	$q_0^2 + \sigma_p^2$	$q_0^2 + Fp_0q_0$

As the reductions in heterozygosity are the same whether derived from inbreeding or drift considerations, we can equate inbreeding and population divergence, as follows (Wright 1969):

$$F = \frac{\sigma_p^2}{p_0q_0} \qquad 5.2$$

This inbreeding coefficient due to non-random mating among populations provides us with a widely used measure of population differentiation (F_{ST}), as detailed later in this chapter.

Many species exhibit some degree of gene flow among population fragments, but less than in pre-fragmentation continuous populations.

What are the consequences of gene flow among fragments?

Gene flow reduces the harmful genetic effects of population fragmentation

The genetic impacts of fragmentation (inbreeding, loss of genetic diversity, and divergence among populations) are inversely related to the rate of gene flow, also referred to as genetic connectivity. For example, resistance to a fungal disease in the *Plantago lanceolata* plant in Finland is higher for populations with greater gene flow (Fig. 5.7).

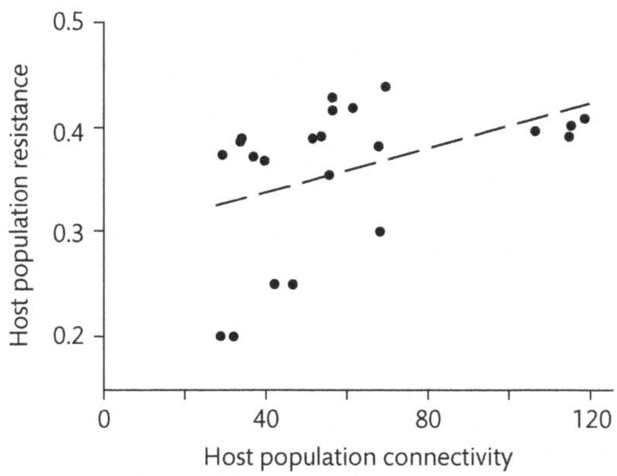

Fig. 5.7 Host population resistance in the lamb's tongue plant (*Plantago lanceolata*) to a powdery mildew fungus (*Podosphaera plantaginis*) is higher in well-connected than in poorly connected populations in Finland (Jousimo et al. 2014). Host resistance was measured from inoculation trials, while connectivity was based on population sizes and spatial locations.

When is gene flow sufficient to overcome the genetic impacts of fragmentation?

A single successful migrant per generation is considered sufficient to prevent complete differentiation of idealized populations, irrespective of their size, but not to prevent damaging inbreeding

Sewall Wright (1931) obtained the surprising result that a single successful migrant per generation, among idealized populations under an island model of migration (see Fig. 5.10), was sufficient to prevent complete differentiation (and fixation), irrespective of population size (Fig. 5.8). This may appear paradoxical, but one migrant that breeds represents proportionally a much higher rate of gene flow in smaller than in larger populations, and the rates of drift differ correspondingly. Importantly, this level of gene flow is not sufficient to prevent damaging levels of inbreeding, as it results in an inbreeding coefficient of 0.2.

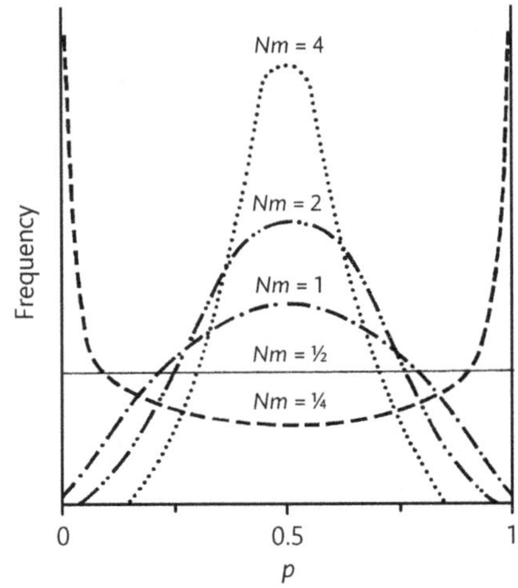

Fig. 5.8 Populations with one or more effective migrants per generation ($Nm \geq 1$) do not differentiate completely (no fixations), while $Nm < 1$ does not prevent fixations. Distributions of allele frequencies at equilibrium between drift and gene flow in idealized populations of size N with different levels of migration (Nm) among them for an "island" model (see later). The rate of gene flow m is the proportion of alleles in the population derived from migrants each generation (Frankham et al. 2010, Fig. 14.9, after Wright 1940).

Fragmentation effects in non-idealized populations

Similar genetic consequences of population fragmentation are expected in non-idealized populations as in idealized ones, except that the rates of change depend on N_e rather than N

If we wish to make predictions for non-idealized populations, we substitute effective population size N_e for N in the appropriate equations, such as 5.1.

Empirical results confirm the theoretical expectations we have described above. For example, a meta-analysis of plant studies showed that fragmented populations are more inbred and have less genetic diversity than non-fragmented populations, and genetic diversity declines with generations of isolation (Fig. 5.9).

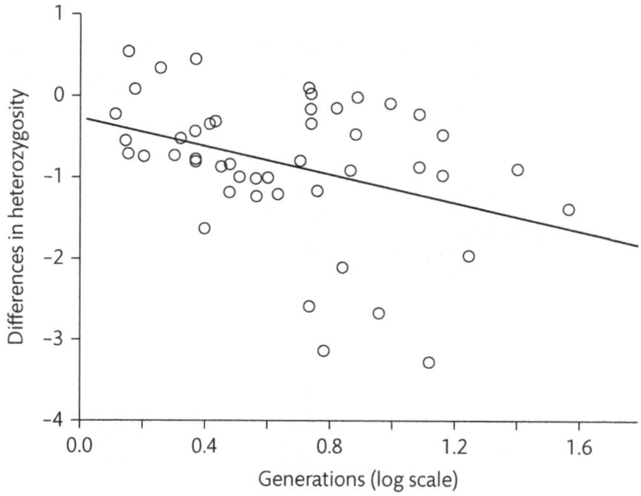

Fig. 5.9 Genetic diversity (H_e) in plant populations declines with generations of isolation, based on data from 47 species (Aguilar et al. 2008). Standardized differences in genetic diversity (heterozygosity in fragmented minus that in the continuous habitat, divided by the pooled standard deviation and adjusted for sample size) are plotted against generations.

In real (non-idealized) populations, more than one migrant per generation is usually required to prevent fixation in population fragments, but one effective migrant has the intended effect

As an immigrant may not breed, or be as successful in producing offspring as residents (Hanski et al. 2000), more than one migrant per generation is typically required to prevent fixation of alleles by drift in real populations. Numbers of 5, 1–10, and >10 migrants per generation have been proposed (Lacy 1987; Mills & Allendorf 1996; Vucetich & Waite 2000). The complications arising from non-idealized population structures are overcome by assessing the effective number of migrants (N_em) where a value of 1 prevents fixation (but not damaging inbreeding) as envisaged by Wright (Wang 2004a).

Selection may alter the impact of fragmentation

Balancing selection typically reduces the genetic differentiation among population fragments, while diversifying selection amplifies it

The theory above assumes selective neutrality of alleles (Falconer & Mackay 1996). If alleles are subject to balancing selection, and have equilibria with allele frequencies in the 0.2–0.8 range, then the rate of diversification will be lower than for neutral loci, but outside this frequency range it may be greater (Robertson 1962). If selection differs among fragments, as for example when one allele is favored in some fragments and harmful in others, then the rate of diversification will be greater than predicted by neutral theory. Accordingly, genetic divergence among populations for quantitative characters is often greater than that for "neutral" molecular markers due to diversifying selection on quantitative traits (Leinonen et al. 2008). However, if populations are small and selection is weak, alleles will often behave as if neutral (effectively neutral).

What are the consequences of different fragmented population structures?

The genetic impacts of population fragmentation depend on the population structure, the geographic location of fragments, and particularly on levels of gene flow among them

The genetic impacts of population fragmentation may range from insignificant to severe, depending upon the resulting population structure and gene flow among fragments. There are several theoretical models of fragmented population structure with gene flow (Fig. 5.10):

- island structure where gene flow is equal among all island populations
- linear stepping-stone structure, where only neighboring populations exhibit gene flow (as in riparian habitat)
- two-dimensional stepping-stone structure, where only nearby populations exchange genes
- source–sink or mainland–island structures
- metapopulations with ongoing extinctions and recolonizations of populations.

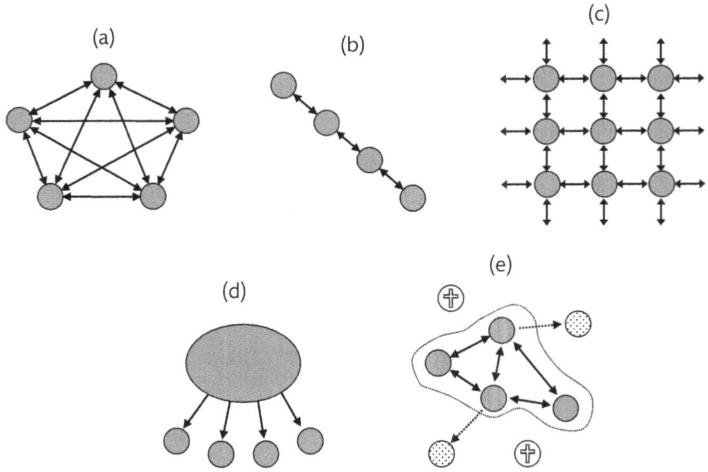

Fig. 5.10 Five different theoretical fragmented population structures (after Frankham et al. 2010, Fig. 14.1): (a) an island structure, where gene flow is equal among equal sized island populations, (b) a linear stepping-stone structure, where only neighboring populations exchange migrants, (c) a two-dimensional stepping-stone structure, where neighboring populations exchange migrants, (d) a source-sink or mainland–island structure, where the source (mainland) population provides all the input to the sink (island) (all after Hedrick 1983), and (e) a metapopulation with extinctions (circles with crosses) and recolonizations (dotted circles) (after Hanski & Gilpin 1997).

Extensive population genetic theory has been developed to model genetic processes for these different population structures, especially for the simple island model that has few assumptions and parameters (Rousset 2004). However, real population structures will often be more complicated than assumed in these simple models.

These structures differ in how harmful their effects are, and in their relationships of genetic divergence with geographic distance (see "Isolation by distance"). For example, the effective size for a source–sink structure is that of the source only, while metapopulations may also have much smaller effective sizes than their total sizes.

In general, the genetic consequences of a metapopulation structure are more harmful than the other population structures (apart from completely isolated fragments)

Metapopulations differ from the other fragmented structures in that there are regular extinction and recolonization events, while no extinctions are assumed for the others. Metapopulations also encompass different sized population fragments and rates of gene flow (Levins 1969, 1970; Hanski & Gaggiotti 2004) (note that we are using the original Levins [1969] definition of a metapopulation). For example, there are ~ 1,600 suitable meadows for Glanville fritillary butterflies on Åland Island in southwest Finland, 320–524 being occupied in 1993–1996, with an average of 200 extinctions and 114 colonizations per year (Saccheri et al. 1998). Figure 5.11 illustrates the dynamics of extinctions and recolonizations in a metapopulation, including bottlenecks during some recolonizations. In the long term, metapopulation structures are often more realistic models for fragmented populations with gene flow than the other structures we have defined.

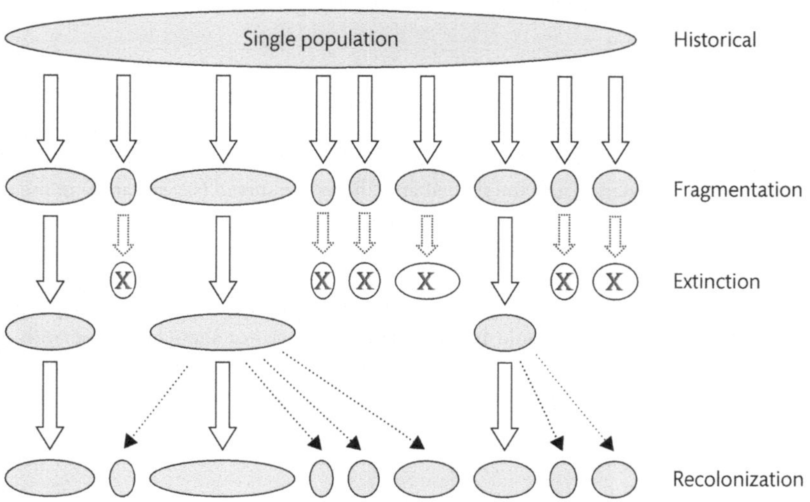

Fig. 5.11 Cycles of extinction and recolonization in a metapopulation lead to reductions in the effective population size of a species. The dotted lines indicate bottlenecks during recolonizations (Frankham et al. 2010, Fig. 14.14).

In general, within population genetic diversity decreases with higher rates of extinction and recolonization in a metapopulation, and as the colonists to a particular location become more related to one another (Barton & Whitlock 1997; Wang & Caballero 1999; Whitlock 2004).

Isolation by distance

Distant population fragments often exhibit reduced gene flow, resulting in genetic differentiation that increases with geographic distance (isolation by distance)

Isolation by distance (Wright 1969) is expected with stepping-stone population structures, but not with the island model (with equal gene flow), or in a continuous population with random mating. For example, Box 5.2 shows the expected patterns of genetic differentiation with distance for populations with different rates of gene flow, matched with three corresponding empirical examples in aquatic insects. Populations of bighorn sheep, gray wolves, and brown bears in North America all show isolation by distance (Fig. 5.12), as do red-cockaded woodpeckers (Box 5.1; Haig et al. 2001), northern spotted owls (*Strix occidentalis caurina*), the plant *Arabidopsis thaliana* (Platt et al. 2010), and many other species. However, isolation by distance is not expected where dispersal rates are high and distances short, as observed in the *Anthyllis vulneraria* plant in Belgium (Honnay et al. 2006).

Box 5.2 Predicted and empirical relationships between genetic divergence and geographic distance with different rates of gene flow

(Phillipsen et al. 2015)

Predicted (upper panel) relationships between genetic divergence (genetic distance) and geographic distance for (a) low, (b) moderate, and (c) high dispersal among populations. The black lines represent regressions of genetic differentiation on geographic distance, and the shaded area shows the spread (i.e. variance) in the pairwise genetic divergences across the geographic distances. (a) When genetic drift is more influential than gene flow, as is predicted for species with low dispersal, the slope of the line does not differ from zero, the intercept is high, and the variance is high. (b) For species with moderate dispersal, a positive slope is predicted, the intercept should be near zero, and variance should increase with increasing geographic distance. (c) Gene flow should be more influential than drift for species with high dispersal. The regression line in this case does not deviate from zero, the intercept is small, and variance is low.

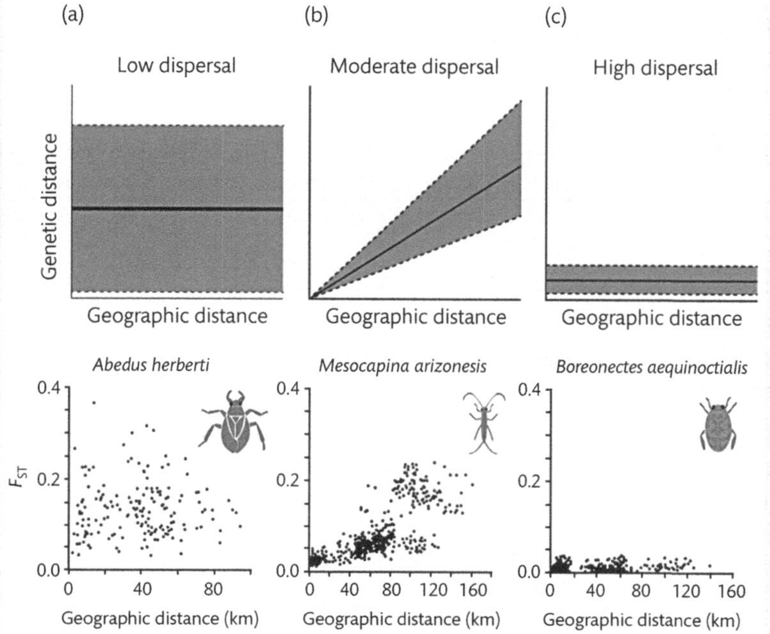

Empirical results (lower panels) of the patterns of genetic divergence (F_{ST}) with geographic distances between populations for three desert aquatic insects *Abedus herberti* (flightless), *Mesocapnia arizonensis* (moderate disperser), and *Boreonectes aequinoctialis* (high disperser) closely matched the predictions for low, moderate, and high dispersal, respectively.

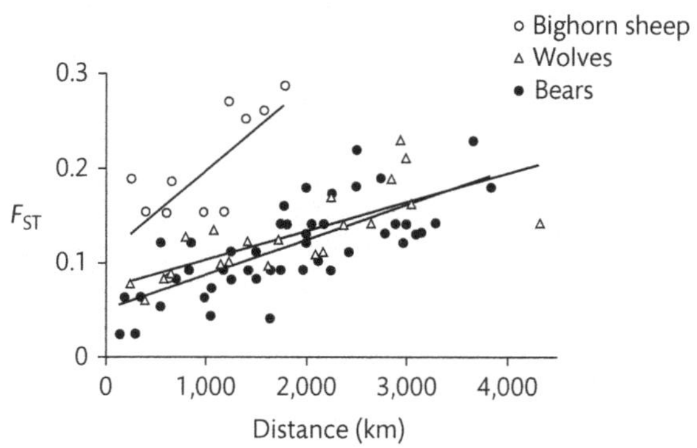

Fig. 5.12 Isolation by distance. Genetic differentiation (F_{ST}) at microsatellite loci increases with geographic distance among bighorn sheep, brown bear, and gray wolf populations in North America (Frankham et al. 2010, Fig. 14.12, after Forbes & Hogg 1999).

Bighorn sheep

Gray wolf

Brown bear

Complex relationships among variables in real fragmented populations

So far, we have typically restricted our considerations to a few variables at a time, but in real populations many variables may be changing simultaneously. Empirical evidence on the white-fronted chat (*Epthianura albifrons*: Box 5.3) illustrates how a complex pattern can result from such effects. Similarly, studies of red campion (*Silene dioica*) plants in a Swedish archipelago revealed a complex story due to the effects of multiple variables (Giles & Goudet 1997).

White-fronted chat (Australia)

Box 5.3 Complex patterns of genetic differentiation among population fragments of an Australian bird with different sizes and degrees of isolation

(Major & Sladek 2012; Major et al. 2014)

The white-fronted chat is widely distributed in salt marshes across the southern half of Australia (dots on small map). Whilst the chats were historically widespread throughout the Sydney region (now a city of ~ 5 million people), they are now restricted to only two isolated and endangered populations at Towra Point (18–24 birds) and Homebush Bay (~ 9 birds).

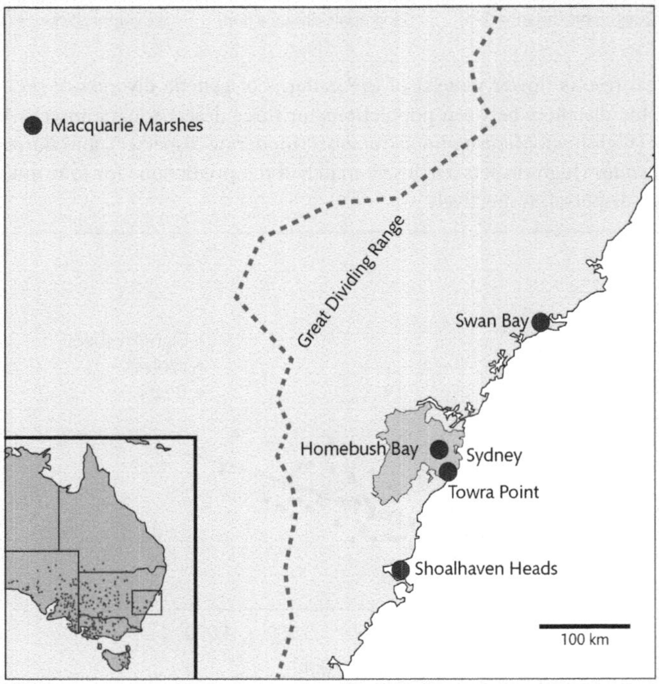

(Major et al. 2014, Fig. 2)

The genetic characteristics of the two small isolated populations were compared with three large connected populations from surrounding areas, up to 500 km distant, based on data for 18 microsatellite loci. Observed heterozygosity was lower than expected heterozygosity (0.652 versus 0.782) across populations, indicating a Wahlund effect. Heterozygosities (and number of alleles) were lower in the two small isolated Sydney populations than in the larger connected populations (bar chart).

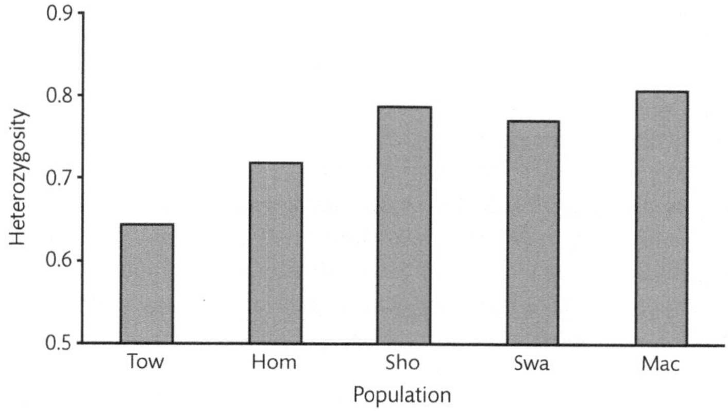

(Major et al. 2014, Fig. 3b)

Population genetic differentiation was greatest between the two small isolated Sydney populations, followed by the Sydney populations versus other populations, and least between the larger connected populations (F_{ST} 0.121, 0.066, and 0.033, respectively), and there was no signal of isolation by distance. Multilocus clustering clearly differentiated individuals from the two small Sydney populations (H1–H6 [dark gray] and T1–T10 [white]) from each other and from individuals from the nearby larger Shoalhaven Heads (S1–S8 [light gray]) population (following graph), and from the other two populations (not shown), using STRUCTURE analyses (described in Chapter 10).

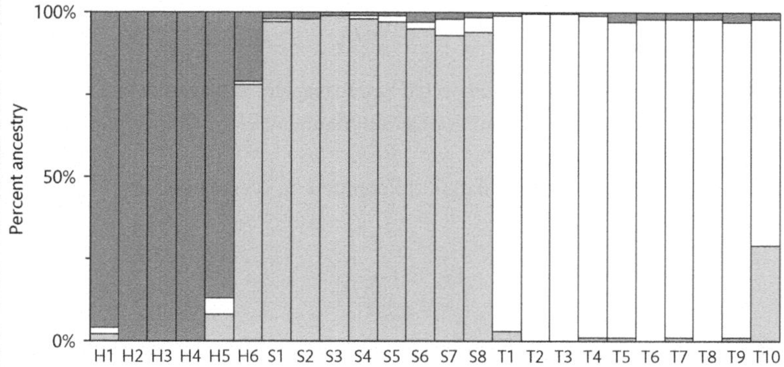

(Major & Sladek 2012, Fig. 3)

To predict the genetic consequences of population fragmentation in real populations with all their complexity, computer simulations are often used (Landguth et al. 2014).

How do we measure genetic differentiation among populations?

F statistics

The degree of differentiation among fragments can be described by partitioning the overall inbreeding into components within and among populations (*F* statistics)

We begin by describing *F* statistics, as these have been the most used measures of population differentiation (Wright 1969; Neigel 2002) and several more recent measures are derivatives of them (see later). Sewall Wright (1951) partitioned inbreeding in the total (*T*) population (F_{IT}) into inbreeding of individuals relative to their population fragment F_{IS}, and inbreeding due to differentiation among population fragments, relative to the total population, F_{ST}. These reflect departures from random mating (Hardy–Weinberg equilibrium) within and among populations, as well as divergence in allele frequencies among population fragments (see later).

F statistics are usually determined from heterozygosity for genetic markers using the following equations (Nei 1987). Thus, the total inbreeding F_{IT} is:

$$F_{IT} = 1 - \frac{H_o}{H_T} \qquad 5.3$$

where H_o is the observed heterozygosity averaged over all fragments (sub-populations), and where H_T is Hardy–Weinberg heterozygosity for the total population.

F_{IS}, the inbreeding of individuals relative to their population fragments, is estimated as:

$$F_{IS} = 1 - \frac{H_o}{H_S} \qquad 5.4$$

where H_S is the Hardy–Weinberg expected heterozygosity averaged over all fragments. A deficiency of heterozygotes within population fragments ($F_{IS} > 0$) indicates inbreeding due to non-random mating.

F_{ST}, the inbreeding due to population differentiation, is estimated as:

$$F_{ST} = 1 - \frac{H_S}{H_T} \qquad 5.5$$

Non-random mating among populations results in genetic differentiation due to genetic drift among them ($F_{ST} > 0$). F_{ST} was originally defined in terms of standardized variance in allele frequencies (Wright 1969; eqn 5.2), which better reflects its use to measure genetic differentiation, but the two definitions are equivalent.

For multiallelic loci, these parameters are also referred to as G_{IS}, G_{ST}, and G_{IT}, and F and G are now often used interchangeably (Allendorf et al. 2013). F_{ST} has a range from 0 (no differentiation among fragments) to 1 (fixation of different alleles in fragments) for loci with 2 alleles (but not with > 2 alleles, as detailed later in the chapter).

Table 5.2 illustrates the application of F statistics using hypothetical examples. In case (A) there is random mating (no inbreeding) within either population fragment so $F_{IS} = 0$. However, the population fragments have different allele frequencies and $F_{ST} > 0$, implying restricted gene flow. Case (B) shows two population fragments with identical allele frequencies, but with inbreeding in fragment 2. $F_{IS} > 0$, but $F_{ST} = 0$, indicating considerable gene flow. In (C) there is both divergence in allele frequencies and inbreeding, F_{IS} and $F_{ST} > 0$, and the total inbreeding, F_{IT}, reflects both effects.

Table 5.2 A hypothetical example demonstrating calculation of F statistics from genotype data (after Frankham et al. 2010, Table 14.3).

Fragment	Genotypes			Allele		
	A_1A_1	A_1A_2	A_2A_2	frequency	F	H_e (= $2pq$)
(A) Random mating within populations, but differentiation in allele frequencies between them						
1	0.25	0.5	0.25	$p = 0.5$ $q = 0.5$	0	0.5
2	0.04	0.32	0.64	$p = 0.2$ $q = 0.8$	0	0.32
Combined		$H_I = 0.41$		$p = 0.35$ $q = 0.65$		$H_S = 0.41$ $H_T = 0.455$
$F_{ST} = 0.099$	$F_{IS} = 0$	$F_{IT} = 0.099$				
(B) Populations with the same allele frequencies, (1) random mating and (2) inbreeding						
1	0.25	0.5	0.25	$p = 0.5$ $q = 0.5$	0	0.5
2	0.4	0.2	0.4	$p = 0.5$ $q = 0.5$	0.6	0.5
Combined		$H_I = 0.35$		$p = 0.5$ $q = 0.5$		$H_S = 0.5$ $H_T = 0.5$
$F_{ST} = 0$	$F_{IS} = 0.3$	$F_{IT} = 0.3$				
(C) Populations with different allele frequencies, (1) random mating and (2) inbreeding						
1	0.25	0.5	0.25	$p = 0.5$ $q = 0.5$	0	0.5
2	0.14	0.13	0.74	$p = 0.2$ $q = 0.8$	0.6	0.32
Combined		$H_I = 0.31$		$p = 0.35$ $q = 0.65$		$H_S = 0.41$ $H_T = 0.455$
$F_{ST} = 0.042$	$F_{IS} = 0.31$	$F_{IT} = 0.312$				

Example 5.1 illustrates the computation of F statistics for the rare Pacific yew tree (*Taxus brevifolia*) in Canada.

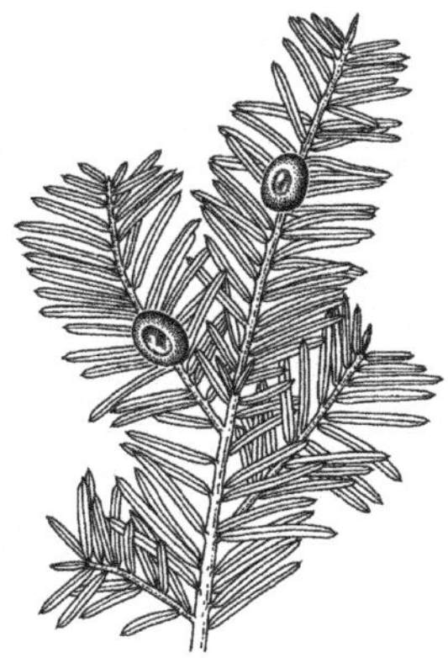

Pacific yew (North America)

Example 5.1 *F* statistics for populations of the Pacific yew
(Frankham et al. 2010, Example 14.4)

Average observed heterozygosity (H_o) for 21 allozyme loci across nine Canadian populations was 0.085, while the average expected heterozygosity for these populations (H_S) was 0.166, and the expected heterozygosity for the nine populations combined (H_T) was 0.18 (El-Kassaby & Yanchuk 1994). Consequently, inbreeding within populations is:

$$F_{IS} = 1 - \frac{H_o}{H_e} = 1 - \frac{0.085}{0.166} = 0.49$$

This inbreeding is not due to selfing because the species is dioecious, but is probably due to offspring establishing close to parents in bird and rodent seed caches, and subsequent mating between close relatives.

Inbreeding due to population differentiation is:

$$F_{ST} = 1 - \frac{H_S}{H_T} = 1 - \frac{0.166}{0.180} = 0.078$$

This indicates only a modest degree of population differentiation.

The total inbreeding across the population fragments due to the combination of the two effects above is:

$$F_{IT} = 1 - \frac{H_o}{H_T} = 1 - \frac{0.085}{0.180} = 0.53$$

Divergence in mtDNA and cpDNA is expected to be more rapid than for nuclear loci

F_{ST} can be estimated from nuclear or organelle DNA. However, in the same populations mtDNA or cpDNA diversities differentiate more rapidly with generations of isolation than those from autosomal markers (Fig. 5.13), because the effective sizes for mtDNA and cpDNA are only about ¼ of that for autosomal loci, due to uniparental inheritance, and even less when selection is accounted for (Frankham 2012). For example, F_{ST} for white campion (*Silene alba*) plants was 0.67 for cpDNA, but only 0.13 for nuclear allozyme loci due to some combination of different effective sizes and sex-biased gene flow (McCauley 1994).

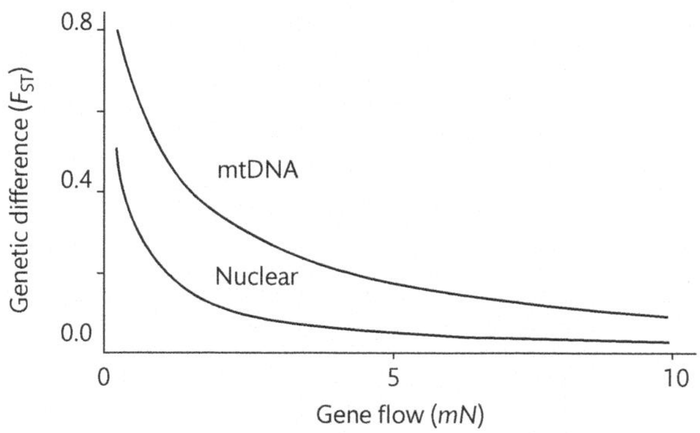

Fig. 5.13 Theoretical values of F_{ST} for different gene flow rates (mN) for nuclear and mtDNA loci (after Allendorf & Luikart 2007). This assumes that the sex ratio is 1:1, gene flow is not sex-biased, and that the mtDNA N_e is ¼ that of nuclear loci.

Other measures of population differentiation

A variety of other parameters are available to measure genetic differentiation among populations (Table 5.3) (see review by Meirmans & Hedrick 2011). Many of these are closely related to F_{ST}, or are analogues of it.

Table 5.3 Methods for estimating genetic diversification among populations, compared to F_{ST}

Estimator	Scaling 0–1	Data	Basis	Comments	References
F_{ST}	No	Allozymes, microsatellites, SNPs	H or σ_p^2	Scales 0–1 with 2 alleles/locus	Wright (1969)
G_{ST}	No	Allozymes, microsatellites, SNPs	Usually H	F_{ST} analogue, devised for use with multiple alleles/locus	Nei (1973)
G'_{ST}	Yes	Allozymes, microsatellites, SNPs	H or σ_p^2	F_{ST} analogue, always scales 0–1	Hedrick (2005a)
K_{ST}	No?	DNA sequences	H	F_{ST} analogue	Hudson et al. (1992)
Φ_{ST}	No	DNA sequences and frequencies	Genealogies	F_{ST} analogue	Excoffier et al. (1992)
Q_{ST}	Yes	Quantitative variation	V	F_{ST} analogue	Wright (1951); Lande (1992)
D_J	Yes	Allozymes, microsatellites, SNPs	n_A	Distinct from F_{ST}	Jost (2008)
D_N	No	Allozymes, microsatellites, SNPs	Standardized genetic distance	Distinct from F_{ST}	Nei (1972)

Estimator	Scaling 0–1	Data	Basis	Comments	References
J_{XY}	Yes	Allozymes, microsatellites, SNPs	Frequencies of shared alleles in two populations	Distinct from F_{ST}, but related to kinship	Nei (1987)
mk_{AB}	Yes	Pedigrees, molecular markers	Relationship (1 – mk_{AB} measures divergence)	Used in endangered species management	Ballou & Lacy (1995)
R_{ST}	Yes	Microsatellites	V number of repeats	Distinct from F_{ST}, requires realistic mutation model	Slatkin (1995)

V = variance, H = heterozygosity, n_A = effective number of alleles.

Because F_{ST}, G_{ST}, and their analogues do not scale 0–1 if there are more than two alleles or haplotypes at a locus, versions scaling 0–1 have been devised, including G'_{ST}. These adjusted estimators allow comparisons to be made between estimates of differentiation obtained from genetic markers with different levels of genetic diversity.

G'_{ST}

G'_{ST} is a measure closely related to F_{ST} that scales 0–1 even with > 2 alleles per locus

A version of G_{ST} that scales 0–1 can be obtained by dividing it by the maximum value for the particular data set ($G_{ST(max)}$; Hedrick 2005a):

$$G_{ST(max)} = \frac{(k-1)(1-H_S)}{k-1+H_S} \qquad 5.6$$

where k is the number of population fragments. Thus, G'_{ST} is

$$G'_{ST} = \frac{G_{ST}}{G_{ST(max)}} \qquad 5.7$$

A similar correction can also be used for all F_{ST} analogues so that they scale 0–1. We recommend the routine use of G'_{ST} rather than G_{ST}, especially for microsatellite data, or when values are being compared across taxa and markers.

Mean kinship

Mean kinship between populations (mk_{AB}) can be used to measure population similarity, so (1 – mk_{AB}) provides a measure of population genetic differentiation. Kinship (k_{ij}, also called coancestry) is based on the relatedness of pairs of individuals and is the inbreeding coefficient of an offspring if they had one, and thus scales 0–1 (see Chapter 13). Mean kinship is simply the mean of all the pairwise kinship values between all individuals in the population, including kinships with self (Chapter 13). Mean kinship is widely used for genetic management of captive populations of threatened species, and in Chapter 13 we show that it is superior to F_{ST} for genetic management of wild and captive populations. Consequently, we recommend its use for genetic management of fragmented wild populations, either at the between population level or the individual level. Kinship can be measured from pedigrees, or estimated from multilocus genetic data (such as microsatellites or SNPs) (Wang 2011; Chapter 13).

There are two primary messages about these different measures of population differentiation. All are able to detect signals of genetic fragmentation and they are often closely correlated, so the choice of which measure to use for this purpose is rarely crucial (Balloux & Goudet 2002; Heller & Siegismund 2009; Meirmans & Hedrick 2011). However, there are crucial differences when they are applied to genetic management of fragmented populations (Chapter 13).

In Section II, we discuss genetic rescue of fragmented populations suffering genetic erosion by the application of augmenting gene flow from one or more other populations (Chapter 6). We subsequently consider the potential risk that such crossing will be harmful, rather than beneficial (Chapter 7), while in Chapter 8 we consider how the risks and rescue prospects are modified in species with diverse mating systems and modes of inheritance.

Summary

1. Most species have fragmented distributions, often with limited gene flow between population fragments.
2. Population fragmentation with cessation of gene flow is harmful in the long term, leading to greater inbreeding, lower fitness, more rapid loss of genetic diversity, less evolutionary adaptation, and elevated extinction risk, when compared to a single unfragmented population of the same total size.
3. Genetic divergence among population fragments is causally associated with reduced genetic diversity within populations and increased inbreeding.
4. The effects of limited gene flow among fragmented populations on genetic diversity, inbreeding, fitness, and evolutionary potential typically lie between those for several small isolated populations and those for a single large population of the same total size. However, metapopulations with high extinction and recolonization rates may suffer particularly harmful genetic consequences.
5. Fragmented populations with limited gene flow diverge in allele frequencies and lose heterozygosity at rates inversely related to effective population sizes and levels of gene flow, and the impacts increase with generations.
6. One effective migrant per generation between otherwise isolated population fragments is sufficient to prevent fixation, but inadequate to avoid substantial inbreeding.
7. *F* statistics and related measures are frequently used to measure genetic differentiation among populations.
8. *F* statistics and related measures are adequate to measure genetic differentiation, but kinship is preferable for genetic management.

FURTHER READING

Aguilar et al. (2008) Meta-analysis on the genetic impacts of habitat fragmentation in plants.

Balkenhol et al. (2015) *Landscape Genetics*: An edited research monograph covering the concepts, methods, and applications of the distribution of genetic diversity across landscapes.

Charlesworth & Charlesworth (2010) *Elements of Evolutionary Genetics*: Authoritative advanced textbook with fine coverage of spatially structured populations.

Meirmans & Hedrick (2011) Review of the characteristics of different estimates of population differentiation.

Phillipsen et al. (2015) Theoretical prediction and empirical data on genetic divergence with geographic distance with different gene flow rates.

Rousset (2004) *Genetic Structure and Selection in Subdivided Populations*: Advanced monograph on the theory of spatial genetic structure.

SOFTWARE

COANCESTRY: a program for estimating kinship, relatedness, and inbreeding coefficients from molecular marker data, as well as doing simulations (Wang 2011). http://www.zsl.org/science/research/software/coancestry

FSTAT: calculates *F* and *R* statistics and does partial Mantel tests (Goudet 2002). https://www2.unil.ch/popgen/softwares/fstat.htm

GenAlEx: computes basic population genetic statistics, including *F* statistics, and conducts Mantel tests (Peakall & Smouse 2006). http://biology-assets.anu.edu.au/GenAlEx/Download.html

GENODIVE: computes measures of genetic divergence that scale 0–1, such as G'_{ST}, G''_{ST}, Jost D, and Φ'_{ST} (Meirmans & Van Tienden 2004). http://www.patrickmeirmans.com

GENEPOP: computes *F* statistics, isolation by distance, and Mantel tests (Rousset 2008). http://kimura.univ-montp2.fr/~rousset/Genepop.htm

SECTION II
Rescue and risk

The adverse impacts of inbreeding and loss of genetic diversity on fitness and ability to evolve in small isolated population fragments described in Section I are usually reversible by gene flow from another population, a process termed genetic rescue.

Despite the expected benefits of gene flow, genetic rescue has rarely been attempted for conservation purposes, with only ~ 30 known cases for all species on the planet, yet there are millions of small isolated populations potentially suffering genetic erosion. Why have there been so few attempts? One reason has been the lack of quantitative information on the benefits and risks associated with interpopulation crosses, but such information is now available. Chapter 6 provides a quantitative assessment of the frequency of beneficial versus harmful fitness impacts of gene flow between populations and the magnitude of such effects across F_1, F_2, and F_3 and later generations. Further, it discusses the variables affecting the magnitude of genetic rescue effects.

In Chapter 7 we consider harmful effects of crossing populations (outbreeding depression) and the variables that predict its risk. A minority of crosses between populations result in decreased reproductive fitness. These are typically cases where strongly diverged populations (potentially on the path to separate species) are crossed. Outbreeding depression is less of a problem than inbreeding depression in outbreeding diploid populations fragmented by human activities. Further, outbreeding depression is typically reversed in time by natural selection. Substantial progress has been made in predicting the risk of outbreeding depression (Frankham et al. 2011), removing an impediment to the use of genetic rescues.

In the final chapter in this section (Chapter 8) we discuss how the magnitude of genetic rescue effects and the risks of outbreeding depression differ in species with mating systems and modes of inheritance other than outbreeding diploids (e.g. self-incompatible, self-fertilizing, asexual, polyploid, haploid, and haplodiploid). Further, we consider their evolutionary genetics attributes, such as levels of genetic diversity, ability to evolve, genetic loads, and genetic differentiation among populations, as background information to inform their genetic management.

Genetic rescue by augmenting gene flow

CHAPTER 6

Inbreeding is reduced and genetic diversity enhanced when a small isolated inbred population is crossed to another unrelated population. Crossing can have beneficial or harmful effects on fitness, but beneficial effects predominate, and the risks of harmful ones (outbreeding depression) can be predicted and avoided. For crosses with a low risk of outbreeding depression, there are large and consistent benefits on fitness from gene flow within outbreeding species that persist across generations. Benefits are greater in species that naturally outbreed than those that inbreed, and increase with the difference in inbreeding coefficient between crossed and inbred populations in mothers and zygotes. However, benefits are similar across invertebrates, vertebrates, and plants. There are also important benefits for evolutionary potential of crossing between populations.

TERMS

Conspecific, evolutionary rescue, heterosis, maternal effect, sub-species

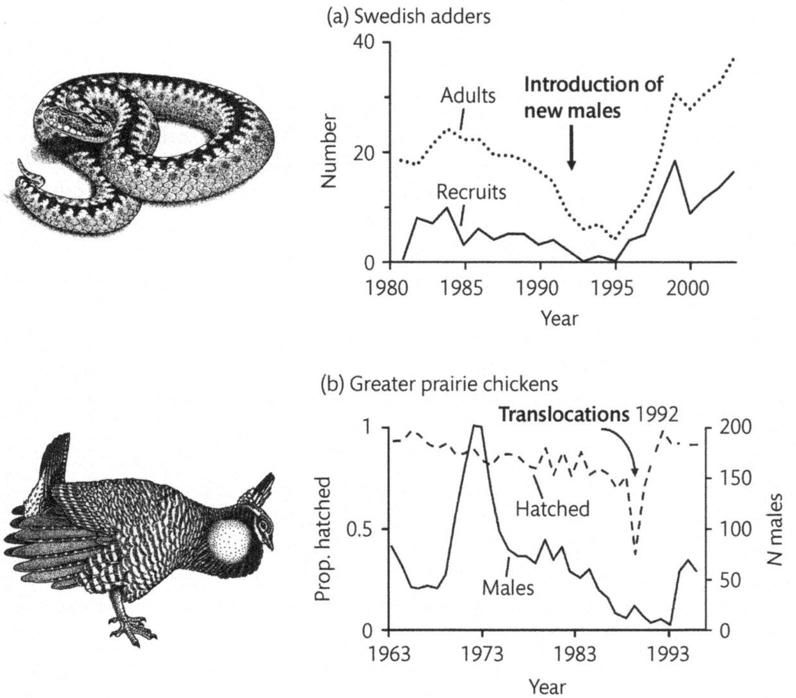

Declines in numbers and fitness due to inbreeding, and genetic rescue following subsequent introduction of immigrants (arrowed) in (a) Swedish adders and (b) greater prairie chickens (North America) (Frankham et al. 2010, Fig. 13.8, after Madsen et al. 2004; and Westemeier et al. 1998, respectively).

Genetic Management of Fragmented Animal and Plant Populations. R. Frankham, J. D. Ballou, K. Ralls, M. D. B. Eldridge, M. R. Dudash, C. B. Fenster, R. C. Lacy & P. Sunnucks. Oxford University Press (2017) © R. Frankham, J. D. Ballou, K. Ralls, M. D. B. Eldridge, M. R. Dudash, C. B. Fenster, R. C. Lacy & P. Sunnucks 2017. DOI 10.1093/oso/9780198783398.001.0001

What are the problems of genetic erosion in small isolated population fragments?

Genetic erosion is a serious problem in many small isolated populations

Small isolated population fragments lose genetic diversity and become inbred over generations, leading to inbreeding depression, reduced evolutionary potential, and increased extinction risk (Chapters 2–5). Many small fragments already have low genetic diversity and are inbred, while others will suffer these problems in the future (Frankham et al. 2010). In this chapter, we consider the prospects for reversing these adverse genetic effects by augmenting gene flow (also referred to as crossing or genetic rescue). For example, Box 6.1 describes genetic rescue of the inbred Florida panther population following gene flow from the Texas population.

Florida panther (USA)

Box 6.1 Genetic rescue of inbred Florida panthers following gene flow

(Roelke et al. 1993; Culver et al. 2000, 2008; Land & Lacy 2000; Driscoll et al. 2002: Mansfield & Land 2002; Hostetler et al. 2010, 2012; Johnson et al. 2010; Benson et al. 2011)

Prior to European settlement, the Florida panther ranged across the entire south-eastern USA, and other sub-species were spread throughout North and South America. By the early 1990s, Florida panthers had declined to a small relict population of ~ 20–25 individuals in southern Florida. Two populations of panthers existed in Florida, one unhybridized remnant and another hybridized with South America panthers. We focus here on the endangered unhybridized population.

The unhybridized Florida panther population had very low levels of genetic diversity for allozymes, DNA fingerprints, microsatellites, and mtDNA compared to earlier museum specimens and other populations of the species (Roelke et al. 1993; Driscoll et al. 2002). They were highly inbred (Chapter 3), and displayed evidence of inbreeding depression, including kinked tails, cardiac defects, a high prevalence of infectious disease, ~ 50% of the males with at least one undescended testis (cryptorchidism), lowered testosterone levels, and low numbers of motile sperm (Roelke et al. 1993).

In 1995, the Florida panther population was augmented with eight wild-caught Texan females, whose population was historically contiguous and genetically connected with it. Five of the eight introduced Texas pumas bred and produced offspring.

The resulting admixed population had ~ 50% higher heterozygosity than the original population, and it had tripled in numbers by 2007 (Johnson et al. 2010). Comparisons of hybrid offspring with unhybridized panthers revealed a 26% benefit in adult and sub-adult survival, 99% advantage in pup survival, and 6.8% higher litter size (Hostetler et al. 2010, 2012; Johnson et al. 2010; Benson et al. 2011). Thus, the overall relative fitness of the crossed individuals is $1.262 \times 1.994 \times 1.068 = 2.688$, i.e. a 169% fitness improvement in the F_1. Admixed offspring also had reduced levels of kinked tails, cardiac defects, and cryptorchidism, improved sperm quality, and increased escape behavior during capture attempts.

Can we reverse inbreeding and loss of genetic diversity?

Inbreeding and loss of genetic diversity can be reversed by crossing to a different unrelated population

Genetic theory predicts that it will be possible to reduce inbreeding and increase genetic diversity of inbred populations by crossing them to another distinct population of the species. For example, if two completely inbred ($F = 1$) populations homozygous for different alleles at a locus are crossed, F_1 individuals have an inbreeding coefficient of zero and all are identically heterozygous (Table 6.1). However, partial inbreeding returns in the F_2: the inbreeding coefficient increases to ½ and the heterozygosity halves. Thus, the reduction in zygotic inbreeding coefficient (ΔF) between the parent and the F_1 is 1, while that between the parent and the F_2 is ½. If only some of the parental inbred individuals are crossed, the inbreeding coefficient will be reduced by a correspondingly smaller proportion in generation 1 and increase at a faster rate over subsequent generations than when all parents are crossed. There are similar benefits of crossing on inbreeding, genetic diversity, and evolutionary potential (see "What variables affect evolutionary rescue?"). But what effects does crossing have on fitness?

Table 6.1 Recovery from inbreeding and low genetic diversity following crossing of distinct completely inbred populations. Inbreeding coefficients (*F*) and genetic diversity in parent populations and in F_1 and F_2 generations are shown, along with the reductions in inbreeding coefficients (ΔF) from parental to crossed generations.

Population	Genotype	F	ΔF	Heterozygosity
Inbred a	A_1A_1	1		0
Inbred b	A_2A_2	1		0
F_1 (a × b)	A_1A_2	0	1	1
F_2 (F_1 × F_1)	¼ A_1A_1: ½ A_1A_2: ¼ A_2A_2	½[a]	½[a]	½[a]

[a] These will remain the same in subsequent generations with random mating in large populations.

Does crossing have beneficial or harmful effects on fitness?

Crossing can have beneficial or harmful effects on fitness, but beneficial effects predominate, and the occurrence of harmful effects is predictable and potentially avoidable

While augmentation of gene flow into inbred populations with low genetic diversity has beneficial effects on inbreeding and genetic diversity, the effects on fitness may be either beneficial (genetic rescue) or, less frequently, harmful (outbreeding depression) (Tallmon et al. 2004; Whiteley et al. 2015). If the populations being crossed are widely diverged genetically, such as horse by donkey species crosses or diploid by tetraploid plant crosses, the progeny may even be sterile. Since the key variables associated with outbreeding depression are known, it is possible to predict the risk that it will occur (Frankham et al. 2011; Chapters 7 and 11).

The remainder of this chapter focuses on the impacts of crossing an inbred population to another population from which it has been isolated for $\leq$ 500 years and where the risk of outbreeding depression is low. These are circumstances relevant to practical conservation attempts to genetically rescue small inbred populations. In what follows, we present the impacts of crossing on fitness as the relative difference of fitness for the crossed progeny compared to the inbred parent (referred to as genetic rescue ΔGR) (Box 6.2).

Box 6.2 Measuring the impact on fitness of interpopulation crossing

A convenient measure of the impacts on fitness of crossing inbred populations is the relative difference caused by genetic rescue (ΔGR), as follows:

$$\Delta GR = \frac{(\textit{fitness of crossed population} - \textit{fitness of inbred population})}{\textit{fitness of inbred population}}$$

Thus, a value > 0 represents beneficial effects of crossing, 0 means equal fitness of crossed progeny and the inbred parent, and negative values indicate outbreeding depression.

For example, inbred African lions in Hluhluwe-iMfolozi Park in South Africa produced an average of 0.465 weaned cubs per female, but this increased to 2.077 following gene flow from Namibian lions. Thus, ΔGR is (2.077 - 0.465)/0.465 = 3.47, a 347% improvement (Trinkel et al. 2008).

Conversely, in the largely selfing poorjoe plant (*Diodia teres*), pollination success was 47.6% for the inbred parent populations, but 44.7% for F_1 population crosses (Hereford 2009b), yielding ΔGR of (44.7 - 47.6)/47.6 = -0.063, a 6.3% decline.

Inbreeding depression is usually reversed by crossing between populations (genetic rescue)

Crossing of populations has frequently been used to successfully reverse inbreeding depression in laboratory and agricultural species, where it is referred to as heterosis, or hybrid vigor (Darwin 1876; Falconer & Mackay 1996; Frankham et al. 2010; Hufbauer et al. 2015a). For example, heterosis has been fundamental to maize production in the USA, and the "green revolution" that has substantially increased yields from several other crop plants (Evenson & Gollin 2003). Similarly, breed or strain crosses are often used to improve productivity of chickens and swine. More recently, gene flow has been used to recover fitness in small inbred populations of several species of wild animals and plants (some of conservation concern), including greater prairie chickens and Swedish adders (chapter frontispiece), bighorn sheep, deer mice (*Peromyscus maniculatus*), Florida panthers (Box 6.1), gray wolves, African lions, Mexican wolves, desert topminnow fish (see Box 6.3), Florida ziziphus, jellyfish trees, and partridge pea plants (*Chamaecrista fasciculata*) (Frankham 2015).

How many genetic rescues have been attempted?

Few genetic rescues have been attempted for conservation purposes

Despite the widespread use of breed and strain crosses and synthetic breeds in domestic plant and animal production, we are aware of only ~ 30 cases where genetic rescue has been attempted for conservation purposes, despite the likely existence of millions of small populations that would benefit from increased gene flow (Chapter 1).

Isolated populations of many species in nature have successfully merged following environmental fluctuations

In contrast to the rare use of genetic rescues in conservation, many mammal, bird, fish, lizard, and plant species throughout the world show evidence of the merging of previously isolated and differentiated populations following climatic cycles, such as glacial advances and retreats (Byrne et al. 2011; Frankham et al. 2011). For example, populations of many European plants and animals have expanded from different glacial refuges, and subsequently hybridized and merged (Hewitt 2000).

How large and consistent are genetic rescue effects for fitness?

Empirical studies reveal large and consistent fitness benefits from gene flow into small isolated inbred populations

One impediment to genetic rescue attempts has been the lack of a quantitative overview of the magnitude of effects. A recent meta-analysis revealed consistently beneficial effects of outcrossing on a wide variety of fitness characters in invertebrates, vertebrates, and plants (92.9% of 156 comparisons beneficial: Frankham 2015). Benefits were reported for total fitness, female and male lifetime reproductive success, fecundity, survival, population growth rate, final population size, gamete quality, and fertilization success.

Median benefits from augmenting gene flow were 148% in stressful (usually wild/field) and 45% in benign (captive/greenhouse) environments for all mating systems combined

Ideally the benefits of crossing inbred populations should be evaluated for total fitness, as this is the definitive measure for conservation purposes. However, such data are challenging to obtain in wild species, and accordingly rare. Nonetheless, the Frankham (2015) meta-analysis encompassed evaluation of composite fitness (comparisons with at least some data on reproduction and survival).

The median increase in composite fitness from augmented gene flow was 57.5% (see selected cases in Table 6.2). For example, there was a 151% genetic rescue effect for total fitness when a small, partially inbred population of the endangered jellyfish tree was crossed to another population in the Seychelles. In the meta-analysis, genetic rescue effects were significantly greater in stressful (usually wild/field; 148%) than in benign (captive/greenhouse; 45%) environments for all mating systems combined. Proxies used in place of total fitness likely underestimate the benefits of crossing, because inbreeding depression is typically greater for total fitness than for its components (Chapter 3).

Table 6.2 Magnitude of genetic rescue effects (ΔGR) for reproductive fitness following gene flow into small, isolated, inbred populations for a sample of species (Frankham 2015).

Taxon	ΔGR (%)	Trait	ΔF	Breeding system[a]
Vertebrates				
African lion (*Panthera leo*)	347	Cubs weaned/female	NA[b]	O
Bighorn sheep (*Ovis canadensis*)	331	Female annual reproductive success	0.25	O
Desert topminnow fish (*Poeciliopsis monacha*)	7,500	Total fitness	NA	O
European tree frog (*Hyla arborea*)	15	Tadpole body mass	0.15	O
Florida panther (*Puma concolor coryi*)	169	Composite fitness	0.58	O
Greater prairie chicken (*Tymphanuchus cupido pinnatus*)	26	Hatching success	0.10	O
Gray wolf (Europe) (*Canis lupus*)	23	Annual population growth	0.14	O
Isle Royale gray wolf (*Canis lupus*)	438	Annual population growth	0.35	O
Mexican wolf (*Canis lupus baileyi*)	184	Composite fitness	≥ 0.17	O
Song sparrow (*Melospiza melodia*)	47	Mean of female and male lifetime reproductive success	0.07	O

South Island robin (*Petroica australis*)	679	Reproductive recruitment/egg	0.21	O
Swedish adder (*Vipera berus*)	233	Male recruitment success	0.75	O
Invertebrates				
Ambrosia beetle (*Xylosandrus germanicus*)	15	Composite fitness	NA	FS/HD
Drosophila fruit fly (*Drosophila melanogaster*)	114	Total fitness	0.19	O
Glanville fritillary butterfly (*Melitaea cinxia*)	211	Egg hatching rate	0.41	O
Mysid shrimp (*Americamysis bahia*)	318	Net increase in *N*	0.31	O
Plants				
Alabama glade cress (*Leavenworthia alabamica*)	64	Total fitness	High	Se
Florida ziziphus (*Ziziphus celata*)	∞	Fertilization success	NA	SI
Italian ryegrass (*Lolium multiflorum*)	43	Flowering heads/plant	0.42	SI
Jellyfish tree (*Medusagyne oppositifolia*)	151	Composite fitness	0.31	SI
Partridge pea (*Chamaecrista fasciculata*)	73	Total fitness	0.06	O
Small scabious (*Scabiosa columbaria*)	114	Composite fitness	0.15	O
Water hyacinth (*Eichhornia paniculata*)	118	No. flowers	0.97	O
White campion (*Silene alba*)	94	Germination %	0.20	O

[a] FS = sib mating; HD = haplodiploid; O = natural outbreeding species; SI = self-incompatible; Se = selfing.
[b] NA = Not available (unknown).

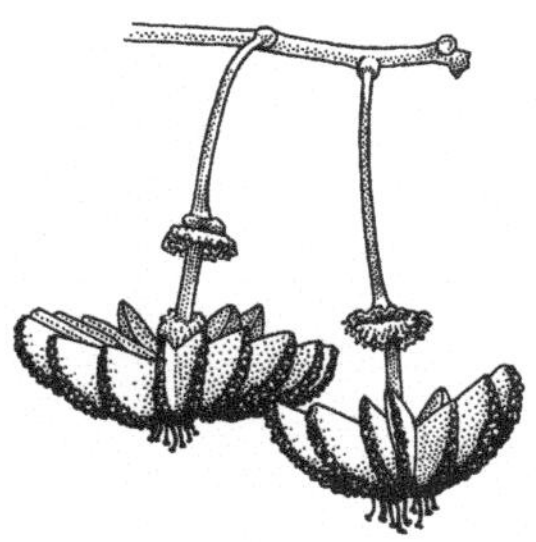
Jellyfish tree (Seychelles)

Changes of outbred to inbred fitness varied from −14% to infinity, the latter being a population of Florida ziziphus, a self-incompatible plant species incapable of fertilization within populations, that only produced seeds following crossing between populations (Weekley et al. 2002; Gitzendanner et al. 2012).

Eight of the nine cases in the meta-analysis where crossing was harmful involved low statistical power and/or base populations that were already highly inbred (Frankham 2015). The only convincing case of outbreeding depression was in a selfing nematode, where both inbreeding depression and genetic rescue effects are expected to be smaller and outbreeding depression more likely than for naturally outbreeding species (Frankham et al. 2011; Chapter 8).

What variables affect the magnitude of genetic rescue effects for fitness?

As genetic rescue (in the absence of outbreeding depression) involves reversal of inbreeding depression (Wright 1977; Vrijenhoek 1994; Falconer & Mackay 1996), the variables predicted to affect it are those that predict inbreeding depression, but with effects in the opposite direction. Thus, our expectations are that genetic rescue will:

- increase with ΔF, the reductions in inbreeding coefficients from parental to crossed generations
- be greater in naturally outbreeding than in habitually selfing or mixed mating species (Byers & Waller 1999; Frankham et al. 2010)
- be larger in wild or stressful environments than in benign ones (Dudash 1990; Armbruster & Reed 2005; Fox & Reed 2011)
- increase with both maternal and zygotic ΔF (Falconer & Mackay 1996)
- be greater when immigrants are outbred rather than inbred (Pickup et al. 2013).

In general, empirical studies support these predictions (Table 6.3).

Table 6.3 Tests for effects of different variables on the magnitude of genetic rescue (ΔGR) (Frankham 2015).

Variable	Median ΔGR (%)	n[a]
Mating system	Outbreeding > selfing	133
Outbreeding	78.8***	
Selfing or mixed mating	16.5	
Environment (outbreeders)	Stressful > benign	114
Stressful/wild	113.9***	
Benign	48.0	
Immigrants	Outbred > Inbred	120
Outbred	113.6***	
Inbred	51.9	
Major taxa (all data)		133
Invertebrates	58.4[ns]	
Vertebrates	94.2	
Plants	59.1	

[a] Number of comparisons.
***$P < 0.001$.
[ns] Non-significant for indicated comparisons.

Box 6.3 documents the combined effects of several of these variables on genetic rescue in the classic study of a small inbred population of desert topminnow fish: there was a 75-fold increase in total fitness, one of the largest benefits recorded, and it involved most of the conditions expected to yield benefits, namely a naturally outbreeding species, a parental population that was highly inbred and exhibiting substantial inbreeding depression, measurement of total fitness in stressful wild conditions, and outbred immigrants.

Box 6.3 Inbreeding depression and genetic rescue in desert topminnow fish in the wild

(Vrijenhoek 1994; after Frankham et al. 2010)

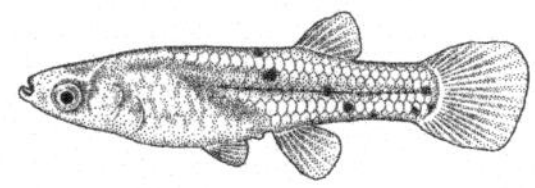

Desert topminnow fish (Mexico)

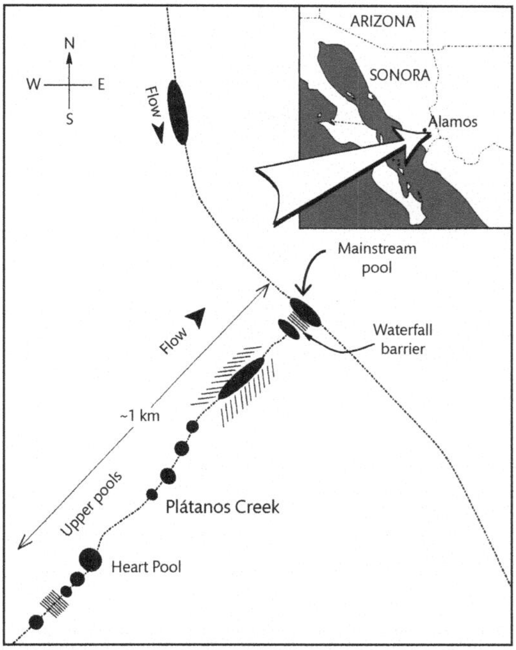

(Frankham et al. 2010, Box 13.1, after Vrijenhoek 1994)

Sexually reproducing populations of the desert topminnow fish in Sonora, Mexico co-exist with a clonal, parthenogenetic fish of the same genus that can be used as a genetically constant control for assessing environmental changes over time. Fish populations inhabiting the upper pools in the Plátanos Creek were extirpated during a severe drought in 1975. Prior to 1975, the sexual and parthenogenetic topminnows represented 76% and 24%, respectively, of the fish density (i.e. the sexual form had a fitness 3.17 times that of the asexual form).

By 1978, these upper pools had been recolonized, but the sexual form was founded by a single gravid female and rapidly became inbred ($F \sim 0.986$, as estimated from allozyme heterozygosity), while the asexual form does not inbreed. After the re-colonization, the inbred sexual form represented only 5% of the topminnow fish, while the re-founded parthenogenetic form represented 95% (i.e. the inbred sexual form now had a fitness only 0.053 that of the asexual form). Thus, inbreeding depression was = 1 – (0.053/3.17) = 0.98.

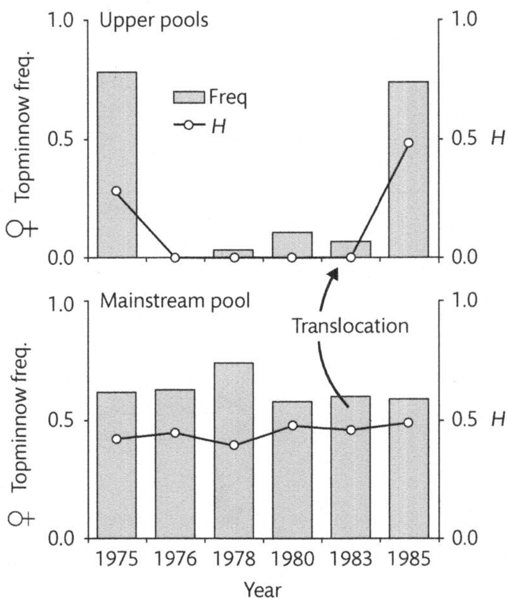

(Frankham et al. 2010, Box 13.1 figure, after Vrijenhoek 1994)

In 1983, 30 genetically variable (outbred) female topminnows from the downstream population were exchanged with 30 inbred topminnow females from the upstream Heart Pool. By 1985, the sexual topminnow had re-established its numerical dominance over the clonal genotype and represented ~ 80% of the fish, indicating that the rescued sexual form had fitness four times that of the asexual form, similar to its fitness prior to the drought. Thus, there has been a 75-fold [(4 – 0.0526)/0.0526] increase in fitness due to gene flow, one of the largest values recorded. Further, heterozygosity in the upstream population recovered following the gene flow.

While temporal changes in fitness may be due to either environmental or genetic differences, or a combination of them, the genetically constant asexual population allows us to conclude that the reduction in fitness was due to inbreeding depression, and the recovery after augmentation to genetic rescue. The lack of change over this time in the frequencies of sexual and asexual fish in the downstream population (where topminnows did not become inbred) supported these conclusions.

Do genetic rescue effects persist across generations?

Benefits of crossing between populations may increase or decrease after the F_1 generation depending on whether the trait is most affected by maternal or zygotic inbreeding, and whether there is outbreeding depression

In commenting on the Frankham (2015) meta-analysis, Waller (2015) worried that harmful effects of crosses may not be manifest until F_2 or later generations. To establish that genetic rescue effects for fitness persist across generations, we need to determine whether they persist until the F_3 generation, because reduction in inbreeding is stable beyond this point for both maternal and zygotic inbreeding effects under large effective population sizes in outbreeding species (Table 6.1). Conversely, in species that regularly self-fertilize, heterozygosity halves in each succeeding generation after the F_1 until there is none, and inbreeding levels increase correspondingly (Chapter 8).

Empirical studies showed that genetic rescue benefits typically persist for many generations in outbreeding species

The percentages of beneficial cases were similar in the F_1, F_2, and F_3, (and later) generations based on a meta-analysis (Table 6.4). Neither the F_2 nor the F_3 differed significantly from the F_1 results. Median fitness benefits of genetic rescue were significantly beneficial in the F_1, F_2, and F_3 generations, respectively, and F_2 and F_3 effects were at least as great as those in the F_1 (Frankham 2016).

Table 6.4 Percentage of beneficial comparisons and median benefits in offspring resulting from crossing inbred populations in the F_1, F_2, and F_3 (and later) generations (Frankham 2016).

Generation	Beneficial %	Median benefits %
F_1	90.5**	42**
F_2	100**	84*
F_3 and later	94.1**	86**

*$P < 0.05$; **$P < 0.01$.

Fitness benefits in outbreeding species persist beyond the F_3 generation, as there was no decline between generations F_3 and F_{16} (R. Frankham, unpublished, based on the Frankham 2016 data set). Further, Bijlsma et al. (2010) reported that genetic rescue benefits did not change between generations F_5 and F_{10}.

Why don't fitness benefits decline after the F_1 generation?

Benefits of crossing between populations due to reductions in zygotic inbreeding are maximal in the F_1 and reduce in the F_2, and are subsequently relatively stable at this level in large populations. However, fitness benefits of crossing due to reduced maternal inbreeding are maximal in the F_2 generation and reduce in the F_3, and are relatively stable thereafter

Some readers may be surprised that the fitness was similar or less in the F_1 than in the F_2 and F_3 (and later) generations in the Frankham (2015, 2016) meta-analyses, because this conflicts with the expectations that zygotically determined traits will show their greatest benefit in the F_1 offspring (Whiteley et al. 2015). In fact, genetic rescue effects in the F_2 and F_3 generations may be either less or greater than those in the F_1, depending on whether traits are determined by the zygotic genotype, or by the maternal genotype (Table 6.5). As some traits are determined by maternal genotypes, others by zygotic genotype, and composite fitness by a combination of these (Roach & Wulff 1987; Falconer & Mackay 1996; Frankham 2015), changes in maternal (ΔF_m) and zygotic (ΔF_z) inbreeding coefficients should both be predictors of the magnitude of genetic rescue effects. In genetic rescues, the reversal of inbreeding depression is delayed by one generation for maternally determined compared to zygotically determined traits (Table 6.5) and the two inbreeding levels are not equal until the F_3 generation. For example, genetic rescue effects on litter size in mice following crossing between populations are only partial when mothers are still inbred, and are not complete until both mothers and offspring have zero inbreeding (Box 6.4); guinea pigs and old-field mice show similar effects (Wright 1977; Lacy, unpublished data).

Table 6.5 Maternal and zygotic inbreeding coefficients in F_1, F_2, and F_3 crosses between (a) different inbred populations, or (b) inbred females and outbred males from different populations (with random mating between multiple individuals in subsequent generations). ΔF_m and ΔF_z are the differences in inbreeding coefficients between inbred parents and crossed generations in mothers and zygotes, respectively (after Frankham 2015).

Generations	Maternal F	Zygotic F	ΔF_m	ΔF_z
(a) Reciprocal crosses between different inbred parental populations with inbreeding coefficients of F_a and F_b with comparisons being made between crossed populations and the mean of the two inbred parents				
F_1	½ $(F_a + F_b)$	0	0	½ $(F_a + F_b)$
F_2	0	¼ $(F_a + F_b)$	½ $(F_a + F_b)$	¼ $(F_a + F_b)$
F_3	¼ $(F_a + F_b)$	¼ $(F_a + F_b)$	¼ $(F_a + F_b)$	¼ $(F_a + F_b)$
(b) Crosses between inbred female and outbred male parent populations with inbreeding coefficients of F_a and 0, respectively, with comparisons being made between crossed populations and the inbred parent				
F_1	F_a	0	0	F_a
F_2	0	¼ F_a	F_a	¾ F_a
F_3	¼ F_a	¼ F_a	¾ F_a	¾ F_a

Box 6.4 Effects of maternal and zygotic inbreeding and crossbreeding on litter size in mice

(Roberts 1960)

Replicate house mice populations from an outbred strain were inbred by full-sib mating for three generations. Crosses were performed between the inbred lines, and litter size recorded each generation. Inbreeding coefficients in mothers and offspring are given for the outbreds, inbreds, and crosses in the following table:

Generation	Inbreeding coefficients	
	Mother	Offspring
Outbred	0	0
Full-sib 1	0	0.25
Full-sib 2	0.25	0.375
Full-sib 3	0.375	0.50
F_1 (inbred × inbred)	0.5	0
F_1 × different F_1	0	0

Litter size declined with inbreeding and recovered following crossing (see graph) due to changes in both maternal and zygotic inbreeding. For example, litter size recovered from 5.69 for inbred offspring from inbred mothers to 6.20 pups/litter due to the reduction in zygotic inbreeding, and to 8.47 when the mothers, as well as the zygotes had inbreeding coefficients of 0, due to reductions in maternal and zygotic inbreeding.

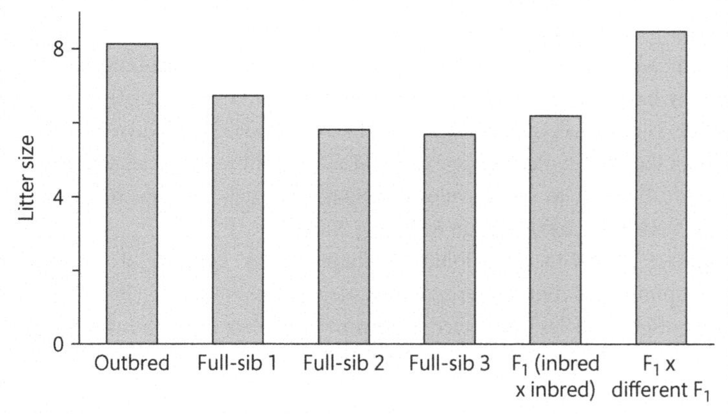

Loss of pups due to inbreeding (2.43) was not significantly different from the gain in pups due to crossing (2.78), supporting the view that genetic rescue (heterosis) represents recovery from inbreeding depression.

Thus, the apparently paradoxical results of the meta-analysis that fitness was similar or less in the F_1 than in later generations indicates that maternal inbreeding is affecting composite fitness, and accordingly maternal ΔF was shown to be a predictor of genetic rescue (Frankham 2015). Different relative effects of maternal and zygotic inbreeding can explain apparently contradictory patterns of relative fitness in F_1 and later generations in multiple meta-analyses. For example, McClelland & Naish (2007) found similar results to Frankham (2015), while Whitlock et al. (2013) found $F_1 > F_2$ for fitness traits across diverse animal and plant taxa (but they avoided traits subject to large maternal effects). Clearly, changes in maternal as well as zygotic inbreeding should be routinely included in considerations of genetic rescue.

There was no statistical support for a zygotic ΔF effect on genetic rescue for fitness in the Frankham (2015) data set, but this was probably an artifact due to truncation of the life cycle in most data sets, thus missing much of the later life where zygotic effects are expected to predominate, and due to the variability of the data. Such ΔF_z effects do exist, as revealed by the significant F_1 genetic rescue effect shown in Table 6.4, based on a different analysis on the same data set (Frankham 2015). Further, they have been documented in extensive maize data sets (Lindstrom 1941; Stringfield 1950; Moll et al. 1965) and they are evident in the mice litter size data presented in Box 6.4.

Mating system effects

Genetic rescue effects are strongly affected by mating system (outbreeders > inbreeders)

Benefits of gene flow into inbred populations were much greater for naturally outbreeding than for inbreeding taxa (median ΔGR of 79% versus 16%, respectively: Table 6.3), as predicted. At a more detailed level, the greatest benefit was seen in self-incompatible species (median ΔGR 78%) > other outbreeders (59%) > mixed mating (16%) < selfers (39%). While the latter two are in reverse order compared to expectations, the number of data points was only small (six for selfing species).

Self-incompatible (SI) plant species suffer an additional impact from loss of genetic diversity (in addition to inbreeding depression and reduced evolutionary potential) because they have genetic systems that reduce matings between relatives (Richman & Kohn 1996; Young et al. 2000; Chapter 8). These self-incompatibility mechanisms are regulated by loci that typically have many SI alleles within a species and loss of these alleles eventually leads to some ovules not being fertilized, and thus to reduced fitness (Castric & Vekemans 2004; Pickup & Young 2008).

The impact of self-incompatibility is illustrated by Lakeside daisies in Illinois, Florida ziziphus, and button wrinklewort daisies in Australia (Box 6.5). Loss of self-incompatibility alleles is expected to become a problem in small isolated population fragments of all self-incompatible plant species.

Box 6.5 Decline in self-incompatible Lakeside daisies, Florida ziziphus, and button wrinklewort daisies due to loss of alleles at self-incompatibility loci, and recovery following gene flow

(Demauro 1993; Young et al. 2000; Weekley et al. 2002; Gitzendanner et al. 2012)

The Lakeside daisy population from Illinois, USA declined to three plants with few self-incompatibility alleles, such that it did not reproduce sexually for 15 years—it was functionally extinct (Demauro 1993). However, plants produced viable seed when fertilized with pollen from a large population in Ohio. A similar dire situation was remedied with augmented gene flow between isolated populations of Florida ziziphus (Gitzendanner et al. 2012). The endangered button wrinklewort daisy in eastern Australia also exhibits reduced fitness in smaller populations with reduced numbers of SI alleles and displayed genetic rescue when populations were crossed (Pickup 2008; Pickup & Young 2008).

Florida ziziphus (USA)

Genetic rescue of fitness in domestic animals and plants

Benefits of genetic rescue on fitness traits have also been found in domestic species

When populations of domestic animals and plants are crossed to conspecific populations, they exhibit similar benefits to those documented above for wild species (Table 6.6). For example, grain yield in F_1 crosses between inbred lines of maize is on average 190% higher than the mean of the inbred parents. Benefits in the other domestic species were lower, because many were inbreeders rather than outbreeders, their environments were relatively benign, and the studies reported fitness components, rather than composite fitness.

Table 6.6 Mean genetic rescues (ΔGR) for fitness traits in a range of domestic plant and animal species.

Species	ΔGR (%)	Trait	Breeding system	Reference
Plants				
Maize	190	Grain yield	Outbreeding	1
Sorghum	100	Grain yield	Mainly selfing	1
Cotton	48	Grain yield	Selfing	1
Wheat	29	Grain yield	Selfing	1
Barley	32	Grain yield	Selfing	1
Tomato	45	Fruit yield	Selfing	2
Animals				
Poultry	22	Eggs to 500 days	Outbreeding	3
Swine	72	Litter weight at 154 days	Outbreeding	4

References: 1. Sinha & Khanna (1975); 2. Williams & Gilbert (1960); 3. Shoffner et al. (1966); 4. Dickerson et al. (1946).

Natural experiments: evolution of invasiveness following population crossing

In some species of animals and plants, crosses of differentiated populations have led to invasive forms

Merging of genetically differentiated colonists within animal and plant species has sometimes been so "successful" that they have become invasive (Ellstrand & Schierenbeck 2000; Kolbe et al. 2007; Sarre et al. 2014). For example, partially differentiated populations of the brown anole lizard (*Anolis sagrei*) from Cuba were introduced into Florida, merged, became invasive, and then spread to several other counties (Kolbe et al. 2007). Further, multiple introductions (and higher genetic diversity) are associated with invasiveness for European starlings (*Sturnus vulgaris*) and house sparrows (*Passer domesticus*) in North America, some house mouse populations, and a range of plant species (Sakai et al. 2001; Frankham 2005; Schierenbeck & Ellstrand 2009).

Can we prevent inbreeding depression with repeated immigration?

Regular low levels of immigration can prevent fitness declines in small isolated populations

Inbreeding depression for fitness in small isolated populations can be avoided by augmentation with a few immigrants each generation. This is expected to greatly reduce inbreeding and has been shown to minimize inbreeding depression in both plants and animals. For example, 30 populations of field mustard (*Brassica campestris*) in the USA were maintained at population sizes of five individuals for six generations, with 0, 1, or 2.5 immigrants (Newman & Tallmon 2001). Inbreeding rose rapidly in the isolated populations, but 2.5 immigrants per generation prevented inbreeding from increasing, whereas populations with a single immigrant per generation had an F of 0.08 at generation 6 (Fig. 6.1a). Both immigrant treatments had much higher fitness than the totally isolated populations (Fig. 6.1b). Similarly, small bottlenecked house fly populations maintained with either 0, 1, or 20 immigrants per generation for 24 generations had larval survivals of 22%, 43%, and 53%, respectively, in the final generations. Further, 66% of the totally isolated populations went extinct, whereas all populations with regular immigration survived (Bryant et al. 1999).

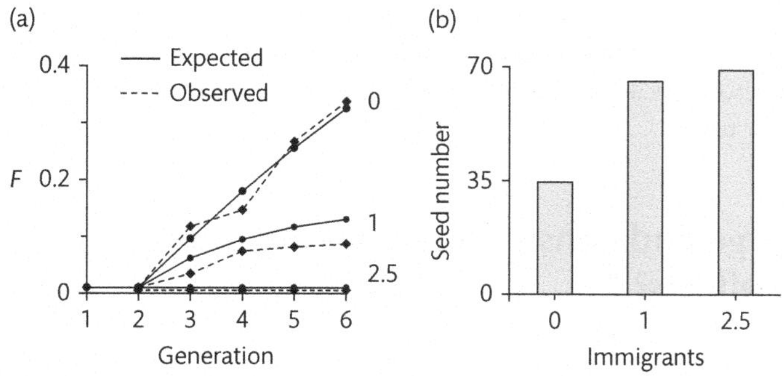

Fig. 6.1 (a) Observed (pedigree) and expected inbreeding coefficients (*F*) for the 0, 1, and 2.5 immigrants per generation treatments in self-incompatible field mustard, and (b) seed numbers for plants of the 0, 1, and 2.5 immigrant treatments at generation 6 in an outdoor common garden (after Newman & Tallmon 2001).

Are immigrant alleles at a selective advantage?

Immigrant alleles often have a selective advantage, so the advantage of crossing will often improve over generations beyond the F_3 level

Rare immigrant alleles are expected to be at a selective advantage and to increase in frequency, when:

- the inbred recipient population is fixed for harmful alleles at loci contributing to inbreeding depression and the immigrants carry advantageous alleles
- a locus exhibits heterozygote advantage and the recipient population is fixed for one allele (contributing to inbreeding depression), and the immigrants carry other alleles whose initial frequencies in the cross are below equilibrium values.

Selective advantages of rare immigrant alleles have been reported in bighorn sheep, mice, wolves, *Drosophila* fruit flies, water fleas (*Daphnia magna*), butterflies (*Bicyclus anynana*), and mustard plants (Scriven 1992; Ball et al. 2000; Newman & Tallmon 2001; Ebert et al. 2002; Saccheri & Brakefield 2002; Miller et al. 2012). For example, a single male immigrant upon entering the inbred Isle Royale gray wolf population (in Lake Superior, North America) represented 14% of the recipient gene pool, but its contribution rose to 56% within 2.5 generations. It fathered all of the next generation of progeny and its genetic contribution increased thereafter (Adams et al. 2011). The mean advantage per generation of immigrant alleles (*s*) over six studies was 38%, with a range from 3.5% to 86% (R. Frankham, unpublished).

Consequently, a small initial genetic contribution from immigrants may amplify across generations, leading to distorted founder representation and to populations becoming inbred again more rapidly than predicted under neutral expectations.

So far, we have focused primarily on the fitness benefits of crossing over the first few generations. However, the ability to evolve, especially in response to environmental change is also expected to be enhanced by augmentation of genetic variation, i.e. evolutionary rescue.

How large and consistent are evolutionary rescue effects?

Crossing restores genetic diversity and the ability to evolve in response to environmental change

If populations with low evolutionary potential are crossed to unrelated inbred or outbred populations, genetic diversity and the ability to evolve are expected to increase (Aitken & Whitlock 2013; Orr & Unckless 2014; Frankham 2015). Empirical studies support this—all six available studies exhibited improvements in evolutionary potential for fitness traits following crossing, the median improvement being 22.4% per generation (Frankham 2015). For example, crosses of replicate *Drosophila* populations previously maintained with N_e = 50 for 50 generations had 79% higher evolutionary potential than their inbred parent populations when all were subjected to increasing concentrations of NaCl (Margan et al. 1998). Further, Stelkens et al. (2014) have observed evolutionary rescue effects for ability to tolerate increasing NaCl concentrations in populations of yeast (*Saccharomyces*). For traits more peripherally related to fitness, 35 of 39 studies exhibited beneficial effects of gene flow on evolutionary rescue, and the few exceptions had low statistical power (R. Frankham, unpublished analysis, based on studies listed in Frankham 2015).

What variables affect evolutionary rescue?

The improvement in evolutionary potential due to gene flow depends on the increase in genetic diversity and the extent to which improved fitness increases the selection differential

The extent of evolutionary genetic adaptation (*GA*) in the short term is predicted by the breeders' equation $GA = h^2Sd$ (Chapter 4) to depend on differences in heritability and selection differential between the parents and offspring of the cross. Thus, the ratio of evolutionary potential in a cross (X) to that of its inbred, low genetic diversity parent (I) (evolutionary rescue ratio *ER*) is:

$$ER = \frac{h_X^2 Sd_X}{h_I^2 Sd_I} \qquad 6.1$$

Since heritabilities are a function of heterozygosity (Falconer & Mackay 1996), the effects of crossing on heritabilities should be related to the ratios of heterozygosities (H_X/H_I) in F_2

and later generations (Frankham 2015). The effect can be predicted as $\text{heterozygosity}_{\text{cross}}/\text{heterozygosity}_{\text{inbred parent}}$, or from inbreeding coefficients as $(1 - F_X)/(1 - F_I)$.

Empirical data for fitness traits support this prediction (Fig. 6.2). All of these data came from invertebrates, but similar results are expected for vertebrates and plants. Accordingly, increased evolutionary potential arising from crossing between populations has also been reported in selection experiments for sternopleural bristle number (a non-fitness trait) in *Drosophila*, body weight in mice, and for nest-building behavior in mice (Falconer & King 1953; Robertson 1969; Eisen 1975; Bult & Lynch 2000; Swindell & Bouzat 2006).

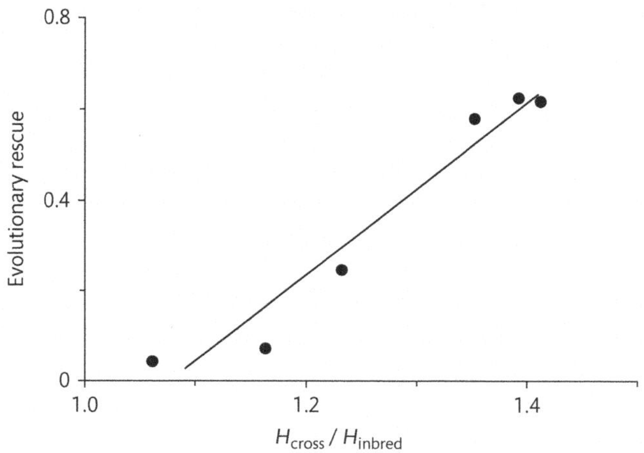

Fig. 6.2 Evolutionary rescue for fitness traits (GR_{Evp}/generation) increases with the ratio of heterozygosities (H) in crossed versus inbred populations, estimated as $(1 - F_X)/(1 - F_P)$, with the line of best fit shown (after Frankham 2015).

Genetic rescue effects for evolutionary potential will also increase if crossing increases the selection differential. Crossing typically increases fitness, and with a constant number of parents per generations, the proportion of the population contributing as parents of the next generation will decrease, thereby increasing the selection differential compared to that in the parent generation (see Box 4.3). However, quantifying the size of this effect requires information that may not be readily available.

Genetic rescue should be routinely considered as a conservation option for small outbreeding populations with limited gene flow. The current reluctance to attempt genetic rescues in conservation settings is not justified scientifically, given the information in this chapter. We recommend a much broader use of augmented gene flow to enhance fitness, population persistence, and evolutionary potential in small inbred populations, and ultimately to reduce species extinctions.

Summary

1. Crossing of an inbred population to an unrelated population reduces inbreeding, and increases genetic diversity.
2. Crossing may have beneficial or harmful effects on fitness, but beneficial effects are more common, and the circumstances associated with harmful effects (outbreeding depression) are understood and predictable.
3. When the risk of outbreeding depression is low, the effects on composite fitness of crossing an isolated inbred population to another population are overwhelmingly beneficial, with a median observed change in composite fitness of 148% under stressful conditions and 45% under benign conditions.
4. Genetic rescue effects typically persist to at least the F_3 generation and are expected to stabilize by this time in outbreeding species, but not in habitually self-fertilizing species.
5. Genetic rescue effects are dependent on mating system (outbreeders > inbreeders).
6. The size of genetic rescue effects increases with both maternal and zygotic ΔF.
7. Augmentation of gene flow between populations also strongly benefits the ability of populations to evolve to cope with environmental change.
8. We recommend use of augmented gene flow between populations of outbreeding species to reverse the detrimental effects of inbreeding and loss of genetic diversity when the proposed crosses have a low risk of outbreeding depression.

FURTHER READING

Demauro (1993) Documents genetic rescue of the Illinois population of self-incompatible Lakeside daisy that had not reproduced for 15 years.

Ebert et al. (2002) Documents a selective advantage of immigrant alleles in water flea (*Daphnia*) populations.

Frankham (2015, 2016) These meta-analyses revealed that gene flow into small inbred populations resulted in large and consistent benefits on fitness and evolutionary potential, and that the fitness benefits persisted until at least the F_3 generation.

Johnson et al. (2010) Describes the beneficial fitness effects of crossing a small inbred population of Florida panthers to Texas individuals.

Newman & Tallmon (2001) Empirical evaluation of the impacts of regular migration into small population fragments on inbreeding and fitness in a plant.

Outbreeding depression is uncommon and predictable

CHAPTER 7

The progeny of crosses between some populations exhibit harmful effects on fitness (outbreeding depression). These are primarily due to populations being different taxa, having fixed chromosome differences, being genetically adapted to different environments, having a long history of isolation, or combinations of these. Even if outbreeding depression occurs as a result of crossing, it is often only temporary, as natural selection acts to remove it, especially in large populations.

TERMS

Centromere, centric fusion, chromosomal translocation, coadapted, coadapted gene complexes, common ancestor, decision tree, epistasis, heterochromatin, heterozygote disadvantage, inversion, linkage disequilibrium, polymorphic, polyploid, telomere, tetraploid

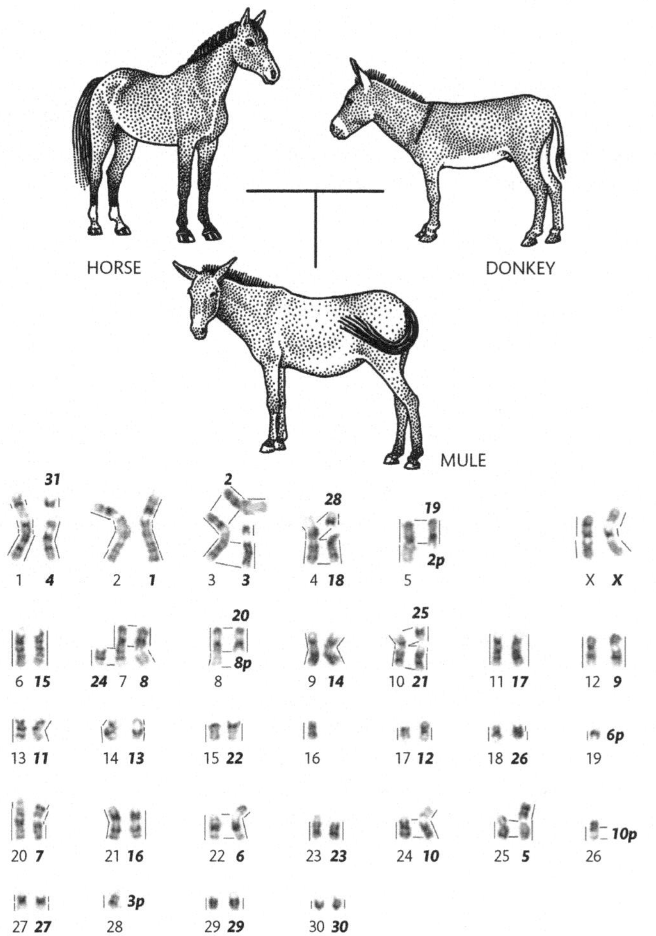

Classic example of outbreeding depression: a horse (*Equus caballus*) × donkey (*Equus asinus*) mating produces sterile F_1 mule offspring. Haploid chromosomes of donkey (numbered in plain type) and horse (numbered in bold and italics) are shown with homologous regions aligned: the former have 31 pairs and the latter 32, with many chromosomal translocations, transpositions, centric repositionings, and inversions. Only nine chromosomes are unchanged between the two species (Yang et al. 2004; Carbone et al. 2006).

Genetic Management of Fragmented Animal and Plant Populations. R. Frankham, J. D. Ballou, K. Ralls, M. D. B. Eldridge, M. R. Dudash, C. B. Fenster, R. C. Lacy & P. Sunnucks. Oxford University Press (2017) © R. Frankham, J. D. Ballou, K. Ralls, M. D. B. Eldridge, M. R. Dudash, C. B. Fenster, R. C. Lacy & P. Sunnucks 2017. DOI 10.1093/oso/9780198783398.001.0001

Why are we concerned about harmful crosses?

Some crosses between populations result in reduced reproductive fitness (outbreeding depression), at least in initial generations

A serious impediment to conservation managers undertaking genetic rescue of small inbred populations is the perceived risk of outbreeding depression (Frankham et al. 2011; Frankham 2015; Banes et al. 2016). But how serious a risk is it, and can that risk be predicted?

If crosses are made between distinct species, or populations adapted to different environmental conditions and/or separated by long periods of times, they may result in harmful effects (Coyne & Orr 2004; Frankham et al. 2011). Species crosses are often harmful—horses crossed with donkeys produce sterile mules, because these species have multiple fixed chromosomal differences, have been isolated for ~ 2 million years, and have evolved on different continents (Chapter frontispiece). However, crosses between species are not always harmful (Arnold & Hodges 1995).

Crosses within species may also be harmful, if the crossed populations are sufficiently distinct (Endler 1977; Templeton 1986; Thornhill 1993; Edmands 2007; Frankham et al. 2011, 2012; Frankham 2015). For example, Box 7.1 illustrates adaptive differences between partridge pea populations from varied geographic distances in the USA and the effects of crossing between different populations—F_1 crosses were beneficial, but F_3 crosses at distances of 1,000 km or more were harmful.

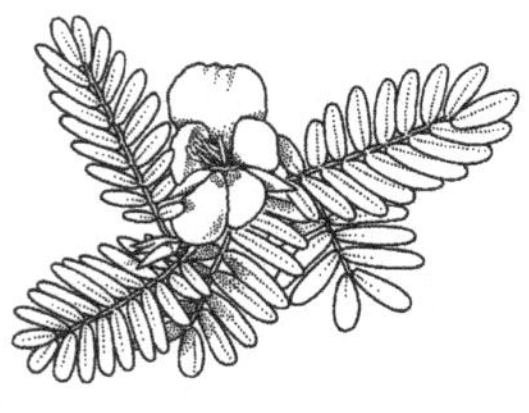

Partridge pea (USA)

Box 7.1 Effects of crossing populations of partridge peas across their range in the USA: adaptation, heterosis, and outbreeding depression

(Fenster & Galloway 2000; Erickson & Fenster 2006)

Fenster and colleagues compared progeny performance from crosses between populations of the annual partridge pea from within Maryland (MD), Illinois and Kansas (KS). The resulting progeny were grown in the three different locations. Populations 1,000 km or more apart showed significant local adaptation, measured as the difference between the fitness of progeny from the home and away parents (home site advantage) (upper panel of the graph). As this species exists in small fragmented populations, F_1 crosses over all distances were beneficial compared to the progeny of the home parent (heterosis/genetic rescue: middle panel). However, F_3 crosses at the two longer distances were harmful compared to the progeny of the home parent (bottom panel), i.e. they exhibited outbreeding depression. However, as we detail later, even these crosses were beneficial by the F_6 generation, due to natural selection acting on the genetic diversity within them. In each case performance is plotted relative to the home parent, with zero reflecting the home parent and values above and below indicating performance better or worse than the home parent. Asterisks indicate performance significantly different from that of the home parent at the $P < 0.05$ level.

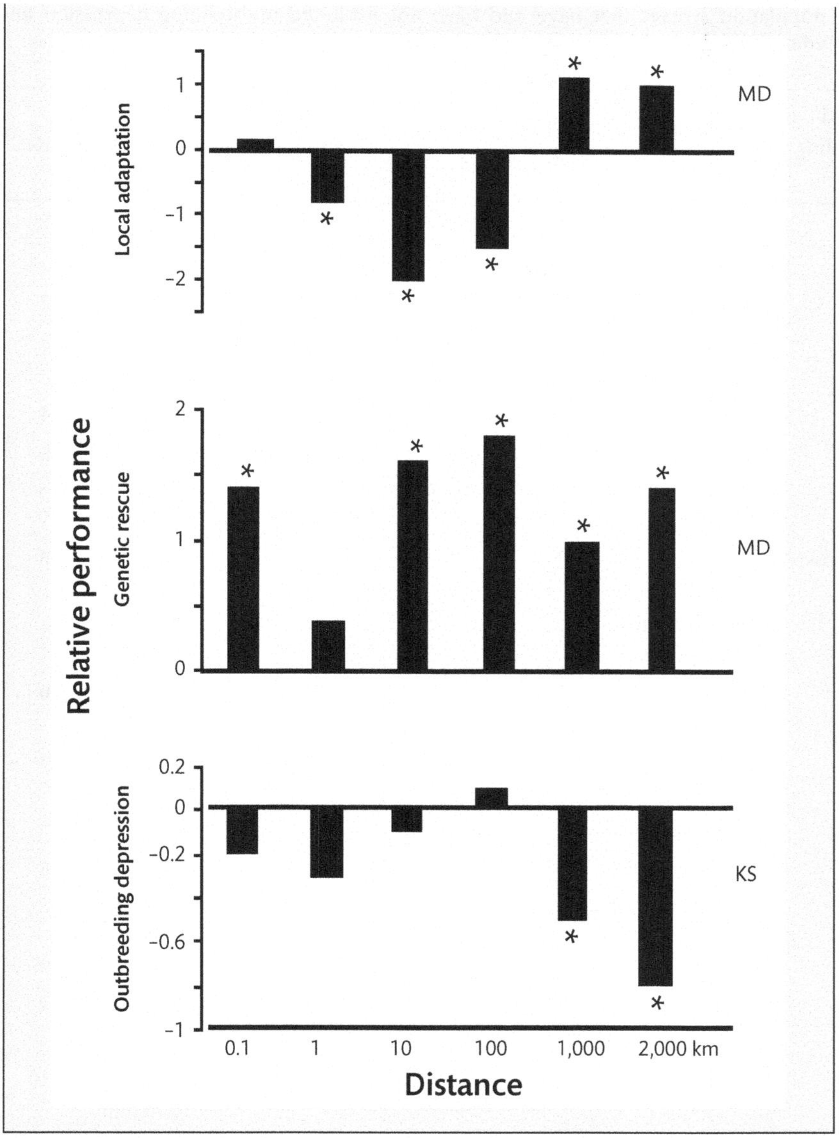

Table 7.1 lists examples of crosses within and between species that exhibit or do not exhibit outbreeding depression, along with information on characteristics that are expected to predict the risk of outbreeding depression (see later). In what follows, we concentrate primarily on the frequency and severity of outbreeding depression within species and its causes.

Table 7.1 Examples of species and population crosses that have and have not exhibited outbreeding depression and associated characteristics likely to predict the risk of adverse consequences (after Frankham et al. 2011).

Taxa	Observed outbreeding depression	Taxonomic status of populations	Fixed chromosome differences	Gene flow in last 500 years	Environments	Ref[a]
Corroboree frog (*Pseudophryne corroboree* and *P. pengilleyi*)	Yes	2 species	?	No	Similar	1
Ibex (*Capra ibex ibex*)	Yes	3 species[b]	Unlikely	No	Different	2
Owl monkey (*Aotus trivirgatus*)	Yes	Some consider them 2 species	Yes	Unlikely		3
Kirk's dik dik (*Madoqua kirki*)	Yes	>1 species	Yes	Unlikely		4
Button wrinklewort (*Rutidosis leptorrhynchoides*)	Yes	Called 1 species	Yes, hexaploid, tetraploid, and diploid	Unlikely	Similar	5
Peromyscus (*P. polionotus leucocephalus* × *P. p. subgriseus* or *P. p. rhoadsi*)	Yes, modest	Beach mouse sub-species × old-field mice sub-species	No	No	Different	6
Peromyscus (*P. p. subgriseus* × *P. p. rhoadsi*)	No	Sub-species of old-field mouse	No	Likely	Similar	6
Partridge pea (*Chamaecrista fasciculata*)	Yes	Crosses between distant populations	No	Unlikely	Different	7
Copepod (*Tigriopus californicus*)	Yes	Likely a species complex	No	No	Some different, some similar	8
Golden lion tamarin (*Leontopithecus rosalia*)	No	1 species	No	Yes	Similar	9
Florida panther (*Puma concolor coryi*)	No	1 species[c]	No	Yes	Moderately different	10
Pink salmon (*Oncorhynchus gorbuscha*)	Unclear	1 species	No (odd and even years have 52 vs 52, 53, and 54 chromosomes)	No	Similar[d]	11

[a] References: 1. Osborne et al. (1996); 2. Turček & Hickey (1951); Templeton (1986); Wilson & Reeder (2005); 3. de Boer (1982); 4. Ryder et al. (1989); 5. Young & Murray (2000); 6. Lacy (1998); 7. Fenster & Galloway (2000); 8. Edmands (1999); Edmands (2002); Lee (2000); 9. Ballou (1995); 10. Culver et al. (2000); Johnson et al. (2010); 11. Phillips & Kapuscinski. (1988); Allendorf & Waples (1996, p. 254); Gharrett et al. (1999); Churikov & Gharrett (2002); Frankham et al. (2011).

[b] Populations used are uncertain, but probable candidates are now classified as separate species.

[c] Previously classified as two sub-species, but subsequently revised based on molecular studies.

[d] Distributions partially different so may be adapted to partially different environments.

Crosses of populations can simultaneously result in beneficial and harmful effects due to different impacts at diverse loci. For example, Lacy (1998) found that crosses between three native mice sub-species in the USA were overwhelmingly beneficial (genetic rescue), but the benefits were smaller than expected in the F_2 and backcross generations due to modest harmful effects (Box 7.2).

Box 7.2 Simultaneous beneficial and harmful effects in crosses between *Peromyscus* sub-species

(Lacy 1998, unpublished data)

Lacy (1998, unpublished data) carried out F_1, F_2, and F_3 crosses, backcrosses, and 3-way crosses between three sub-species of *Peromyscus polionotus*—two of which are old-field mice (*P. p. subgriseus* or *P. p. rhoadsi*), and one of which is a beach mouse (*P. p. leucocephalus*) that occupies a different habitat, is more divergent genetically, has lower genetic diversity, and is inbred due to long isolation on an island.

All of the F_1, F_2, and F_3 crosses had greater fitness (total mass of offspring weaned) than the mean of the respective parental populations: the respective means for parents, F_1, F_2, and F_3 were 33.3, 51.4, 55.5, and 51.0 g. The higher fitness in the F_2 than the F_1 was due to maternal effects—F_1 females had greater reproductive success than did non-hybrid dams.

After adjusting for maternal effects, the backcrosses and 3-way crosses (overall mean reproductive success of 47.9 g and 47.3 g) showed modest reductions in fitness compared to what would have been predicted from the heterosis shown in the F_1, F_2, and F_3 crosses. Thus, there were some deleterious consequences of mixed ancestry, but these were much less than the beneficial effects. Moreover, this weakly harmful effect was observed only in the crosses between the beach and old-field sub-species inhabiting different environments.

Old-field mouse (USA)

Frequency and magnitude of outbreeding depression

The precise frequency and severity of outbreeding depression in crosses within species is not known, but it is less frequent and, on average, of a lesser magnitude than inbreeding depression

Surprisingly, there is limited information on the frequency and the magnitude of outbreeding depression, and reports in the literature are conflicting, including those from three meta-analyses. In fish, population crosses within species varied in effect, but had on average beneficial fitness effects in F_1 and F_2 generations (McClelland & Naish 2007). Conversely, Whitlock et al. (2013) found that fitness did not differ significantly from the means of parents in the F_1 (1.3% benefit), but was 8.8% lower in the F_2 for plant and animal studies. However, Frankham (2015, 2016) found that the effects on fitness of crossing inbred populations to another population adapted to similar environments and with the same karyotypes were overwhelmingly beneficial, and of large magnitude in the F_1, F_2, and F_3 and later generations. The variation in results from the Frankham studies and the others is probably because the other authors did not evaluate whether effects were related to chromosomal or adaptive differences between populations.

Edmands (2002) reported that outbreeding depression across a range of species averaged 9.0% in F_1 and 12.5% in F_2 crosses, while inbreeding depression within populations was 27.3%. Rollinson et al. (2014) also found that outbreeding depression was less frequent and less severe than inbreeding depression in Atlantic salmon (*Salmo salar*).

What mechanisms generate outbreeding depression?

In crosses between populations within species, outbreeding depression arises primarily when there are fixed chromosomal differences between populations, and/or they are adapted to different environments, with the risk typically increasing with generations of isolation

Four primary mechanisms have been proposed for generating outbreeding depression in within species crosses (Frankham et al. 2011):

1. Fixed chromosomal differences
2. Adaptive differentiation among populations
3. Coadapted gene complexes
4. Population bottlenecks and genetic drift.

As we shall see, empirical evidence provides little support for the hypothesis that genetic drift is an important factor in rapid development of outbreeding depression.

We use the terms outbreeding depression and reproductive isolation interchangeably to encompass reductions in reproductive fitness in the progeny of population crosses due to any combination of pre-zygotic and/or post-zygotic effects (Frankham et al. 2011).

Mechanism 1: Fixed chromosomal differences

Populations with fixed chromosomal differences have elevated probabilities of outbreeding depression when crossed

Fixed chromosomal differences between populations are well-established causes of reproductive problems when such populations are crossed, as we saw in the horse by donkey cross (White 1978; Rieseberg 2001; Severns & Liston 2008; Frankham et al. 2011; Charron et al. 2014). The major types of chromosomal differences involved in causing outbreeding depression are (Table 7.2):

- ploidy
- translocations
- inversions
- centric fusions.

When populations with such differences are crossed, the F_1 progeny produce some gametes with portions of the genome missing and other parts duplicated (unbalanced chromosomal constitution). The severities of outbreeding depression are usually greatest with ploidy mismatches, next most severe in the presence of chromosomal translocations or complex centric fusions, followed by simple centric fusions and inversions, as detailed below. Adverse effects typically increase with the number of fixed chromosomal differences (White 1973; Lai et al. 2005).

Ploidy differences

Polyploid populations have chromosome numbers that are multiples of the haploid number (n) beyond diploid (2n), such as tetraploid (4n), hexaploid (6n), etc. The progeny of crosses between populations with different ploidy levels are typically sterile

Many plant species have populations with different ploidy levels, yet are still referred to as the same species, and crosses between them are harmful (Frankham et al. 2011). For example, the endangered button wrinklewort daisy in eastern Australia exhibits diploid, tetraploid, and hexaploid populations with similar phenotypes (Fig. 7.1). Progeny of crosses between diploid and tetraploid forms are triploid and usually sterile because meiosis produces gametes with a great diversity of chromosome numbers. Crosses between other ploidy levels result in similar problems.

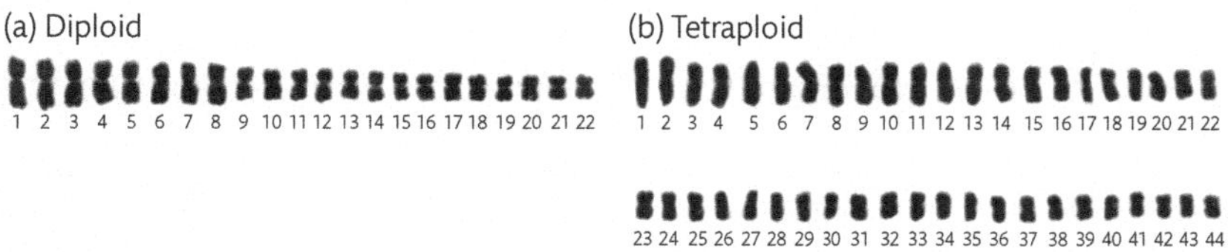

Fig. 7.1 Karyotypes in (a) diploid (2n = 22) and (b) tetraploid (2n = 44) button wrinklewort daisies (after Murray & Young 2001, Fig. 3).

Button wrinklewort daisy (Australia)

Chromosomal translocations

Translocations are chromosomal rearrangements where segments of non-homologous chromosomes are swapped between chromosomes. Meiosis in translocation heterozygotes results in ~ 50% of unbalanced gametes

Chromosomal segregation during meiosis in reciprocal translocation heterozygotes (see Table 7.2) results in about 50% of the gametes being chromosomally unbalanced, causing about one-half of the gametes to be inviable in plants and about one-half of the zygotes to be inviable in animals. A direct causal relationship between chromosomal translocations and pollen sterility has been reported in *Mimulus* plant species hybrids (Stathos & Fishman 2014).

Table 7.2 Associations of fixed chromosomal differences with outbreeding depression when populations are crossed (Frankham et al. 2011).

Chromosomal variant	Examples of population cross	Genetic effect	Fitness effect
Polyploid	4n[a] (tetraploid) × 2n (diploid)	3n (triploid)	High F_1 sterility
Translocation	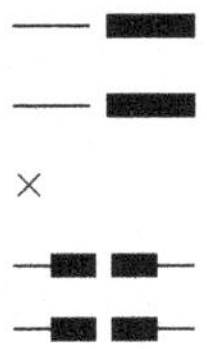	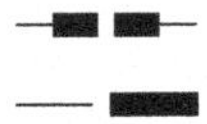 F_1 normal, but meiosis results in ~ ½ gametes with unbalanced chromosomal constitution	~ ½ reduction in gamete (plants) or zygote (animals) viability
Centric fusion (chromosome formed by fusion at the centromere of two previously acrocentric chromosomes)	V V[b] × I I I I[c] (2n – 2) × 2n	V I I F_1 normal, but meiosis results in some unbalanced gametes	Modest reduction in gamete (plants) or zygote (animals) viability
Inversion	ABCDE[d] / ABCDE × ADCBE / ADCBE	ADCBE / ABCDE F_1 normal, but meioses with crossovers in the inverted region yield ~ ½ gametes with unbalanced chromosomal constitution (a minority of meioses)	Modest reduction in gamete (plants) or zygote (animals) viability
Duplication (or deletion) of heterochromatin		F_1 and F_2 have normal complement of functional loci	Little or no change in reproductive fitness of F_1 or F_2

[a] n is number of haploid sets of chromosomes.
[b] Two meta-centric chromosomes.
[c] Four acrocentric chromosomes with the same gene content as the two meta-centric chromosomes.
[d] Gene order within a chromosome.

Inversions

Inversions are chromosomal rearrangements that result from two breaks within a chromosome, with the segment in the middle flipping through 180 degrees and the ends rejoining. When there are single crossovers in the region of the inversion during meiosis in inversion heterozygotes, half of the gametes have unbalanced genetic constitutions

The harmful effects of inversion polymorphisms arise because single (and other odd-numbered) crossovers in the inverted region result in half the gametes being unbalanced (duplicated and deficient gene contents) (Table 7.2). Since only a minority of meioses have crossovers in the region of the inversion, only a modest proportion of the gametes typically have unbalanced genetic constitutions. As probability of single crossovers increases with the size of inversions, the harmful effects increase correspondingly (Coyne et al. 1993). Not all inversion heterozygotes produce elevated frequencies of unbalanced gametes, presumably due to lack of chromosomal pairing and crossing over in the inverted region (Coyne et al. 1993).

Centric fusions

A centric fusion involves the joining of two non-homologous rod shaped (acrocentric) chromosomes at their terminal centromeres, creating a V- or J-shaped chromosome with a medial centromere. Meiosis in simple centric fusion heterozygotes may result in a low frequency of unbalanced gametes

The impacts on reproductive fitness of crossing populations that differ by a simple centric fusion may be small or undetectable, as meiosis in centric fusion heterozygotes typically results in only a low proportion of unbalanced gametes (Baker & Bickham 1986; Table 7.2). For example, introduction of house mice from the Orkney Island of Eday into the genetically depauperate Isle of May population was successful despite the populations differing by three fixed centric fusions (Scriven 1992).

Crosses between populations differing by complex centric fusions yield progeny that produce ~ 50% unbalanced gametes

By contrast, crossing between populations fixed for centric fusions involving one common and one different chromosome, such as a fusion between chromosomes 1 and 2 in one population and between chromosomes 1 and 3 in another (termed monobrachial fusions), are harmful because of the difficulties they pose in meiosis. They yield only ~ 50% balanced gametes (Gropp et al. 1972; Baker & Bickham 1986; Nunes et al. 2011). Such monobrachial centric fusion differences are relatively common, including examples in bats (*Rhogeesa*), beavers (*Castor*), rock-wallabies (*Petrogale*), rodents (*Mus musculus* and *Rattus*), shrews (*Sorex araneus*), grasshoppers (*Dichroplus pratensis*), and velvet worms (*Planipapillus*) (Baker & Bickham 1986; Bidau 1991; Ward et al. 1991; Rockman & Rowell 2002; Potter et al. 2015).

Other chromosomal differences

Duplications and deletions of heterochromatin (dense blocks of differentially staining chromatin containing few functional loci, often concentrated near centromeres and at telomeres; Biscotti et al. 2015) are likely to have small or no fitness effects.

How do fixed chromosomal differences arise?

Populations polymorphic for translocations, inversions, or centric fusions exhibit heterozygote disadvantage and unstable equilibria. Consequently, fixation of chromosomal rearrangements can result from genetic drift overwhelming heterozygote disadvantage, and/or from natural selection favoring diverse gene arrangements in different populations (White 1973; Lande 1979; Coyne 1984; Rieseberg 2001).

How can chromosomal differences be detected?

Cytological methods for chromosomal analyses in animals and plants are reviewed by Rowell et al. (2011) and Kirov et al. (2014), respectively, while Lysak & Mandáková (2013) provide details of chromosomal painting in plants. Genome sequencing with long-reads can also be used to detect chromosomal variants, as done in thale cress (*Arabidopsis thaliana*) (Zapata et al. 2016). However, this is currently practical only in species with small chromosomes and excellent reference genomes.

Mechanism 2: Adaptation to different environments

Rapid development of reproductive isolation and outbreeding depression in crosses between allopatric populations (with the same karyotypes) is primarily associated with their adaptation to different environments

Darwin (1859) concluded that natural selection caused reproductive isolation as a secondary consequence of genetic adaptation to different environments (ecological speciation). There is now compelling evidence from many species that rapid development of reproductive isolation evolves primarily by this mechanism (see Frankham et al. 2011; Sexton et al. 2011; Thompson 2013). The lines of evidence are:

- field studies
- laboratory studies
- molecular analyses of speciation loci
- theoretical comparisons of the feasibility of alternative mechanisms.

Funk et al. (2006) reported positive associations between ecological divergence and reproductive isolation from field studies for > 500 species across several major taxa. For example, stickleback fish (*Gasterosteus* spp.) from three isolated lakes in British Columbia, Canada that had independently evolved benthic and limnetic forms showed low spawning rates in crosses between benthic and limnetic forms, but normal rates in crosses between the same forms, both within and across lakes (Box 7.3). Modest reproductive isolation between isolated laboratory populations has evolved in *Drosophila* adapting to different environmental conditions, but not for populations maintained in the same environment (Kilias et al. 1980; Dodd 1989).

Box 7.3 Outbreeding depression in crosses between populations of stickleback fish adapted to different environments, but not in those adapted to similar environments

(Rundle et al. 2000)

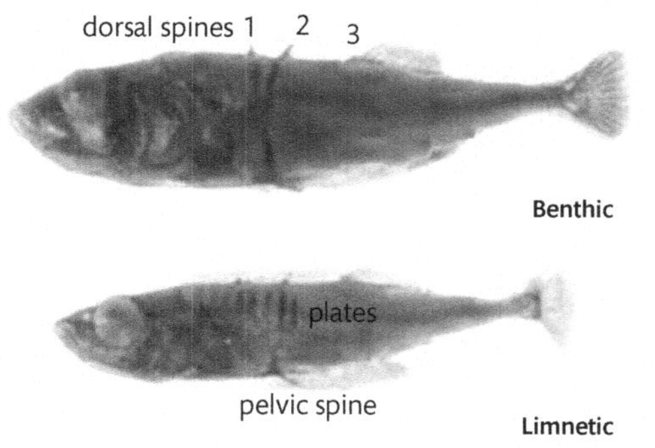

(Reprinted by permission from Macmillan Publishers Ltd: Nature. From Figure 1 in Peichel, C. L., Nereng, K. S., Ohgi, K. A., Cole, B. L. E., Colosimo, P. F., Buerkle, C. A., Schluter, D., and Kingsley, D. M. (2001). The genetic architecture of divergence between threespine stickleback species. *Nature*, **414**(6866), 901–905. Copyright 2001)

Three-spined stickleback fish (*Gasterosteus* spp.) were trapped in three isolated lakes in British Columbia by falling sea levels in the Pleistocene ~ 6,000 generations ago. Benthic (bottom feeding) and limnetic (water column) forms evolved independently in each lake. Crosses were carried out between similar and different forms in the same and different lakes (figure). Progeny of crosses between benthic × benthic and limnetic × limnetic forms showed normal spawning rates, while crosses between the forms adapted to different environments (benthic × limnetic) showed much reduced spawning rates for both crosses within and across lakes, i.e. there was outbreeding depression in crosses between populations adapted to different environments, but not in crosses between populations adapted to similar environments.

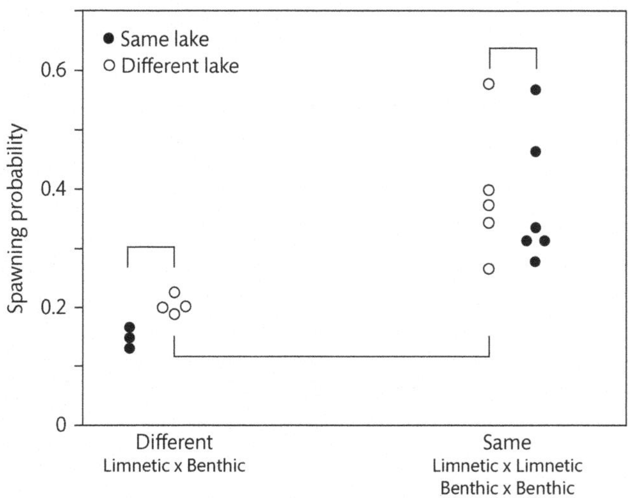

(From Figure 2 in Rundle, H. D., Nagel, L., Boughman, J. W., & Schluter, D. (2000). Natural selection and parallel speciation in sympatric sticklebacks. *Science*, **287**(5451), 306–308. Reprinted with permission from AAAS)

A speciation gene is defined as any locus contributing to the evolution of reproductive isolation (Nosil & Schluter 2011). Most investigated speciation genes involved in pre-zygotic isolation (outbreeding depression) show molecular signals of positive selection (McCartney & Lessios 2004; Orr et al. 2007; Maheshwari et al. 2008). Finally, theory indicates that reproductive isolation is more likely to arise and occurs more rapidly with selection than genetic drift (Gavrilets 2004).

Whether adaptive differentiation is associated with all cases of reproductive isolation remains controversial (Coyne & Orr 2004; Templeton 2008; Presgraves 2010), but adaptation is involved in most cases where reproductive isolation has rapidly evolved in isolated populations (excluding chromosomal differences) (Frankham et al. 2011; Castillo et al. 2015).

How is adaptive differentiation converted into reproductive isolation?

Pre-zygotic reproductive isolation arises from either pleiotropic effects on reproductive traits of alleles involved in adaptive differentiation, or from linkage disequilibrium between such alleles and ones affecting reproductive isolation (Rice & Hostert 1993). Adaptive changes in timing or location of reproduction in plants and animals are associated with pre-zygotic reproductive isolation (Hall & Willis 2006; Savolainen et al. 2006; Nosil 2007; Binks et al. 2012).

Outbreeding depression also occurs if the F_1 and later generation progeny of crosses are approximately intermediate in fitness between the two parents in each parental environment (as often occurs) (Rundle & Whitlock 2001; Hereford 2009b). Such outbreeding depression has been found in *Drosophila*, stickleback fish, and leaf beetles (*Neochlamisus bebbianae*) (de Oliveira & Cordeiro 1980; Rundle, 2002; Egan & Funk 2009). Post-zygotic isolation may also result from Dobzhansky–Muller incompatibilities, as detailed later.

Mechanism 3: Coadapted gene complexes

Coadapted gene complexes may develop in isolated populations within the same environment and result in outbreeding depression when the populations are crossed, but this is more likely for populations adapted to different environments

Coadapted gene complexes refer to beneficial combinations of alleles at different loci whose overall fitness effects are better than expected from the average effects of the constituent alleles. This is a form of epistasis (interaction between loci: Falconer & Mackay 1996). For example, coadapted gene complexes involving different combinations of mitochondrial and nuclear DNA genotypes (mitonuclear interactions) in different populations have been shown to result in outbreeding depression in *Drosophila*, copepods, wasps (*Nasonia*), newts (*Triturus*), nematodes (*Caenorhabditis briggsae*), plants, and yeast (Galloway & Fenster 1999, 2001; Sackton et al. 2003; Lee et al. 2008; Arntzen et al. 2009; Ellison & Burton 2010; Koevoets et al. 2012; Chang et al. 2016). However, these mitonuclear coadaptations seem more likely to develop between populations adapting to different, rather than similar environments.

Drift plus natural selection may lead to the evolution of different coadapted gene complexes in completely isolated populations within the same environment (Fig. 7.2; Whitlock et al. 1995). However, drift will not be strong enough to produce coadapted gene complexes unless populations are small or have experienced bottlenecks. Such populations become inbred, have reduced genetic diversity and adaptation, and show genetic rescue effects upon crossing that may obscure the effects of coadaptive differences (Fenster & Galloway 2000). Similarly, drift plus sexual selection in the same environments between isolated populations may also lead to reproductive isolation and speciation, but again genetic rescue effects will tend to mask them (Coyne & Orr 2004; Sobel et al. 2010; see also Oneal & Knowles 2013).

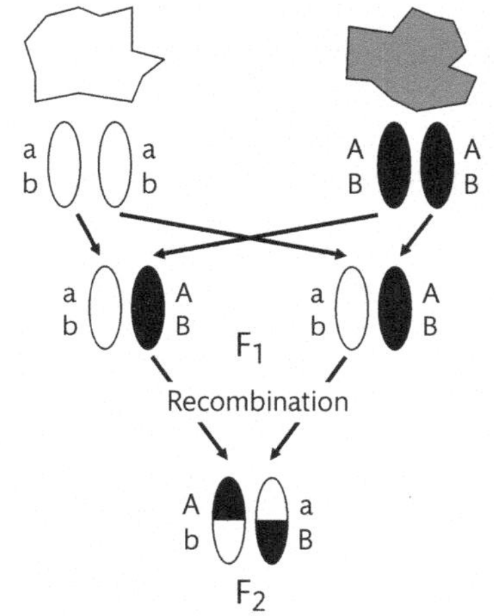

Fig. 7.2 Outbreeding depression due to coadapted gene complexes (for populations with the same karyotypes) (Frankham et al. 2010, Fig. 16.5). Coadapted gene complexes arise when the combinations of alleles at two loci, ab and AB, have high fitness, while aB and Ab have reduced fitness. The ab and AB are favored in different populations and go to fixation. However, crossing the two populations results in lowered fitness, especially in the F_2 because segregation and recombination break up the coadapted gene complexes, yielding gametes Ab and aB.

Molecular evidence indicates that the same loci and alleles are often involved in evolutionary change in large isolated populations located in similar environments (Colosimo et al. 2005; Hoekstra 2006, but see Ferchaud & Hansen 2016). For example, the same insecticide resistance mutations are often observed across populations and species (McKenzie & Batterham 1994; Hartley et al. 2006). Thus, the evolution of different coadapted gene complexes in replicate populations in the same environment is improbable unless there is a significant drift component (Frankham et al. 2011).

Empirical data on speciation (above) in similar environments and the genetic rescue meta-analyses by Frankham (2015, 2016) both indicate that outbreeding depression in crosses between populations adapted to similar environments is rare, or of a lesser magnitude than genetic rescue effects. Different coadapted gene complexes are more likely to evolve in populations in different rather than similar environments, because the selective forces are likely to favor different genetic variants (Pickup 2008; Frankham et al. 2011).

Other coadaptive mechanisms

Dobzhansky–Muller incompatibilities describe post-zygotic reproductive isolation (outbreeding depression) due to interactions between two loci. However, they are likely to take very many generations to evolve, and there is little empirical support for them between populations within species that are adapted to the same environments

Dobzhansky–Muller incompatibilities are widely viewed as a common coadaptive mechanism leading to the evolution of reproductive isolation. They initially involve a two locus system, with homozygous genotype aabb (Fig. 7.3). Independent mutations in the A and B loci occur in two allopatric populations and rise to fixation, yielding AAbb and aaBB genotypes in the respective populations (Dobzhansky 1937; Muller 1940; Coyne & Orr 2004; Presgraves 2010). Under the Dobzhansky–Muller model, crossing of these genotypes is difficult or impossible, due to incompatibilities of some allelic combinations at the A and B loci, resulting in reproductive isolation.

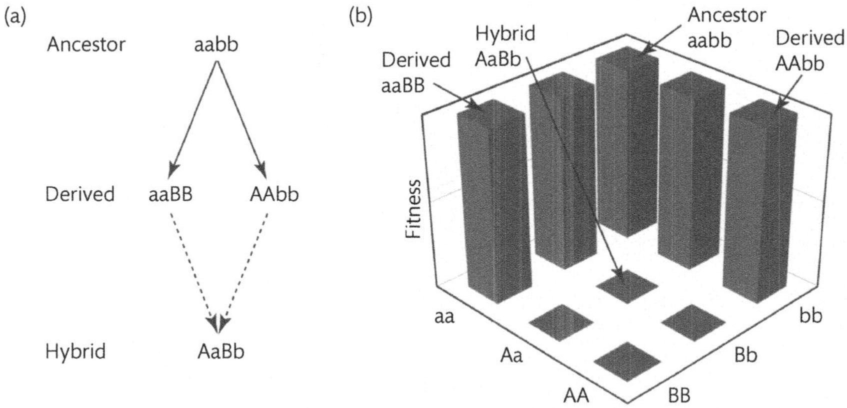

Fig. 7.3 Development of Dobzhansky–Muller incompatibilities. (a) Genotypes of ancestral and derived populations and their hybrid. (b) Theoretical fitnesses of the different genotypes (Cutter 2012).

However, development of these incompatibilities involves the fixation of new mutations by drift, such that their occurrence is improbable in the time frame of recent human habitat fragmentation (Frankham et al. 2011). Further, empirical evidence for them in populations adapted to similar environments is weak. Nevertheless, if the alleles involved in incompatibilities are initially polymorphic in populations, as reported by Cutter (2012) and Corbett-Detig et al. (2013), a variation of this mechanism cannot be ruled out. However, Kao et al. (2010) failed to find simple Dobzhansky–Muller incompatibilities causing post-zygotic isolation in crosses between yeast species.

Other mechanisms, such as genetic conflicts, host–pathogen conflicts, and mutation pressure have also been proposed as means for the evolution of reproductive isolation between populations. However, the near absence of outbreeding depression following crossing between populations from similar environments indicates that these mechanisms are probably of minor importance when the populations reside in similar environments (Frankham et al. 2011; Frankham 2015, 2016).

Mechanism 4: Genetic drift and bottlenecks

While genetic drift and population bottlenecks were proposed as major factors in the development of reproductive isolation, both theory and empirical studies point to the predominant role of natural selection

Mayr (1963) and Templeton (2008) promoted the role of genetic drift as a major force in the development of reproductive isolation. However, theory and empirical studies favor selection over drift as the predominant mechanism producing reproductive isolation, apart from the role of drift in chromosomal differentiation (Coyne & Orr 2004; Gavrilets 2004; Templeton 2008; Frankham et al. 2011; Castillo et al. 2015). For example, divergence due to natural selection, rather than genetic distance (due to genetic drift), predicted reproductive isolation among populations of walking stick insects (*Timema cristinae*), sea urchins (*Arbacia, Echinometra, Heliocidaris, Lytechinus*, and *Strongylocentrotus* genera), cichlid fish (*Pseudocrenilabrus philander*), anole lizards (*Anolis roquet*), and poison-dart frogs (*Dendrobates pumilio*) (Nosil et al. 2002; Zigler et al. 2005; Stelkens & Seehausen 2009; Thorpe et al. 2010; Wang & Summers 2010).

How many generations does it take to develop outbreeding depression?

For populations adapted to similar environments, it takes at least thousands of generations of evolution in isolation before their crosses are likely to exhibit outbreeding depression. Conversely, the first signs of outbreeding depression have been observed after a few dozen generations in crosses of populations adapting to different environments

Given the importance of similar versus different environments in developing outbreeding depression, we discuss the estimates for the two circumstances separately. The generation length for a species is the average age of parents when they have offspring (O'Grady et al. 2006).

Similar environments

For isolated populations evolving in similar environments, we did not find data sets where crosses between populations resulted in outbreeding depression (Table 7.3). These data indicate that it takes at least several thousand generations before outbreeding depression will be observed. For example, stickleback fish populations isolated in similar niches in different lakes ~ 10,000 years ago (> 6,000 generations of isolation) do not show reproductive isolation (Rundle et al. 2000), yet they had sufficient genetic diversity and adequate population sizes to evolve benthic and limnetic forms in each lake.

Table 7.3 Generations without development of outbreeding depression in similar environments and characteristics of the populations studied.[a]

Species	Generations of isolation[b]	Isolation	Population sizes (parents/generation)	Adaptive changes possible	References
Three-spined stickleback fish (*Gasterosteus* spp.)	~ 6,000	Yes	Very large (lakes)	Yes (evolved benthic and limnetic forms)	Rundle et al. (2000)
Thousands of stocks from several species of *Drosophila*	Up to ~ 2,400	Yes	Variable often modest, possibly ≥ 50	Likely	Coyne & Orr (2004)
Caribou (*Rangifer tarandus*)	~ 750		Unknown, but probably relatively large	Yes	McDevitt et al. (2009)
Drosophila pseudoobscura	280	Yes	Several thousand	Yes. High initial genetic diversity	Ehrman (1969)
Cowpea weevil (*Callosobruchus maculatus*)	230–250	Yes	400	Probable. Likely high initial genetic diversity	Bieri & Kawecki (2003)

[a] The use of these data sets to estimate generations to develop outbreeding depression in similar environments depends upon them being isolated populations and having sufficient genetic diversity and large enough effective population sizes to evolve. In most cases these conditions are satisfied.
[b] Generations since the two populations separated from the common ancestral population.

Different environments

In different environments, signs of reproductive isolation between isolated populations of invertebrates and vertebrates can commence within only dozens of generations (Hendry et al. 2007). For example, sockeye salmon introduced into Lake Washington, USA developed differentially adapted river and beach spawning populations with partial reproductive isolation within 14 generations (Hendry et al. 2007).

Can we predict the risk of outbreeding depression?

The risks of outbreeding depression are predictable, as they are associated with crossing of populations that are different species, exhibit fixed chromosomal differences, are adapted to different environments, or some combination of these

It is critical for genetic management of fragmented population that we have methods to predict the risk of outbreeding in population crosses. As we have seen above, outbreeding depression is caused in most cases by crossing of populations that belong to different species, exhibit fixed chromosomal differences, or are adapted to different environments (especially for extended periods). In what follows, we elaborate on these threat factors, and in Chapter 11 we use them in a decision tree for predicting the risk of outbreeding depression.

Fixed chromosomal differences

Fixed chromosomal differences can be assessed by karyotyping populations, so new theory is not required to predict when they will occur.

Adaptive differentiation

We predict that the degree of adaptive differentiation between isolated populations and thus their risk of exhibiting outbreeding depression when crossed is an increasing function of five factors: selection applied, genetic diversity for reproductive fitness, effective population sizes, generations of isolation, and the difference between their environments

Quantitative genetic theory has been used to predict the extent of outbreeding depression for crosses between diploid populations with the same karyotypes by focusing on adaptive differentiation (Frankham et al. 2011). As we saw in Chapter 4, the cumulative genetic adaptation (ΣGA_t) over t generations in a population due to pre-existing quantitative genetic diversity (*PD*) depends upon the heritability of reproductive fitness, the selection differential, and the effective population size.

New mutations also contribute to genetic variation and adaptation (*MU*), and this contribution depends upon the increase in additive genetic variation due to mutation in each generation, the phenotypic variation, the selection differential, N_e, and the number of generations. Thus, the cumulative total adaptive genetic change is the sum of contributions from pre-existing genetic diversity and new mutations, as follows (Frankham et al. 2011):

$$\Sigma GA_t \approx PD + MU \qquad 7.1$$

In Table 7.4 we consider several adaptive scenarios, all of which predict that similar variables will affect differential adaptation and the development of outbreeding depression. In scenario 1, where one population remains in the original environment to which it is adapted (i.e. it is no longer showing directional evolution in fitness), and a subsample of substantial N_e moves to a new environment, adaptive differentiation is given by eqn 7.1. In scenario 2, both populations originate from the same source population, but move to new, different environments and the source population is extirpated. Genetic differentiation between them is the sum of adaptation in the two environments, multiplied by *E*, the proportion of the adaptation of the new populations that is to different new features in their environments. *E* is 0 when the two environments are the same and 1 when they are entirely different.

In scenario 3, where populations from a cline (or otherwise with initial adaptive divergence) become isolated in different environments, the total genetic differentiation will be as in scenario 2, plus the initial adaptive difference (GA_0).

Table 7.4 Equations to predict adaptive differentiation under a range of scenarios (Frankham et al. 2011).

Scenario	Prediction equations[a]
1. Two completely isolated populations from the same source, one in the original environment and the other adapting to a new one	$\Sigma GA_t \approx PD + MU$
2. Populations from the same source move into two new and different environments (a, b) and undergo adaptive divergence (source now extinct)	$\Sigma GA_t \approx E(PD_a + MU_a + PD_b + MU_b)$
3. Partly diverged populations from once connected range with gene flow (cline), isolated in different habitats and locally adapting	$\Sigma GA_t \approx GA_o + E(PD_a + MU_a + PD_b + MU_b)$

[a] ΣGA_t is the cumulative adaptive genetic divergence between the populations over t generations, PD from pre-existing genetic diversity, and MU due to new mutations, E is the proportion of the adaptations that are to different new features of their environments, and GA_0 is the initial adaptive genetic differentiation when the population was fragmented.

Thus, the degree of adaptive differentiation between two populations and the risk of outbreeding depression in crosses between them is an increasing function of five factors (Frankham et al. 2011):

- selection differentials
- heritabilities (a function of genetic diversity for reproductive fitness)
- effective population sizes
- number of generations since the populations were isolated
- differences between the environments to which the populations are adapting.

Empirical evidence confirms the roles of the first four variables in predicting genetic adaptation (Weber 2004; Frankham 2008; Leimu & Fischer 2008). Further, the degree of adaptive differentiation between populations increases with the environmental difference between populations, as predicted (Hereford 2009a). Alterations in predictions for species with other mating systems and modes of inheritance are discussed in Frankham et al. (2011) and Chapter 8.

At what generation after crossing is outbreeding depression evident?

Outbreeding depression may be first observed in the F_1, F_2, or F_3 generation depending on the mechanism causing it, but it should be detectable (if present), and at relatively stable levels by the F_3, apart from that due to coadapted gene complexes

Outbreeding depression is often not manifest until F_2 or F_3 generations (Fenster & Galloway 2000; Tallmon et al. 2004; Frankham 2015, 2016). Different generations are informative for a range of causative factors, as shown in Table 7.5. For example, adverse effects of differential adaptation will typically be evident in the F_1 generations, but their impacts reduce to about half this level in the F_2 for zygotic effects, with maternal contributions delayed by a generation. Fixed chromosomal effects due to translocations, inversions, or centric fusions are expected to show a similar pattern. Effects of coadapted gene complexes will not be evident until the F_2 offspring for zygotic effects and the F_3 offspring for maternal effects. Conversely, crosses of populations differing in ploidy (e.g. diploid versus tetraploid) typically result in viable F_1 zygotes that are sterile, such that there are few F_2 and F_3 progeny.

Table 7.5 Patterns of occurrence of outbreeding depression (OD) across F_1, F_2, and F_3 generations for different causative factors in offspring resulting from crossing between populations (after Frankham 2016).

Effects	F_1	F_2	F_3
Differential adaptation			
– zygotic effects	Maximum OD	~ ½ the F_1 level (for additive effects)	~ ½ the F_1 level (for additive effects)
– maternal effects	Nil	Maximum OD	~ ½ the F_2 level (for additive effects)
Fixed chromosomal differences: ploidy			
– zygotes and maternal	Nil (soma)	Present[a]	Present[a]
Fixed chromosomal differences: translocations, inversions, and centric fusions			
– zygotes and maternal	Nil (soma)	OD present	~ ½ that in F_2
Coadapted gene complexes			
– zygotes	Nil	OD present	Worsening
– maternal	Nil	Nil/minimal	OD present and worsens with subsequent generations

[a] F_1 individuals are typically sterile, so there may be no F_2 or F_3 offspring.

Of the causes of outbreeding depression, only coadapted gene complexes are expected to result in progressively worsening fitness over generations beyond the F_3. Fitness is expected to decline in an approximately exponential decay until alleles at different loci contributing to outbreeding depression reach linkage equilibrium, with the rate of this decay depending on the recombination rate between the loci. If the screen against outbreeding depression used in the Frankham (2015) study (see Fig. 11.4) was not excluding such effects, the data sets should exhibit a progressive decline in fitness beyond the F_3, but this was not observed (R. Frankham, unpublished analyses). Further, Bijlsma et al. (2010) did not find a decline between the F_5 and F_{10} fitness in their study.

Can populations recover from outbreeding depression?

Even if crosses between populations result in immediate outbreeding depression, natural selection will usually lead to recovery and, sometimes, to higher eventual fitness than either parental population

Crosses between populations that exhibit outbreeding depression will usually recover their fitness over subsequent generations due to natural selection acting upon the extensive genetic variation in the hybrid population. This has been observed in all reported cases of which we are aware, both for crosses between species and for crosses between locally adapted populations within species (Lewontin & Birch 1966; Rieseberg et al. 1996; Carney et al. 2000; Edmands et al. 2005; Erickson & Fenster 2006).

For example, three diploid sunflower species (*Helianthus anomalus*, *H. deserticola*, and *H. paradoxus*) have each evolved from crosses between the same two diploid parents *H. annuus* and *H. petiolaris* that differ by 10 fixed chromosomal rearrangements (Lai et al. 2005). F_1 crosses between the two parent species have pollen viability of < 5% and 2.1% viable seed, but all three derivative species have pollen viability of 90–95% (similar to that in the parent species) and > 80% seed set. Experimental studies of crosses between the same two parental species exhibited recovery of pollen viability from < 10% in the F_1 to > 90% over five generations (Rieseberg et al. 1996). For crosses between distant populations within the partridge pea that showed strong F_3 hybrid breakdown, fitness in the field had returned to the level of the F_1 performance by the F_6 generations, which was superior to the parents (Erickson & Fenster 2006).

Summary

1. Crosses between populations within species sometimes result in reduced fitness, especially in F_2 and later generations (outbreeding depression).
2. The primary mechanisms causing outbreeding depression in crosses between populations are fixed chromosomal differences and adaptive genetic differences, especially for long-isolated populations.

3. Outbreeding depression is usually observed after crossing populations with ploidy differences or fixed differences for translocations, inversions, or centric fusions: the magnitudes are usually ploidy > translocations and monobrachial centric fusions > inversions and simple centric fusions.
4. Populations adapted to different environments (but with the same karyotype) often exhibit outbreeding depression when crossed, especially in the F_2 and later generations.
5. For populations adapted to similar environments, at least thousands of generations of evolution in isolated populations is required before their crosses are likely to exhibit outbreeding depression (in the absence of fixed chromosomal differences). Conversely, for populations adapting to different environments, their crosses may exhibit outbreeding depression within dozens of generations.
6. Even if outbreeding depression occurs, it is often only temporary, as natural selection acts to remove it, especially in large populations.

FURTHER READING

Erickson & Fenster (2006) Found outbreeding depression in intraspecific crosses between adaptively divergent populations of a native legume, followed by recovery in fitness by the F_6 generations.

Frankham (2015, 2016) Meta-analyses in which crosses screened as having a low risk of outbreeding depression were found to show very low frequencies of outbreeding depression across F_1, F_2 and F_3 generations.

Frankham et al. (2011) Reviewed mechanisms leading to the evolution of outbreeding depression and proposed means to predict its occurrence.

Rundle et al. (2000) Classic study on the role of adaptive differentiation in outbreeding depression, based on studies in stickleback fish.

Whitlock et al. (2013) Meta-analysis on outbreeding depression in diverse taxa.

CHAPTER 8 Modified rescue and risk expectations for species with diverse mating systems and modes of inheritance

TERMS

Allopolyploid, amphidiploid, asexual, autopolyploid, biparental inbreeding, clone, disomic, gynodioecious, haplodiploid, haploid, hermaphrodites, mixed mating, outcrossing, parthenogenesis, selfing, tetrasomic

The risks of inbreeding and outbreeding depression, and the prospects for genetic rescue are often altered in species with different mating systems or modes of inheritance (compared to outbreeding diploids), such as self-incompatible, self-fertilizing, mixed mating, non-diploids, and asexual.

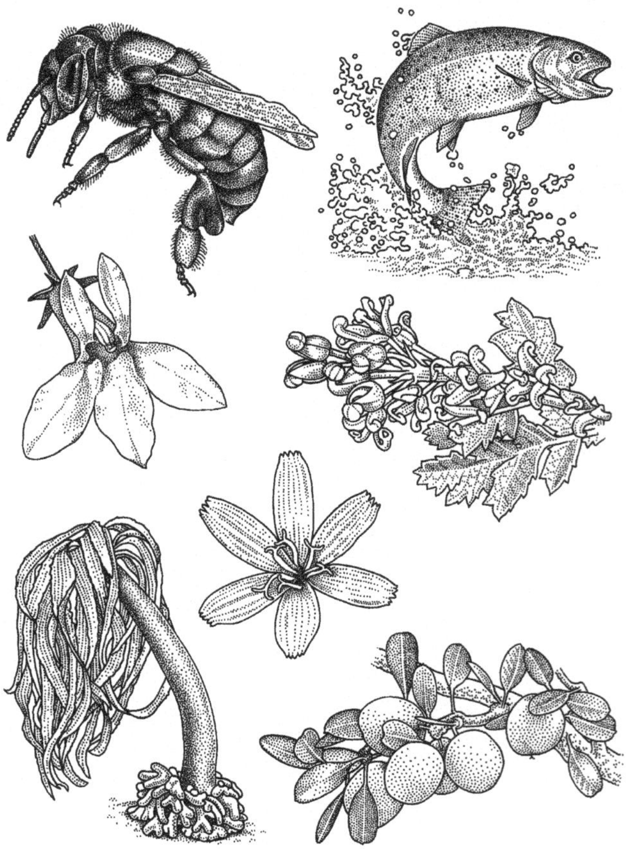

Top left: haplodiploid orchid bee (*Euglossa imperialis*, Central and South America); top right: polyploid brown trout (*Salmo trutta*, Europe); middle left: gynodioecious pale-spike lobelia (*Lobelia spicata*, USA); middle right: asexual King's lomatia (*Lomatia tasmanica*, Australia); bottom left: predominantly haploid sea palm kelp (*Postelsia palmaeformis*, North America); bottom middle: self-fertilizing Malheur wirelettuce (*Stephanomeria malheurensis*, USA); bottom right: self-incompatible Florida ziziphus (*Ziziphus celata*, USA).

Genetic Management of Fragmented Animal and Plant Populations. R. Frankham, J. D. Ballou, K. Ralls, M. D. B. Eldridge, M. R. Dudash, C. B. Fenster, R. C. Lacy & P. Sunnucks. Oxford University Press (2017) © R. Frankham, J. D. Ballou, K. Ralls, M. D. B. Eldridge, M. R. Dudash, C. B. Fenster, R. C. Lacy & P. Sunnucks 2017. DOI 10.1093/oso/9780198783398.001.0001

Why do we need to consider diverse mating systems and modes of inheritance?

Species that are not outbreeding diploids often have different risks of inbreeding and outbreeding depression, and prospects for genetic rescue, and thus may require altered genetic management regimes

Thus far, we have concentrated on the conservation genetics of outbreeding diploid species. While most mammal, bird, reptile, and amphibian species of conservation concern fit this category, many species of plants and invertebrates have other mating systems and modes of inheritance, with different rates of loss of genetic diversity, increases in homozygosity under inbreeding, genetic rescue following crossing, and ability to adapt evolutionarily. For example, haploids do not experience inbreeding depression, while diploids and polyploids do, and failure to account for the distinction has led to erroneous conclusions (Box 8.1).

Box 8.1 Results of inbreeding in predominantly haploid sea palm kelp cannot be extrapolated to make inferences about outbreeding diploids

(Barner et al. 2011; Johansson et al. 2013; Wootton & Pfister 2013, 2015)

An apparent exception regarding the links between inbreeding and extinction was found in a study by Wootton & Pfister (2013) on experimental populations of sea palm kelp, a species that spends much of its lifecycle as a haploid gametophyte. They stated "The minor effects of genetic treatment on extinction risk were unexpected for several reasons . . . " and "the attributes of sea palms that facilitated our experiments do not necessarily predispose this species to strong demographic effects relative to genetic effects."

However, their assertions are incorrect. First, **inbreeding depression cannot occur in the haploid gametophyte** because it requires heterozygous alleles to become homozygous (Chapter 3). Consequently, this kelp has much lower susceptibility to inbreeding depression than species that are predominantly diploid. By contrast, Johansson et al. (2013) reported inbreeding depression in the **sporophytic** stage (diploid) for the giant kelp (*Macrocystis pyrifera*). Second, natural selection against harmful recessives is much more effective in haploids than diploids, so this species will have a very low frequency of harmful alleles (Frankham et al. 2010; Hufbauer et al. 2015b).

Contrary to the authors' assertions, this study is not a valid guide to the relationship between inbreeding and extinction in outbreeding diploid and polyploid species.

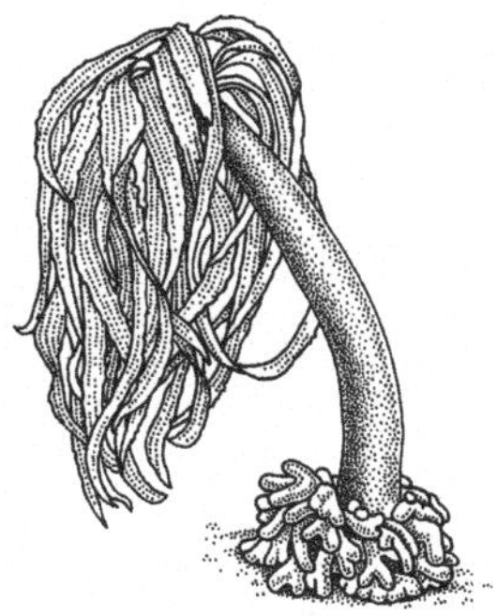

Sea palm kelp (USA)

We now consider the risk and rescue expectations and other relevant evolutionary genetic characteristics for taxa with alternative mating systems and modes of inheritance, based largely on reviews by Dudash & Murren (2008), Frankham et al. (2011), and Weeks et al. (2011). We consider only a selection of the most common departures from outbreeding diploidy; readers are referred to Dudash & Murren (2008) for details of others.

How are genetic rescue and risk expectations modified for different mating systems?

The mating system refers to the proportion of mating events resulting from selfing and/or mating between related individuals (Lloyd & Webb 1986; Webb & Lloyd 1986). Before turning to the risk and rescue expectations, we consider what taxa exhibit these systems.

How are species with diverse mating systems distributed across major taxa?

Plants have the greatest diversity of mating systems, but deviations from random-mating diploidy are also evident in invertebrates and in a minority of lower vertebrates

Table 8.1 lists the frequency of diverse mating systems in major taxa. Mating systems sometimes differ between species within the same genus (Tsuchimatsu et al. 2010), or among populations of the same species (Schemske & Lande 1985; Barrett & Charlesworth 1991; Murray & Young 2001). For example, Caliente clarkia (*Clarkia temblorensis*) in California has populations with outcrossing rates of 16% and 74% (Holtsford 1996).

Table 8.1 Occurrence of different mating systems in major taxa (Richards 1997; Goodwillie et al. 2005; Jarne & Auld 2006; Renner 2014; Sawada et al. 2014).

Taxa	Mating system
Animals	
Vertebrates	
Mammals and birds	Predominantly outbreeding dioecious diploids
Reptiles, fish, and amphibians	Mostly outbreeding dioecious diploids
Invertebrates	Mostly outbreeding or mixed mating diploids
	Hermaphrodites exist in some species (e.g. snails, marine invertebrates, and nematodes)
	Some selfing species (e.g. snails, marine invertebrates, and nematodes)
Plants	
Vascular	
Angiosperms	~ 85% hermaphrodites, 5–6% dioecious
	~ 50% obligate outbreeders (self-incompatible)
	~ 40% mixed mating species
	~ 10–15% species habitually selfing
Gymnosperms	Primarily outcrossing
Non-vascular	In mosses, ferns, and their allies, it ranges from dioecy to intragametic selfing (where the haploid gametophyte produces gametes of both sexes, resulting in complete homozygosity)

Characteristics of species with diverse mating systems

Different mating systems often result in different levels of inbreeding, resulting in modified levels of genetic diversity, risks of inbreeding and outbreeding depression, and prospects for genetic and evolutionary rescue

We begin by considering the most extreme outbreeding system, followed by selfing mating systems.

Self-incompatible species

Self-incompatible species are obligate outcrossers due to self-incompatibility loci

Self-incompatibility (SI) systems are controlled by one or more loci that, in large populations, typically have many SI alleles (Richman & Kohn 1996; Young et al. 2000; Castric & Vekemans 2004; Charlesworth 2006; Sawada et al. 2014). Pollen carrying an allele also present in the stigma will often fail to germinate and fertilize an ovule (Fig. 8.1). Strong natural selection maintains SI allele diversity, due to the advantage of new or rare alleles (Wright 1939).

Plants may have either gametophytic or sporophytic self-incompatibility, as illustrated in Fig. 8.1 (Charlesworth 2006; Briggs & Walters 2016). Angiosperms exhibit a positive relationship between presence of self-incompatibility and outcrossing rate, as expected (Raduski et al. 2012).

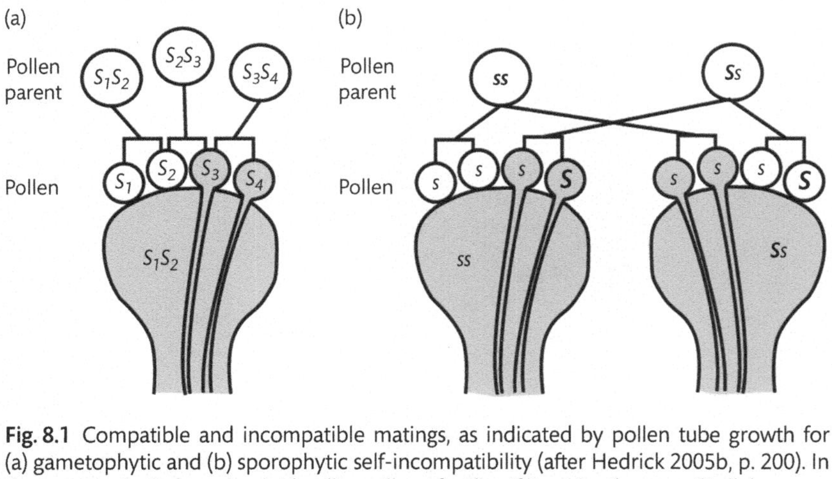

Fig. 8.1 Compatible and incompatible matings, as indicated by pollen tube growth for (a) gametophytic and (b) sporophytic self-incompatibility (after Hedrick 2005b, p. 200). In the gametophytic form, haploid pollen will not fertilize if it carries the same SI allele as one of the two alleles in the female plant. In the sporophytic form, the diploid genotype at the SI locus of the pollen parent determines whether fertilization will occur—it will not occur if it is the same as the style genotype, irrespective of the haploid genotype of the pollen.

Self-incompatible species are expected to have the following characteristics, compared to outbreeding diploids:

- loss of SI allele diversity reduces fitness
- more susceptible to small population size
- suffer normal or higher levels of inbreeding depression when deliberately inbred
- higher genetic diversity
- suffer outbreeding depression, possibly less than in self-compatible species
- larger genetic rescue effects
- larger evolutionary rescue effects.

Despite balancing selection on self-incompatibility loci, small populations of self-incompatible species lose SI alleles. This reduces mate availability, lowers the proportions of fertilized ovules, and consequently increases extinction risk (Box 8.2; Demauro 1993; Young et al. 2000; Pickup & Young 2008; Morgan et al. 2013). Thus, the need to re-establish gene flow among fragmented populations is more critical in self-incompatible species than in self-compatible ones (Willi & Fischer 2005).

Button wrinklewort daisy (Australia)

Box 8.2 Loss of SI allele diversity in small populations reduces reproductive fitness in the endangered self-incompatible button wrinklewort daisy

(Young et al. 2000; A. G. Young *pers. comm.*; Pickup & Young 2008)

The endangered button wrinklewort daisy is found in southeastern Australia (map). In Victoria it has both diploid and tetraploid (T) populations, but in this box we consider only the diploid populations.

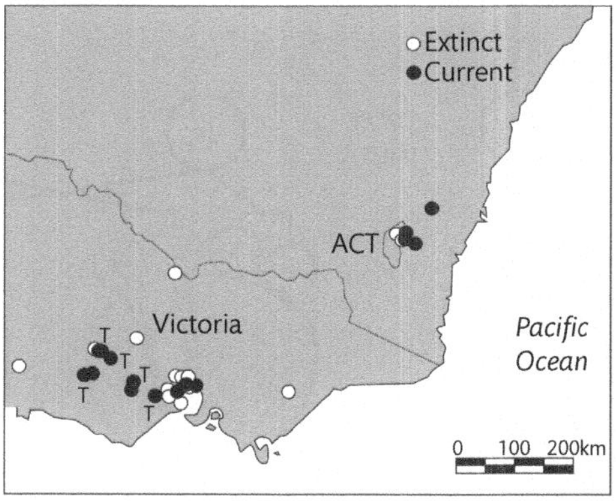

SI allelic diversity (and allozyme diversity) declined with population size in eight population fragments of the daisy with sizes from 5 to 70,000 plants. Further, the number of filled seeds per plant increased with population size and this was not due to insufficient pollen. Use of pollen from other populations increased seed set in small populations, confirming that the small number of SI alleles in small populations limited mating opportunities, and reduced the seed set.

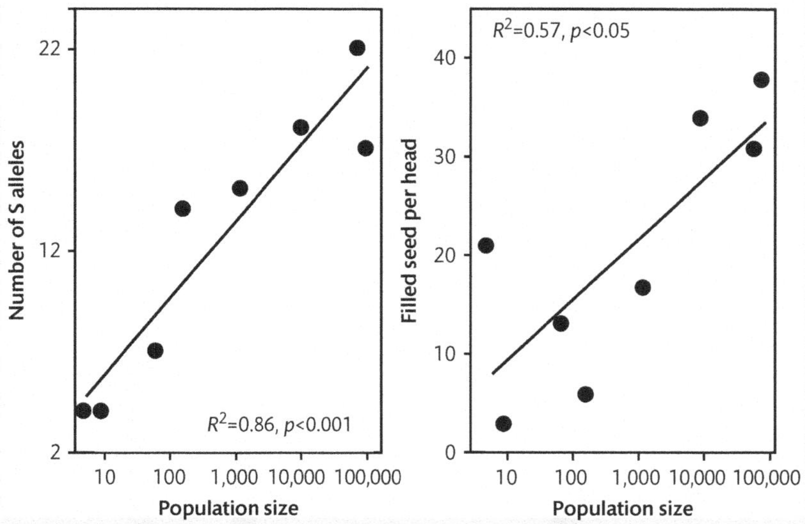

(Frankham et al. 2010, Box 11.1 figure)

With natural pollination, self-incompatibility prevents most inbreeding, so populations often do not exhibit serious inbreeding depression. However, if self-incompatible species are deliberately selfed (by bud pollination that can avoid self-pollen rejection), as done in *Arabidopsis lyrata* (Sletvold et al. 2013), they exhibit greater inbreeding depression than in other outbreeding self-compatible angiosperms. Genetic rescue effects are greater in self-incompatible species than in outbreeding or selfing ones, based on a meta-analysis (Frankham 2015). For example, crosses between 12 populations of the self-incompatible wrinklewort daisy from similar environments revealed genetic rescue effects and no outbreeding depression (Pickup et al. 2013). We are not aware of a comparison of the frequency and severity of outbreeding depression in self-incompatible and self-compatible species. Some self-incompatible species do not exhibit outbreeding depression (*Arnica montana*, two species of aster [*Eupatorium*], and *Ranunculus reptans*: Byers 1998; Luijten et al. 2002), but it was reported in *Ipomopsis aggregata* (Waser et al. 2000).

If populations become sufficiently small, they may evolve loss of their self-incompatibility and become highly selfing, as has occurred many times in plant evolution, but the risk of extinction is typically high during the transition (Porcher & Lande 2005; Tsuchimatsu et al. 2010; Sletvold et al. 2013; Vekemans et al. 2014).

Selfing species

Self-fertilizing species have, by definition 80% or more of fertilizations from self-pollen or sperm

The critical characteristic of inbreeding mating systems is that they lead to increased homozygosity (Chapter 2). This exposes harmful recessive alleles to selection, reducing the genetic load and inbreeding depression (Chapter 3). Some self-fertilizing species of animals and plants are obligate, while most have occasional outcrossing (Fig. 8.2).

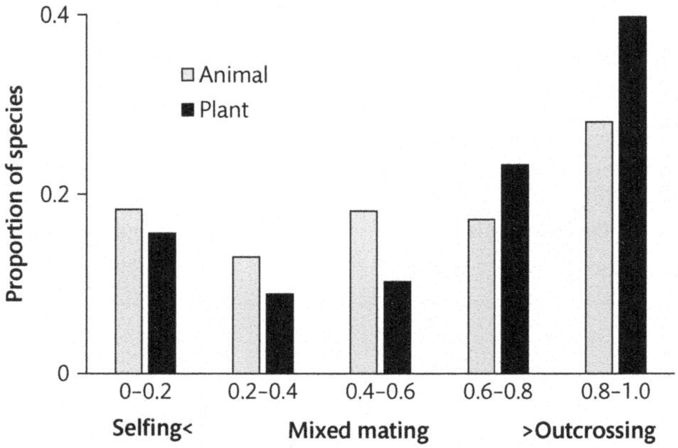

Fig. 8.2 Distribution of mating systems in animals and plants, defined by the proportion of self-fertilizations. Primarily selfing species have outcrossing rates < 0.2, mixed mating species > 0.2 and < 0.8, and primarily outcrossing species > 0.8 (Jarne & Auld 2006).

Obligate selfing species

Obligate selfing species compared to outbreeding diploids are expected to have the following attributes:

- exhibit little inbreeding depression
- lower levels of genetic diversity
- do not outcross, and consequently do not exhibit outbreeding depression or genetic rescue
- greater genetic differentiation among populations
- suffer from higher extinction rates.

Malheur wirelettuce (USA)

For example, highly selfing endangered water howellia (*Howellia aquatilis*) and Malheur wirelettuce in North America both have low genetic diversity (Lesica et al. 1988; Karron 1991). It is unclear whether these two species are obligate selfers, or exhibit very low levels of outcrossing. Selfing lineages suffer from higher extinction rates than outcrossing ones (Goldberg et al. 2010).

Selfing species with occasional outcrossing

These species are expected to have the following characteristics, compared to outbreeding diploids:

- reduced inbreeding depression
- lower heterozygosity and allelic diversity
- higher risk of outbreeding depression (Fenster & Dudash 1994; Dudash & Fenster 2000)
- genetic rescue exhibited in the early generations after crossing, but it is progressively lost with continuous selfing
- greater differentiation among populations.

As inbreeding increases homozygosity, it exposes harmful recessive alleles to natural selection, resulting in lower genetic loads in inbreeding populations than in random mating ones (Chapter 3). Thus, species and populations with only occasional outcrossing (selfing, or mixed mating) are typically less affected by inbreeding depression (Table 8.2). Byers & Waller (1999) reported inbreeding depression of 23% in selfing species and 53% in outcrossing ones, and a negative correlation between inbreeding depression and selfing rate, based upon a meta-analysis. For example, selfing *Mimulus micranthus* exhibited less inbreeding depression than the mixed mating *M. guttatus* (Dudash & Carr 1998).

Table 8.2 Inbreeding depression for selfing, mixed mating, and outcrossing species (number of species in parentheses) across major plant taxa. Inbreeding is due to selfing (ΔF = 0.5), except for intragametic inbreeding where ΔF = 1.

Taxa	Selfing	Mixed mating	Outcrossing	References
Angiosperms and gymnosperms	0.31 (6)	0.49 (13)	0.75 (15)	Husband & Schemske (1996)[a,b]
	0.26 (9)	0.58 (33)	0.54 (17)	Winn et al. (2011)[b]
Pteridophytes	Intragametic			
	0.22 (1)		0.61 (1)	Flinn (2006)
	0.44 (1)		0.91 (1)	Klekowski (1982)[c]
Bryophytes	0.04 (1)		0.62 (1)	Taylor et al. (2007)[d]

[a] Values re-calculated using cut-offs for selfing, mixed mating, and outcrossing defined in this Chapter.
[b] Impact on total fitness, typically measured in the greenhouse or common garden.
[c] Our calculation from data in the source.
[d] Combined impact on four traits in a temperature controlled room.

Selfing species have lower genetic diversity on average for near neutral markers, and greater differentiation among populations than outcrossing species (Li 1976; Hamrick & Godt 1989; Charlesworth & Charlesworth 1995; Frankham et al. 2014a). Further, heritabilities are lower in inbreeding than outbreeding species (0.15 vs 0.29) (Geber & Griffen 2003).

There is a perception that outbreeding depression is higher in selfing than outbreeding species (Dolgin et al. 2007; Frankham 2015), and this was observed in orchid species (*Epipactis*: Brys & Jacquemyn 2016). Selfing and mixed mating (see later) species show less genetic rescue than outbreeders, based upon a meta-analysis (Frankham 2015). For example, beneficial effects of crossing (usually modest) have been reported in selfing species such as cut-leaved monkey flower (*Mimulus laciatus*: Sexton et al. 2011), Alabama glade cress (*Leavenworthia alabamica*: Busch 2005, 2006), and Alsinidendron (*Schneidea viscosa*: Weller et al. 2005), but Barrett & Charlesworth (1991) reported essentially no genetic rescue effects in selfing forms of water hyacinth (*Eichhornia paniculata*), but considerable amounts in an outbreeding form.

Genetic rescue in selfing species that are capable of outcrossing decays over generations of selfing and is eventually lost, in contrast to the situation in outbreeding species. For example, if two homozygous populations (A_1A_1 and A_2A_2) are crossed, all F_1 individuals are heterozygous, but the heterozygosity is halved with each subsequent generation of selfing until all individuals are homozygous. The inbreeding coefficient increases as the heterozygosity declines. Natural selection may cause some increase in fitness in the crossed population, but it has limited effectiveness within selfing lineages, though among-lineage selection may be more effective. We are not aware of relevant empirical evidence on this issue.

Local adaptation is expected to be more prominent in selfing than outcrossing species, as there is less inflow of non-adapted alleles. However, selfers typically have less genetic diversity than outbreeders, reducing their ability to adapt. Hereford (2010) found no relation between selfing rate and frequency of local adaptation, based on a meta-analysis.

Mixed mating species

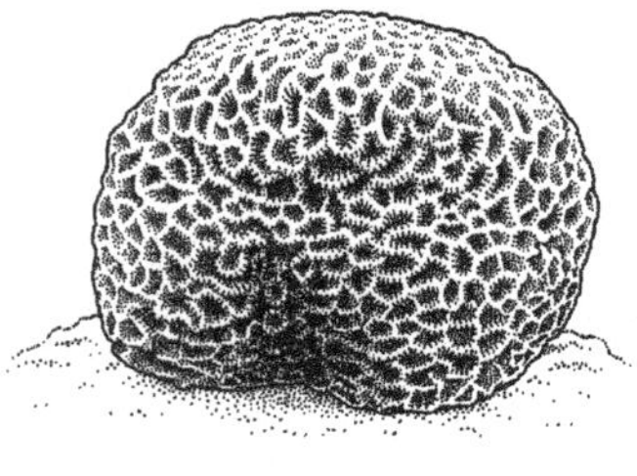

Brain coral (Australia)

Mixed mating species are typically characterized as having selfing rates of 20–80%, with ~ 42% of all plant species falling into this category, as do many invertebrates (Fig. 8.2). Animal examples of mixed mating include some corals and some snails (Stoddart et al. 1988; Brazeau et al. 1998; Tsitrone et al. 2003; Jarne & Auld 2006).

Mixed mating species are expected to have the following characteristics compared to outbreeding diploids:

- reduced inbreeding depression (Table 8.2)
- stronger differentiation among populations
- higher outbreeding depression
- less potential for genetic rescue, and reduced persistence over generations
- exhibit evolutionary rescue.

The evolutionary genetic characteristics of mixed mating species generally lie between those for selfing and outcrossing species. Mixed mating species have similar or less genetic diversity within species, and greater differentiation between populations than outbreeders (Loveless & Hamrick 1984; Hamrick & Godt 1996a).

Whitlock et al. (2013) reported no significant difference in outbreeding depression among inbreeding, mixed mating, and outbreeding species, based on a meta-analysis. In contrast, Frankham (2015) found significantly lower genetic rescue in mixed mating and selfing than outbreeding species, based on a meta-analysis, consistent with theoretical expectations. Persistence of genetic rescue effects for fitness under mixed mating is predicted to be less than for outbreeding species, because heterozygosity (H) is expected to reach a non-zero equilibrium value that depends on the selfing rate (Sf), as follows (Crow & Kimura 1970):

$$\frac{H_t}{H_0} = \frac{2(1 - Sf)}{(2 - Sf)} \qquad 8.1$$

By substituting the selfing values for the lower and upper boundaries of selfing for mixed mating species (0.2 and 0.8), we obtain equilibrium heterozygosities of 0.89 and 0.33, compared to random mating species (relative heterozygosity = 1.0). Relative fitness levels should approximate these proportions. Natural selection will likely enhance these levels, but we are unaware of any empirical evidence of the long-term persistence of genetic rescue effects in mixed mating species. Mixed mating species are expected to exhibit evolutionary rescue, but we are not aware of empirical data on the topic.

Other outcrossing mating strategies

Many other less common outcrossing mating strategies exist for plants and animals, but they cannot be covered in detail in this book. Some highly outcrossing plant systems include dioecy, monoecy, tristyly, and distyly (Dudash & Murren 2008; Dalton et al. 2013; Renner 2014). Many animals have behavioral avoidance of mating between close relatives, as for example in some mammals, birds, reptiles, and invertebrates (Ralls et al. 1986; Pusey & Wolf 1996; Stow & Sunnucks 2004; Bilde et al. 2007; Reid et al. 2015). Conversely, some animal species exhibit lekking, polyandry, harems, and other forms of promiscuous behavior that typically reduce N_e and may lead to higher levels of inbreeding.

Gynodioceous plant species are dimorphic, with some individuals producing only female flowers (male sterile) and others producing hermaphroditic flowers (Richards 1997; Dalton et al. 2013; Renner 2014). They exhibit increased proportions of individuals with female flowers when inbred, reducing seed set and increasing extinction risk (Byers & Meagher 2005; Vilas & García 2006). In general, they are more susceptible to the adverse genetic effects of fragmentation, and likely to show greater genetic rescue effects than outbreeding hermaphroditic species, but are similarly susceptible to outbreeding depression.

Species with permanent translocation heterozygosity over much of the genome behave in a somewhat similar manner to clonal species in that heterozygotes mostly do not segregate to produce homozygotes. They exhibit excess heterozygosity compared to Hardy–Weinberg expectations, linkage disequilibrium, limited reduction in heterozygosity upon inbreeding, and consequently little inbreeding depression or genetic rescue, but are at risk of outbreeding depression when crossed (Levy & Winternheimer 1977; James et al. 1983; Heiser & Shaw 2006).

Given that mating systems affect risk and rescue, we now turn to means for determining them.

How can we identify mating systems?

Mating systems can be determined from the genotypes of parents and offspring

The mating system can be inferred by following the transmission of genetic markers from parents to offspring (Table 8.3). If offspring contain only alleles present in the mother, but have a diversity of genotypes, then they are the result of self-fertilization or meiotic parthenogenesis. Offspring containing alleles not found in the mother are the result of outcrossing (except for new mutations).

Table 8.3 Determination of mating systems using genetic markers (after Frankham et al. 2010, Table 21.2).

Mating systems	Parent genotypes at a locus		Offspring genotypes
Selfing	AB	⇒	AA AB BB
Outbreeding	AB × CD	⇒	AC AD BC BD
Mixed selfing and outcrossing	AB × (AB & CD)	⇒	AA AB BB AC AD BC BD (Combination yields heterozygote deficiency compared to outcrossing)

Selfing rates can be determined indirectly from the reduction in heterozygosity compared to Hardy–Weinberg expectations

The most commonly used model to estimate selfing and outcrossing is the mixed mating model (Ritland 1989; Barrett & Kohn 1991). This assumes that there are only two types of matings, self-fertilization and random mating, so it attributes other inbreeding, such as full-sib, half-sib, and cousin matings (**biparental inbreeding**), to selfing.

Deviations of genotype frequencies from Hardy–Weinberg equilibrium can be used to estimate selfing rates (Sf) (Example 8.1).

Example 8.1 Estimating selfing rate from genotype frequencies

The ratio of observed to expected heterozygosity (H_o/H_e) is 0.68 in endangered round-leaf honeysuckle (*Lambertia orbifolia*) plants from Western Australia (Coates & Hamley 1999). Using eqn 5.4 for F_{IS}, the inbreeding coefficient (F) is:

$$F = 1 - \frac{H_o}{H_e} = 1 - 0.68 = 0.32$$

The selfing rate can be determined from the inbreeding coefficient, as follows:

$$Sf = \frac{2F}{(1+F)} = \frac{2 \times 0.32}{(1+0.32)} = 0.48$$

Thus, the selfing rate in the round-leaf honeysuckle is 48% (assuming that all inbreeding is due to selfing).

In plants with mixed selfing and outcrossing, selfing rates can be determined directly by genotyping maternal parents and progeny

A more direct and precise method estimates selfing rates from maternal and progeny genotypes (Hedrick 2005b). This method also provides a basis for understanding more complex methods described below. Selfing of homozygous maternal plants results in only homozygous progeny, while outcrossing yields heterozygotes (H) at a rate dependent upon the frequency of alleles not found in the homozygote (q). Thus, the frequency of selfing (Sf) is:

$$Sf = 1 - \frac{H}{q} \qquad 8.2$$

Estimates of selfing rates can also be obtained from the genotypes of the progeny of heterozygous maternal plants (Table 8.4; Brown 1989; Hedrick 2005b). Here, the difference between a single-locus and multi-locus estimate yields the degree of biparental inbreeding (Ritland 2002). For example, the Pacific yew is dioecious (separate sexes), but has an F of 47% that must all be due to biparental inbreeding (El-Kassaby & Yanchuk 1994). A selfing rate of 0.11 was estimated for hummingbird-pollinated native *Silene virginica* using this method, and it closely corresponded to the predicted value based on breeding system and pollinator behavior (Dudash & Fenster 2001).

Table 8.4 Proportions of progeny genotypes expected from homozygous (A_1A_1) and heterozygous (A_1A_2) maternal genotypes as a result of self-fertilization (*Sf*) and outcrossing (*T*): *p* and *q* are the frequencies of alleles A_1 and A_2 in the population (Hedrick 2005b).

Maternal genotype	Frequency of matings	Progeny genotypes		
		A_1A_1	A_1A_2	A_2A_2
A_1A_1	*Sf*	*Sf*		
	T	*Tp*	*Tq*	
A_1A_2	*Sf*	¼ *Sf*	½ *Sf*	¼ *Sf*
	T	½ *Tp*	½ *T*	½ *Tq*

The maximum likelihood estimation of mating systems (MLTR; Ritland 2002) assumes that there are multiple types of mating events, self-fertilization, biparental inbreeding, and random mating. For example, *Caryocar brasiliense*, an endangered hermaphroditic tree species in Brazil was shown to be outcrossing, but to be experiencing considerable biparental inbreeding within population fragments, based on analyses using the MLTR method on genotypes of both maternal tissue and seed progeny at 10 microsatellite loci (Collevatti et al. 2001).

A recently derived and more powerful method, Bayesian Outcrossing Rate and Inbreeding Coefficient Estimation (BORICE; Koelling et al. 2012) is more robust in estimating individual family level outcrossing rates without maternal level information.

Mating systems can also be inferred from pollinator behavior, or determined from mating experiments

If genetic marker information is unavailable, the mating systems of plants can be inferred by multiple approaches (Dudash & Murren 2008), as follows:

- observing plant pollinator movements, aided by use of fluorescent dyes, visual inspection, and video recorded observations
- through the use of common garden experiments that compare progeny performance following selfing, outcrossing, and open-pollination
- performing hand-pollinations and bagging flowers or inflorescences to determine if pollinators are required for seed set, and if the plant species is self-compatible (Pinto-Torres & Koptur 2009)
- inference from the flower size, color, perfume, and nectar: small flowered species with white flowers and no nectar or perfume may indicate a selfing species, especially if other related species have larger colored flowers with nectar or perfume.

How are genetic rescue and risk expectations modified for different modes of inheritance?

We begin this section by considering the distribution of species with modified modes of inheritance.

How are species with diverse modes of inheritance distributed across major taxa?

Plants have the greatest diversity of modes of inheritance, but deviations from random-mating diploidy are also evident in invertebrates, and in a minority of lower vertebrates

The modes of inheritance refer to alternatives to independent assortment of chromosomes and transmission of haploid gametes from both parents to offspring, including polyploidy, haploidy, haplodiploidy, and asexual reproduction. Table 8.5 lists the distribution of these in major taxa.

Table 8.5 Modes of inheritance other than diploid with haploid gametes exhibited by major taxa (Richards 1997; Coghlan et al. 2005; Jarne & Auld 2006; Wood et al. 2009; Husband et al. 2013).

Mode of inheritance	Taxa	
Polyploid	Vertebrates	Rare, except for some frogs and salmonid fish
	Invertebrates	Hundreds of polyploid invertebrate species are known, including arthropods, nematodes, molluscs, and flatworms
	Plants	Many angiosperms
		Most gymnosperms
		Many pteridophytes (ferns)
		Frequent in red and green algae, unclear in mosses, rare in liverworts, and absent in hornworts
Haploid		Uncommon except in algae, fungi, and bryophytes (liverworts, mosses, and hornworts), where the diploid phase of the life cycle is short and the haploid phase predominates
		Pteridophytes (ferns) have a free living haploid stage
		Most prokaryotes are haploid
Haplodiploid		~ 15% of animal species (most Hymenoptera and a few other species)
Asexual	Vertebrates	Rare, except for some parthenogenetic fish and lizards
	Invertebrates	Rare
	Plants	Obligatory asexuality rare, but sprinkled across species from diverse major taxa
	Prokaryotes	Typically asexual, but often move genetic material between species, populations, and individuals by transformation, transduction, and mobile elements

Characteristics of species with alternative modes of inheritance

Species with different ploidies have different levels of genetic diversity, inbreeding depression, rates of adaptation, and possibly risks of outbreeding depression

For random mating species with different ploidies, Table 8.6 details the predicted loss of genetic diversity per generation, equilibrium Hardy–Weinberg heterozygosity for neutral loci due to mutation-drift equilibrium, efficiency of selection, and mutation-selection equilibria, as background for discussing the expectations for evolutionary genetic characteristics of such species. The models are based on two alleles at a locus with frequencies of p and q in both sexes, random mating, effective population size of N_e, mutation rate of u per generation to deleterious alleles, and selection coefficient of s against the harmful homozygote (or haploid) and hs against heterozygotes for partial recessives.

Table 8.6 Characteristics of species with different ploidies, in relation to loss of genetic diversity per generation (ΔH), heterozygosity due to mutation-drift equilibrium ($\hat{H}$), rates of allele frequency change due to selection (Δq), and mutation-selection equilibrium ($\hat{q}$).

Ploidy	ΔH	$\hat{H}$	Selection (Δq)		Mutation-selection equilibrium $\hat{q}$	
			Additive	Recessive	Recessive	Partial recessive
Haploid (n)	$\frac{1}{N_e}$	$\approx \frac{2N_eu}{2N_eu+1}$	$\frac{-spq}{1-sq}$	$\frac{-spq}{1-sq}$	$\frac{u}{s}$	$\frac{u}{s}$
Haplodiploid (n/2n)	$\frac{1}{1.5N_e}$	$\approx \frac{3N_eu}{3N_eu+1}$	$\frac{-\frac{2}{3}spq}{1-sq}$	$\frac{-\frac{1}{3}spq}{1-sq}$	$\frac{3u}{s}$	$\frac{1.5u}{s}$
Diploid (2n)	$\frac{1}{2N_e}$	$\approx \frac{4N_eu}{4N_eu+1}$	$\frac{-\frac{1}{2}spq}{1-sq}$	$\frac{-spq^2}{1-sq^2}$	$\sqrt{\frac{u}{s}}$	$\frac{u}{hs}$
Tetraploid (4n)	$\frac{1}{4N_e}$	$\approx \frac{8N_eu}{8N_eu+1}$	$\frac{-\frac{1}{4}spq}{1-sq}$	$\frac{-spq^4}{1-sq^4}$	$\approx \left(\frac{u}{hs}\right)^{\frac{1}{4}}$	$\frac{u}{hs}$

Sources: Crow & Kimura (1970), Birky et al. (1989), Moody et al. (1993) and Frankham et al. (2010).

Polyploid species

Polyploid species have more than two copies of the haploid chromosome complement

Two forms of polyploids occur, autopolyploids (where all copies of the chromosomes derive from one species) and allopolyploids (where the chromosomes derive from more than one species) (Chenuil et al. 1999). For example, tetraploid *Rutidosis leptorrhynchoides* (Australia) is an autopolyploid, while the tetraploid Spanish monocotyledon *Borderea chouardii* is an allopolyploid (Brown & Young 2000; Segarra-Moragues et al. 2005). These exhibit tetrasomic and disomic chromosomal segregation, respectively. Tetrasomic inheritance involves the four copies of each homologous chromosome associating and segregating, while disomic inheritance involves chromosomal pairing and segregation as in diploids (amphidiploid) (Brown & Young 2000; Segarra-Moragues et al. 2005; Catalán et al. 2006).

Sexually reproducing polyploids have similar evolutionary genetic characteristics to outbreeding diploids, but the magnitude of some effects is expected to differ (Table 8.6; Lande & Schemske 1985; Dudash & Fenster 2000; Pickup & Young 2008; Madlung 2013), as follows:

- suffer lesser inbreeding depression initially, but it is more persistent
- lose genetic diversity at a slower rate for the same effective population size (Table 8.6)
- higher genetic diversity than diploids for the same N_e (Table 8.6)
- exhibit outbreeding depression (Fenster & Dudash 1994; Dudash & Fenster 2000)
- benefit from genetic rescue for fitness and evolutionary potential
- evolve at a slower rate than equivalent diploids with similar genetic diversity (Table 8.6).

There is a perception that polyploids have higher genetic loads than otherwise similar diploids (e.g. Otto 2007), based on the equilibrium frequencies for recessive harmful alleles (Table 8.6). However, most harmful alleles are partially recessive (Simmons & Crow 1983), so tetraploids and diploids are expected to have similar equilibrium frequencies of harmful alleles (Table 8.6). Consequently, polyploids are expected to be less susceptible than diploids to inbreeding depression (other things being similar), as homozygosity is achieved more slowly in polyploids (Bever & Felber 1994; Frankham et al. 2010), and empirical evidence trends in this direction (Husband & Schemske 1997; Hardy & Vekemans 2001; Ramsey & Schemske 2002; Birchler et al. 2003; Ozimec & Husband 2011). For example, inbreeding depression was lower in polyploid than diploid populations of *Clarkia* plants (Barringer & Geber 2008).

Higher genetic diversity in autopolyploid populations than in diploid populations has been reported in several plant species, as predicted (Soltis & Rieseberg 1986; Hokanson & Hancock 1998; Brown & Young 2000; Mahy et al. 2000; Luttikhuizen et al. 2007; Ng et al. 2004; Frankham et al. 2010). Polyploids show genetic rescue (Pickup & Young 2008). Further, polyploids exhibit outbreeding depression, but whether it is more or less than in diploids is unclear (Grindeland 2008).

The ability of polyploids to evolve in response to environmental change may be similar to equivalent diploids, as they should have higher average genetic diversity (if this also exists for quantitative genetic variation), and slower loss of diversity for the same N_e, but a lower rate of selection response (Table 8.6). We are not aware of empirical evidence on the issue.

Haploid species

Haploidy throughout the life cycle is found in many prokaryotic microbes. Most fungi, bryophytes, and many marine algae have typical haploid gametophytic and diploid sporophytic phases of their life cycle, but the haploid phase predominates. Haploids are expected to have the following characteristics compared to outbreeding diploids:

- no inbreeding depression
- lower genetic diversity for similar sized populations
- experience outbreeding depression
- no genetic rescue for fitness
- experience evolutionary rescue if new genotypes are added.

For example, kelp exhibits a life cycle that is predominantly haploid and shows little inbreeding depression, as predicted (Box 8.1).

Haplodiploid species

Many invertebrate species, primarily Hymenoptera (ants, bees, and wasps), have haplodiploid reproduction systems, and are expected to have the following characteristics compared to outbreeding diploids (Table 8.6; Hedrick & Parker 1997):

- lower genetic loads
- less inbreeding depression and only in diploid females
- more rapid loss of genetic diversity (for the same N_e)
- lower genetic diversity for the same N_e
- experience outbreeding depression
- exhibit genetic rescue for fitness in females, but not males
- experience evolutionary rescue in both sexes
- lower N_e/N ratios for eusocial species with sterile workers
- loss of sex alleles will lead to diploid males, reduced population fitness, and increased extinction risks.

Harmful recessives are effectively removed by natural selection in haploid males, leading haplodiploid species to be less sensitive to inbreeding depression than outbreeding diploids, and expressing it only in diploid females (Werren 1993; Hedrick & Parker 1997; Henter 2003; Peer & Taborsky 2005; Frankham et al. 2010). However, mild inbreeding depression in haplodiploid females has been reported in a rotifer (*Brachionus plicatilis*) and a parasitoid wasp (*Venturia canescens*) (Tortajada et al. 2009; Vayssade et al. 2014).

Haplodiploids are susceptible to outbreeding depression when populations are crossed, but it is unclear whether more or less is found than in outbreeding diploids. Outbreeding depression, but little or no inbreeding depression was reported in haplodiploid ambrosia beetles (*Xylosandrus germanicus*) and fig wasps (*Platyscapa awekei*) that practice substantial sib mating (Peer & Taborsky 2005; Greeff et al. 2009). Conversely, the inbreeding parasitic wasp *Nasonia vitripennis* exhibited modest inbreeding depression, but no outbreeding depression (Luna & Hawkins 2004). Genetic rescue effects have been reported for female fitness for haplodiploid honeybees (*Apis mellifera*), ambrosia beetles, and a rotifer (Cale & Gowen 1956; Peer & Taborsky 2005; Tortajada et al. 2010).

The haplodiploid system has only ¾ the N_e compared to diploids with the same number of adults and an equal sex-ratio. Consequently, haplodiploids have less genetic diversity than diploids with similar N_e (Hedrick & Parker 1997; Frankham et al. 2010). Haplodiploids with eusociality have very low ratios of N_e/N due to the existence of numerous sterile workers.

Sex in haplodiploid Hymenoptera is controlled by a nuclear locus (complementary sex determining, CSD) that in large populations has many alleles. This results in heterozygous diploid females and haploid males (Box 8.3). However, small populations lose alleles at the sex locus, resulting in some diploids being males (rather than females), with consequent reduced reproductive rates for populations or species, reduced N_e, and elevated extinction risk (Box 8.3).

Box 8.3 Loss of genetic diversity at the sex locus reduces fitness and increases extinction risk in Hymenoptera

(Cook & Crozier 1995; Pamilo & Crozier 1997; Zayed et al. 2004; Zayed & Packer 2005)

Until recently, Hymenoptera were considered less susceptible to the impacts of small population size, as they typically suffer less inbreeding depression than diploids. However, they may be more susceptible than diploids, due to the loss of alleles at the sex locus and the production of diploid males (Zayed & Packer 2005).

Sex is determined by the genotype at the CSD locus, with heterozygous diploid females, haploid males, and homozygous diploid males. Large populations maintain many CSD alleles (commonly 9–20) by rare-advantage selection, resulting in diploid females and haploid males, as illustrated:

Female ♀		Male ♂
a/b	×	c
	↓	
Diploid progeny		Haploid progeny
½ a/c ♀		½ a ♂
½ b/c ♀		½ b ♂

However, in small populations, CSD alleles are lost due to drift, leading to matings between heterozygous females and males sharing one of their CSD alleles, resulting in half of the diploid progeny being males, rather than females, as follows:

Female ♀		Male ♂
a/b	×	b
	↓	
Diploid progeny		Haploid progeny
½ a/b ♀		½ a ♂
½ b/b diploid ♂		½ b ♂

For example, the orchid bee (*Euglossa imperialis*) from Panama has an average of only 3.1 sex alleles and ~ 34.3% of males are diploid (Zayed et al. 2004; but see Souza et al. 2010). Diploid males are inviable in some species, and where viable and successful in fertilizing females, the progeny are sterile triploid females. Further, lower genetic diversity across the genome for "neutral" markers is associated with elevated frequencies of diploid males (Collet et al. 2016).

Loss of CSD alleles and diploid male production is associated with reduced population growth rate in Hymenoptera and stochastic computer modeling indicates it greatly elevates extinction risk (Zayed & Packer 2005). Field studies have confirmed these predictions (Whitehorn et al. 2009).

Orchid bee (Central America)

Obligate asexual or clonal species

King's lomatia (Australia)

Asexual lineages have higher extinction risk than sexual ones (Maynard Smith 1978; Normark et al. 2003; Tucker et al. 2013). Consequently, obligate asexual species are rare, but asexual populations are found in many plant species, especially in aquatic, rare and endangered species, and ones in geographically marginal habitats (Ellstrand & Roose 1987; Eckert 2002; Silvertown 2008). Several rare plant species persist as a single clone per population with no known sexual reproduction, including King's lomatia, Meelup mallee (*Eucalyptus phylacis*), *Santalum lanceolatum*, *Grevillea infecunda*, Grampians pincushion lily (*Borya mirabilis*), and *Acanthocladium dockeri*, all in Australia (Lynch et al. 1998; Rossetto et al. 1999; Warburton et al. 2000; Coates et al. 2002; Kimpton et al. 2002; Jusaitis & Adams 2005). Asexual species also exist in fungi, invertebrates (e.g. *Timema* walking stick insects), lizards, fish, and many other animals (Law et al. 2002; Vrijenhoek & Parker 2009; Schwander et al. 2011; Fisher & Henk 2012; Bogart & Bi 2013; Neiman et al. 2014).

Asexual species do not undergo segregation or recombination, and are expected to have the following characteristics, compared to sexual diploids:

- no inbreeding depression
- higher genetic loads
- no outbreeding depression
- no genetic rescue for fitness
- evolutionary rescue if new clones are added
- limited capacity to evolve in single clones, unless population sizes are very large
- susceptibilty to extinction from mutational meltdown.

Asexual eukaryotic species and populations without clonal genetic diversity have limited potential to show adaptive evolution, as they rely on recent mutations for genetic change. For example, Hersch-Green et al. (2012) reported slower evolution of a chitinase locus involved in defense against plant pathogens in an asexual than in a sexual lineage of evening primrose (*Oenothera biennis*). Further, evolutionary change was slower in asexual than sexual populations of the green alga *Chlamydomonas reinhardtii* subjected to increasing concentrations of NaCl (Lachapelle & Bell 2012). Asexual populations with multiple clones evolved faster than those with single clones (Lachapelle & Bell 2012; Stelkens et al. 2014).

Species capable of both sexual and asexual reproduction often exhibit outbreeding depression when long-isolated clonal populations are crossed. In diploid yeast, 500 generations of asexual divergence in disparate environments, followed by crossing resulted in outbreeding depression of 46%, while hybrids of parallel adapted populations in similar laboratory environments experienced a 28% reduction in fitness (Dettman et al. 2007).

Mutational meltdown appears to be of more importance in asexual and selfing species (Gabriel et al. 1993; Lynch et al. 1993; Paland et al. 2005; Paland & Lynch 2006). However, even then it may be only a modest threat (Judson & Normark 1996; Normark et al. 2003). For example, Zeyl et al. (2001) observed no fitness decline or extinction in asexual yeast populations (N_e of 250) with normal mutation rates over 2,900 generations.

Sexual species that exhibit extensive clonality have characteristics between those of asexual and sexual species (Vallejo-Marin et al. 2010).

How do we identify modes of inheritance?

Chromosome counts or DNA content are used to determine ploidies

Haploid, diploid, and polyploid individuals and populations can be distinguished using chromosomal counts (Rowell et al. 2011). Alternatively, DNA content can be compared to other related taxa with known ploidies. For example, the endangered tetraploid *Dioscorea trifida* has approximately double the DNA contents of its presumed diploid wild Amerindian yam progenitor (Fig. 8.3).

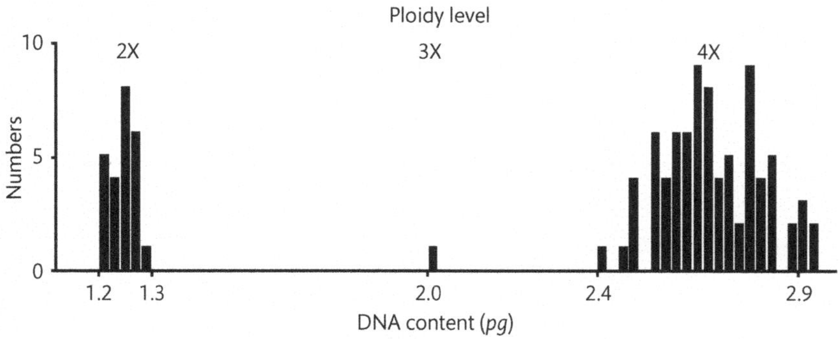

Fig. 8.3 DNA contents of endangered tetraploid (4X) *Dioscorea trifida* and its presumed diploid (2X) wild Amerindian yam progenitor, and a triploid hybrid between them (after Bousalem et al. 2010). The tetraploid form has 80 chromosomes and the diploid 40.

Autopolyploids and allopolyploids can be distinguished by genotyping individuals using multiple microsatellite loci or SNPs (Chenuil et al. 1999). Autotetraploid individuals will only contain a maximum of two alleles per locus, while allotetraploid individuals may contain up to four alleles.

How can we identify asexually reproducing species?

If all offspring contain the same genotype as the parent, then reproduction is asexual (including asexual parthenogenesis)

For example, the mode of inheritance (sexual or asexual) was unclear in grain aphids (*Sitobion miscanthi*) that have been introduced into Australia. Genotyping with molecular markers established that they were reproducing asexually (predominantly or exclusively parthenogenetic), as there were only seven multilocus genotypes among

555 individuals, and no evidence of regular segregation (Fig. 8.4). Related observations using allozyme markers identified clonality and asexual reproduction in the endangered shrub *Acanthocladium dockeri* in South Australia (Jusaitis & Adams 2005). If all individuals in a species are genetically identically homozygous, as in the Wollemi pine (*Wollemia nobilis*), then the mode of reproduction can be either asexual or sexual (Peakall et al. 2003).

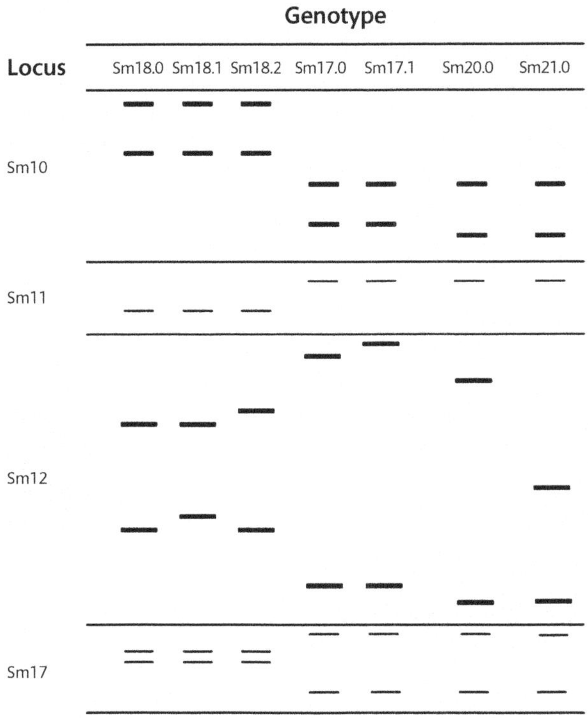

Fig. 8.4 Identifying parthenogenetic reproduction in grain aphids. Only seven Australian *Sitobion miscanthi* genotypes were detected among 555 individuals, based on data for four microsatellite loci (Sunnucks et al. 1996), whereas 24,750 genotypes are possible under random meiotic segregation. Further, all individuals are heterozygous for the Sm10 and Sm17 loci, so there is no meiotic segregation.

We have now completed discussion of essential background material and next turn to Section III 'Developing management decisions'.

Summary

1. Levels of genetic diversity, risk of inbreeding and outbreeding depression, prospects for genetic rescue, and adaptive potential often differ among species with diverse mating systems and modes of inheritance, compared to those in outbreeding diploids.

2. Self-incompatible species are more susceptible than comparable self-compatible species to the effects of small population size due to the fitness effects of loss of SI alleles, in addition to normal inbreeding depression, but similarly or less susceptible to outbreeding depression. They have higher potential for genetic rescue than similar self-compatible species.
3. Obligate selfing species typically exhibit low genetic diversity, and less inbreeding depression than sexual species. They do not outcross and so show no outbreeding depression, or genetic rescue for fitness, but may show evolutionary rescue.
4. Mixed mating and selfing species with occasional outcrossing typically exhibit less inbreeding depression, but have stronger isolation among populations than outbreeding species, and an increased likelihood of outbreeding depression. They have, on average reduced potential for genetic rescue for fitness that persists less over generations, and exhibit evolutionary rescue.
5. Polyploids lose genetic diversity at a slower rate than similar sized diploid populations. They are expected to be less susceptible to inbreeding depression and to suffer outbreeding depression. They benefit from genetic rescue for fitness and evolutionary adaptation.
6. Haploid species do not exhibit inbreeding depression, but are susceptible to outbreeding depression, and exhibit evolutionary rescue, but not genetic rescue for fitness.
7. Haplodiploid species exhibit less inbreeding depression in females than otherwise similar diploid species, and none in males. They may exhibit outbreeding depression when crossed. Benefits of genetic rescue are restricted to females, but both sexes benefit from evolutionary rescue.
8. Obligate asexual species do not inbreed or outcross, and so do not exhibit inbreeding or outbreeding depression, or genetic rescue for fitness, but may show evolutionary rescue if novel clonal genotypes are added.

FURTHER READING

Byers & Waller (1999) Meta-analysis showing that inbreeding depression is less in selfing than outcrossing plant species.

Dudash & Murren (2008) Review of the influence of breeding system and mating system in plant conservation genetics.

Frankham (2015) Compared genetic rescue effects for self-incompatible, outbreeding, selfing and mixed mating species, based on a meta-analysis.

Hereford (2010) Reported no relationship between selfing rate and frequency of adaptation, based on a meta-analysis.

Young et al. (2000) Wide ranging study on the fitness consequences of loss of self-incompatibility alleles in an endangered daisy.

Zayed & Packer (2005) Evaluated the impacts of loss of sex alleles at the CSD locus on the persistence of haplodiploid species, using computer modeling.

SOFTWARE

MLTR: Free software for estimating mating systems and selfing rates from genetic data (Ritland 2002). http://www.genetics.forestry.ubc.ca/ritland/programs.htm

BORICE: Free software for estimating outcrossing and selfing rates from genetic data including robust family level estimates (Koelling et al. 2012). https://eeb.ku.edu/john-kelly

SECTION III
Developing management decisions

Having described genetic problems associated with reduced gene flow and identified augmentation of gene flow as a remedy, we are now ready to consider the genetic management of fragmented populations. Additionally, we have identified the important risk factors for outbreeding depression that allow managers to minimize the chance of experiencing this problem.

Objectives are fundamental to decision-making in conservation. Our objective in genetic management of threatened species is to minimize the risk of population and species extinction due to genetic and associated factors.

The questions we need to answer to achieve this objective are outlined in this Section, and means for addressing them detailed in the following chapters.

What is the management unit?

The first question is whether the taxonomy is reliable and appropriate for conservation purposes (Chapter 9). We provide guidelines for determining whether current taxonomy is adequate for conservation purposes. If not, we recommend appropriate definitions of species for conservation purposes and means to implement species delineations with them.

How can we determine the number and location of genetically differentiated population fragments?

Chapter 10 addresses methods for determining the number and boundaries of population fragments within species. One commonly used method is based on deviations from Hardy–Weinberg equilibrium for genetic markers to identify non-random mating. In general, the number of populations and their boundaries can be delineated by mapping genetic clusters onto the landscape.

If gene flow is limited between population segments due to physical or reproductive isolation, it is common to conclude that the segments should be managed separately—even though this is typically a recipe for extinction for small inbred populations with low genetic diversity. Consequently, evidence for genetically structured populations should trigger the question (Chapter 11):

Are there isolated populations suffering genetic erosion?

Thus, we are seeking to know if there are populations with low genetic diversity that are inbred and have an elevated risk of extinction. If so, we next ask in sequence:

- Would they benefit from augmented gene flow?
- If yes, are there any populations that can be used to genetically rescue them?
- If yes, would crossing between populations be beneficial or harmful?

How can we manage gene flow?

If crossing between populations is likely to be beneficial, then we consider the details of managing gene flow among isolated population fragments (Chapters 12 and 13). Worthwhile genetic management of fragmented populations can usually be conducted with minimal population information, or information obtained at minimal cost (Chapter 12). However, management efficiency can typically be improved with more detailed information, particularly by using mean kinship (Chapter 13). However, we emphasize that **any augmentation of gene flow is usually better than none**, when the risk of outbreeding depression is low.

How should we modify genetic management under global climate change?

In Chapter 14 we address genetic management of fragmented populations under global climate change, applying the principles we developed earlier to the even more trying situation of a persistently changing environment. Global climate change increases the need for genetic management, as genetic diversity is needed to cope with it by adapting *in situ*.

We conclude this section by emphasizing the need to integrate genetic management with other demographic and ecological considerations across populations, disciplines, institutions, and political boundaries to develop a comprehensive plan for the conservation of a species, such as the "One Plan approach" developed by Byers et al. (2013). Details of that approach are beyond the scope of our book. However, there may be the need to compromise on favored genetic management options to address other competing demands. For example, a habitat corridor that allows genetic mixing and prevents local inbreeding can also be the route by which diseases, predators, and invasive species spread through a metapopulation (Haddad et al. 2014). There might be intermediate levels of population connectivity that are adequate to prevent genetic decay while still inhibiting spread of diseases and other threats (Lacy et al. 2013).

Is the taxonomy appropriate? Delineating species for conservation purposes

CHAPTER 9

The first step in conservation management is to delineate groups for separate versus combined management. However, there are many problems with species delineation, including diverse species definitions, a lack of standardized protocols, and poor repeatability of delineations. Definitions that are too broad will lead to outbreeding depression if populations are crossed, while those that split excessively may preclude genetic rescue of small inbred populations with low genetic diversity. To minimize these problems, we recommend the use of species concepts based upon reproductive isolation (in the broad sense) and advise against the use of Phylogenetic and General Lineage Species Concepts. We provide guidelines as to when taxonomy requires revision and outline protocols for robust species delineations.

TERMS

Allopatric, Biological Species Concept, conspecific, chloroplast DNA, Differential Fitness Species Concept, Evolutionary Species Concept, exchangeability, fixed gene differences, General Lineage Species Concept, gene tree, integron, introgression, lineage sorting, parapatric, Phylogenetic Species Concept, phylogenetic tree, reciprocal monophyly, selective sweep, sympatric, Taxonomic Species Concept, transposon

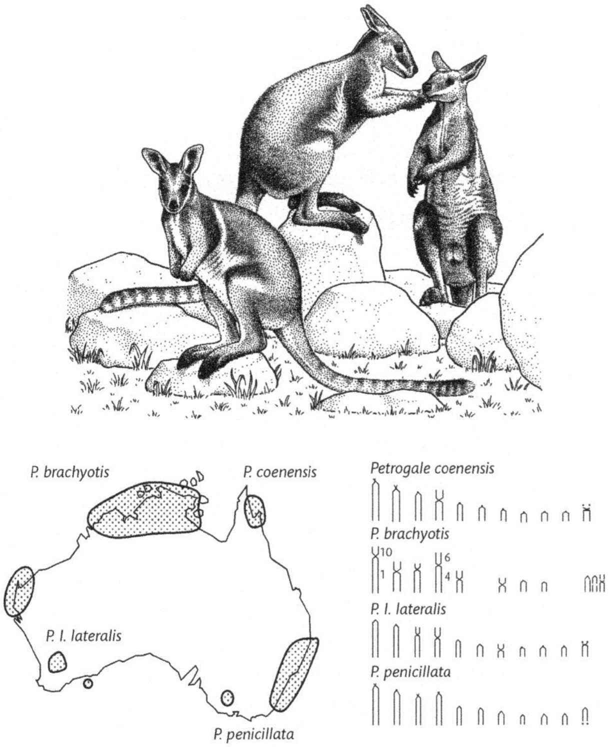

The taxonomy of rock-wallabies in Australia was controversial, but was resolved using extensive geographic sampling, chromosomal, molecular genetic and morphological analyses, and crossing between putative taxa. One species is illustrated, along with the distributions and chromosomes for four species (Frankham et al. 2010, p. 362)

Genetic Management of Fragmented Animal and Plant Populations. R. Frankham, J. D. Ballou, K. Ralls, M. D. B. Eldridge, M. R. Dudash, C. B. Fenster, R. C. Lacy & P. Sunnucks. Oxford University Press (2017) © R. Frankham, J. D. Ballou, K. Ralls, M. D. B. Eldridge, M. R. Dudash, C. B. Fenster, R. C. Lacy & P. Sunnucks 2017. DOI 10.1093/oso/9780198783398.001.0001

What is the role of species delineation in conservation?

The crucial first step in conservation management is to delineate groups (species, etc.) for separate versus combined management

If populations and species are not appropriately delineated, genetically differentiated populations (including distinct species) may be inappropriately crossed, leading to outbreeding depression. Alternatively, populations that are inbred and depleted of genetic diversity may be inappropriately classified as distinct, causing genetic rescue of small populations to be blocked by regulatory and legal hurdles (O'Brien & Mayr 1991; Haig et al. 2006; Ellstrand et al. 2010).

What is our objective?

We seek to delineate species such that the probability of persistence is maximized, especially in the face of changing environments

The probability of species persistence will be maximized if they have high fitness, ample genetic diversity to evolve, and high population sizes (Chapters 2–7 and 14). Thus, species delineations for conservation purposes need to facilitate achievement of these conditions.

As species are delineated based on heritable differences in some combination of morphology, behavior, reproduction, or genotypes, we now consider how these are characterized.

How do we characterize population genetic differentiation in taxonomy?

Genetic differentiation of populations can be characterized in terms of reproductive isolation, or indirectly as differentiation for "neutral" markers (indicating reproductive or physical isolation)

If populations are in contact but exhibit no gene flow, they are reproductively isolated. Allozymes, microsatellites, SNPs, DNA sequences, chromosomes, or heritable morphological differences can provide evidence of isolation. Deliberate attempted crosses can be used, if feasible.

Completely diverged populations are described as having non-overlapping distributions for heritable quantitative characters, or exhibiting no shared alleles, fixed gene differences, or exhibiting reciprocal monophyly at genetic loci

If populations have non-overlapping heritable differences in quantitative characters (e.g. morphology, behavior, or life-history) they are often delineated as distinct species. Since such characters may differ in response to differing environmental conditions, rather than representing genetic differences, taxonomists are relying increasingly on "near neutral" molecular genetic markers, such as allozymes, microsatellites, SNPs, and DNA sequence data.

In what follows, we consider how genetic marker differences are characterized. Closed populations derived from the same ancestral population drift apart and eventually become homozygous, some for the same ancestral allele and some for different ancestral alleles (lineage sorting: Fig. 9.1) or a new mutation. In taxonomy, diverged populations are described as having no shared alleles, fixed gene differences, or exhibiting reciprocal monophyly in phylogenetic trees. Reciprocal monophyly occurs when the members of a group are more closely related to each other than they are to any members of an alternative group.

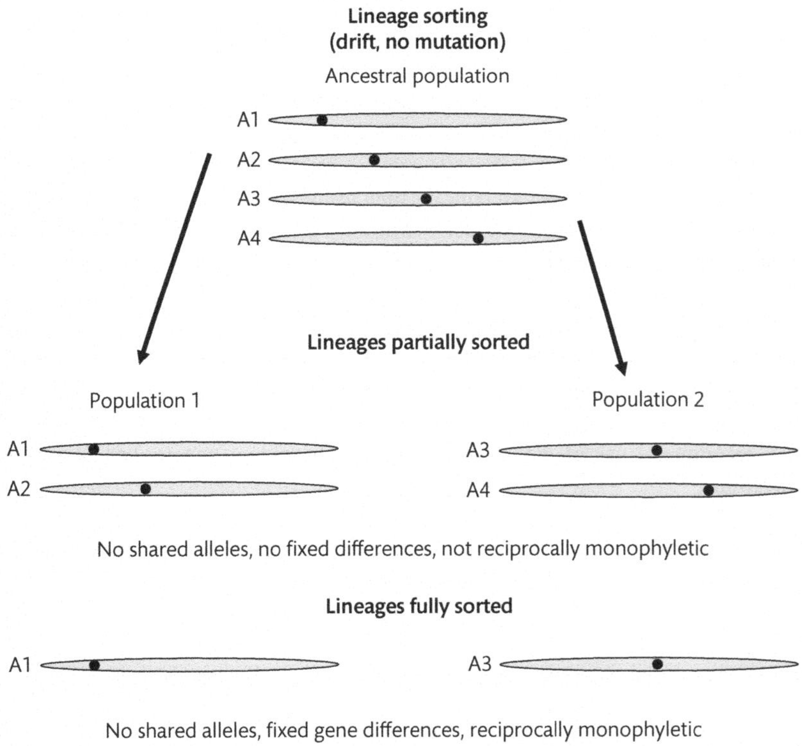

Fig. 9.1 Divergence of two populations from a polymorphic ancestral population (with four haplotypes A1–A4 differing in sequence, as indicated by the black circles) due to drift (lineage sorting) over generations to the point where the populations share no alleles, have fixed gene differences, and exhibit reciprocal monophyly. Populations 1 and 2 are isolated and drift independently, with population 1 losing alleles/haplotypes A3 and A4 in the initial phase and then A2, while population 2 by chance loses A1 and A2 in the first phase and then A4 (after Frankham et al. 2012, Fig. S1).

Taxonomists frequently represent evolutionary similarities among groups of individuals in phylogenetic trees, where increasing levels of differentiation progress from a star phylogeny, through polyphyly, paraphyly to reciprocal monophyly (Fig. 9.2).

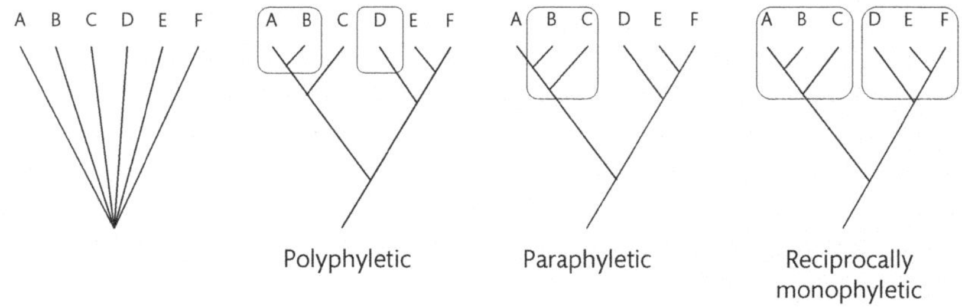

Fig. 9.2 Phylogenetic trees illustrating increasing divergence of populations from left to right: random mating (star phylogeny–all offspring similarly related to recent common ancestor), polyphyly, paraphyly, and reciprocal monophyly.

Phylogenetic trees are typically inferred from gene trees, but different loci often yield different trees, due to genetic drift

Phylogenetic trees were in the past typically built using morphological data, but they are now usually inferred from genetic data. Critically, **gene trees inferred for the same populations using different loci are often different, due to genetic drift.** Consequently, only combined information from multiple loci is likely to yield a reliable phylogenetic tree (Nichols 2001). Coalescent methods have recently been devised to build species trees from data on multiple loci (Fujita et al. 2012). They are superior to prior methods when populations are genetically isolated, or have low levels of gene flow (Carstens & Dewey 2010; Zhang et al. 2011).

We recommend use of data from multiple independent genetic loci in species delineations.

This chapter defines problems with delineating species, and recommends methods for avoiding them in delineations for conservation purposes.

What problems occur with species delineations?

There are frequent controversies about species delimitations, as there are many different species concepts and no standardized protocols, often leading to disparate species delineations

Species delineation as currently practiced is often deeply flawed, with serious consequences for conservation of biodiversity (Mace 2004; Frankham et al. 2012). The main problems are:

- no standardized sampling regimes, list of characters to use, or analyses to perform
- widespread use of diverse methods
- instability of delineations to technological change, especially with some species concepts
- poor repeatability of delineations, especially those done using different approaches
- many disparate definitions of species that often lead to different delineations
- over-lumping is common (especially in older delineations)
- over-splitting is common currently, and worsening
- widespread use of markers (e.g. sole reliance on mtDNA) that have little statistical robustness
- delineations using neutral molecular markers that fail to detect recent adaptive differentiation that has resulted in partial reproductive isolation
- defining morphologically similar plant populations with different chromosomal numbers (ploidy levels) as conspecific, despite crosses between them resulting in sterility
- failure to check chromosomes in many delineations, meaning that chromosomally caused reproductive isolation may be missed (e.g. those due to differences in ploidy, translocations, inversions, and centric fusions)
- morphological differences relevant to taxonomy must reflect heritable differences, but this is rarely evaluated in common garden experiments.

Mace (2004) concluded that "taxonomists and conservationists need to work together to design some explicit rules to delimit the units included as species for the purposes of conservation planning and assessment." Many of the above issues are illustrated in Box 9.1 by the controversial and changing taxonomy of the Bornean and Sumatran orangutans (*Pongo pygmaeus* and *P. abelii*).

Box 9.1 Controversial and changing taxonomy of recently diverged Bornean and Sumatran orangutans, based on conflicting information and different species concepts

(after Frankham et al. 2010)

Orangutans on the islands of Borneo and Sumatra in Southeast Asia have been designated as sub-species or separate species (Janczewski et al. 1990; Xu & Arnason 1996; Zhi et al. 1996; Kanthaswamy et al. 2006; IUCN 2016) as they differ in morphology (but see Groves et al. 1992), behavior, a fixed chromosomal inversion (Ryder & Chemnick 1993), mtDNA sequences (but see Kanthaswamy et al. 2006 and later in Box 9.1), and nuclear loci. Since they differed genetically by at least as much as chimpanzees and bonobos, and their divergence was dated at 10.5 million years, using mtDNA (Arnason et al. 1996), full species status for the two forms was suggested by Xu & Arnason (1996) and Zhi et al. (1996). In addition, the Bornean form has been separated into three sub-species (Wilson & Reeder 2005).

However, hybrids between Bornean and Sumatran orangutans are viable and fertile in the F_1 and F_2 generations (Muir et al. 1998), meaning they are not distinct species according to the Biological Species Concept. Further, a broader geographic sampling of individuals revealed that the two forms do not have fixed differences in mtDNA (Kanthaswamy et al. 2006), and new molecular dating yielded a divergence time of only 334,000 years, based on whole genome sequences (Mailund et al. 2011). However, the fixed differences at nuclear loci and the chromosomal inversion are sufficient to designate the two forms as separate species under the Phylogenetic Species Concept (see later). This illustrates the confusion created by use of different species definitions, different characters, and inadequate sampling regimes.

The taxonomic revision created a management dilemma, as the two orangutan species have been hybridized through multiple generations in zoos. The hybridized individuals are now prevented from breeding, filling valuable zoo breeding space that might be used for breeding endangered species (Frankham et al. 2010). Given the declining numbers of orangutans, gene flow in the wild will likely be needed in the future to preserve the genetic health of the two island populations.

Bornean and Sumatran orangutans

In what follows, we seek to minimize these problems by advocating:

- appropriate species concepts for conservation purposes
- use of multiple characters and loci (integrative taxonomy)
- wide sampling of species distributions
- use of scientifically robust analyses
- development of standardized protocols for robust species delineations for conservation purposes.

In this Chapter we concentrate on the first of these items and defer most consideration of the remaining issues to Appendix 3. To help focus the topic, Box 9.2 provides examples of what we consider thorough versus inadequate taxonomic determinations.

Box 9.2 Thorough versus inadequate taxonomic delineations

Thorough taxonomic delineations: extensive sampling and use of multiple lines of evidence

Thorough taxonomic delineations are typically based on wide geographic sampling of individuals along with evidence from multiple characters, as illustrated by rock-wallabies (*Petrogale*) in Australia (Chapter frontispiece). Rock-wallabies are small macropodid marsupials < 1 m tall, with fragmented distributions on rocky outcrops (see Chapter frontispiece), and they have diverged relatively recently. Their taxonomy was controversial, with between 5 and 11 species (many threatened) recognized by different authors. In 1976 studies were begun at Macquarie University to resolve their taxonomy using extensive geographic sampling, chromosomal analyses, allozymes, mtDNA, DNA-DNA hybridization, nuclear DNA sequences, morphology, and crosses between putative taxa (Sharman et al. 1990; Eldridge & Close 1992, 1993, 1997; Campeau-Péloquin et al. 2001; Eldridge et al. 2001; Metcalfe et al. 2002; Potter et al. 2012a, 2012b). These studies revealed 20 sharply delineated chromosomal races (four shown on the frontispiece map) whose distributions were typically concordant with one or more other attributes (morphology, results of crossing, nuclear DNA, and mtDNA; Eldridge et al. 2001).

Inadequate: diploid and polyploid forms within the same "species"

Endangered button wrinklewort daisies (*Rutidosis leptorrhynchoides*) in south-eastern Australia contain morphologically similar diploid (2n), tetraploid (4n), and hexaploid chromosomal forms (6n) (Murray & Young 2001) that when crossed result in outbreeding depression, as happened during a recovery program (Young & Murray 2000). These forms should be designated as distinct species for conservation purposes and managed separately (Frankham et al. 2012). This is a widespread problem throughout plant taxonomy.

Button wrinklewort daisy (Australia)

What are the main species concepts and how do they differ?

There are ~ 27 definitions of species. Use of different concepts on geographically isolated (allopatric) populations often results in different species delineations, some inappropriate for conservation purposes

Conservation biology is bedeviled by the existence of at least 27 definitions of biological species (see Wilkins 2009; Hausdorf 2011). Use of different species concepts leads to lack of repeatability, controversy, and sometimes to inappropriate delineation of species for conservation purposes (Hey et al. 2003; Frankham et al. 2012).

Table 9.1 defines the four concepts most widely used by the systematic and conservation communities: the Biological Species Concept (Wallace 1865; Mayr 1942, 1963), the Evolutionary Species Concept (Simpson 1951, 1961; Wiley 1978), the Phylogenetic Species Concept (Eldredge & Cracraft 1980; Cracraft 1997), and the General Lineage Species Concept (de Queiroz 1998, 2007). We also define the Differential Fitness

Species Concept (Hausdorf 2011), and the concept of ecological and genetic exchangeability, as applicable to species delineation (Crandall et al. 2000), as these are relevant for conservation purposes. An alternative to the use of defined species concepts is to rely upon the judgment of taxonomists, sometimes referred to as the Taxonomic Species Concept (Mayden 1997). This is widely practiced: papers on species delineations or revisions usually fail to specify what species concept has been used (see McDade 1995).

Table 9.1 Species definition according to different concepts.

Species concept	Species definition	Reference
Biological (BSC)	"groups of actually or potentially interbreeding natural populations, which are reproductively isolated from other such groups"	Mayr (1942)
Evolutionary (ESC)	"a species is a lineage of ancestral descent which maintains its identity from other such lineages and which has its own evolutionary tendencies and historical fate"	Wiley (1978)
General lineage (GLSC)	"species are (segments of) separately evolving metapopulation lineages"	de Queiroz (2007)
Phylogenetic (PSC) (diagnostic)	"a species is the smallest diagnosable cluster of individual organisms within which there is a parental pattern of ancestry and descent"	Cracraft (1983)
Differential fitness (DFSC)	"species can be defined as groups of individuals that are reciprocally characterized by features that would have negative fitness effects in other groups and that cannot be regularly exchanged between groups upon contact"	Hausdorf (2011)
Exchangeability	Species are populations that are not ecologically or genetically exchangeable on the recent time scale and at least one of those attributes is not exchangeable on the historical timescale.	Crandall et al. (2000)
Taxonomic (TSC)	"whatever a competent taxonomist chooses to call a species"	Wilkins (2009)

Despite the disparate definitions, species concepts typically indicate that species are cohesive clusters of individuals that have at least partially different evolutionary paths and represent different lineages (see Avise & Ball 1990; de Queiroz 1998; Hausdorf 2011). All serious concepts recognize that populations inherently incapable of gene exchange are distinct species, while those exhibiting random mating in sympatry are conspecific. However, there are major differences in the treatment of fragmented (allopatric) populations capable of gene flow without adverse fitness consequences, or with beneficial consequences.

All of the commonly used species concepts must be considered in light of the following empirical observations (Hausdorf 2011):

1. "reproductive barriers are often semipermeable to gene flow"
2. "species can differentiate despite ongoing interbreeding" (sympatric speciation)
3. "parallel speciation can occur due to parallel adaptation or recurrent polyploidizations"
4. "uniparental organisms are actually organized in units that resemble species of biparental organisms"

In addition, we note that:

5. development of reproductive isolation between allopatric populations usually accompanies genetic adaption to different environments via natural selection, as proposed by Darwin (1859), and/or fixed chromosomal differences (reviewed by Sexton et al. 2011; Frankham et al. 2012).

Points 1 and 2 cause severe difficulties for the BSC; furthermore, it does not apply to asexual organisms. For PSC and GLSC, points 1, 2, and 3 cause difficulties. Point 3 may cause difficulties for ESC, but ESC copes with the other points (but means for diagnosing species with it are not well established).

Use of PSC results in more splitting than BSC. For example, PSC yielded 49% more species than BSC on the same group of organisms (Agapow et al. 2004), and revisions using PSC approximately doubled the number of primate and ungulate species (Groves 2001; Groves & Grubb 2011; see critique by Heller et al. 2013).

Changed taxonomy due to application of different species concepts can lead to altered management, with adverse consequences for biodiversity conservation, as illustrated by black and white ruffed lemurs (*Varecia variegata*) (Box 9.3).

Box 9.3 Confused management resulting from conflicting taxonomies of critically endangered black and white ruffed lemurs from Madagascar

Prior to 2001, black and white ruffed lemurs were recognized as a single species. However, Groves (2001) using PSC delineated three sub-species based upon pelage pattern differences (map a). Vasey & Tattersall (2002) disagreed, because they interpreted the different pelage forms as a color polymorphism without clear geographic patterning. Further, Wyner et al. (1999) delineated distinctive populations north and south of the Mangoro River (map b) that are discordant with Groves' three sub-species: they also used PSC, but based their delineation upon mtDNA control region sequences. Baden et al. (2014) supported the Wyner distinction, based on mtDNA and microsatellite data (map b and the STRUCTURE plot), and found isolation by distance in both populations.

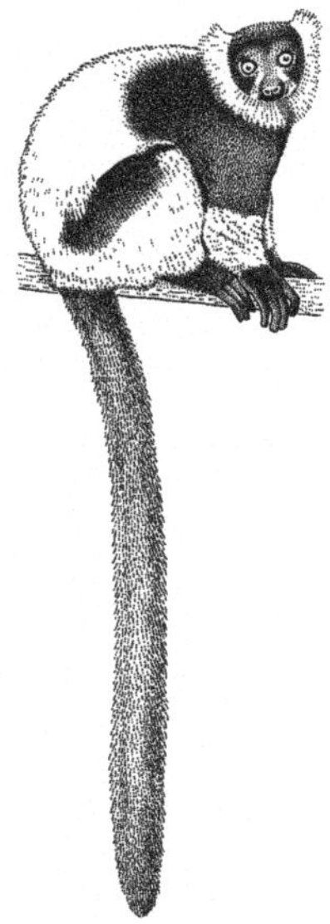

Black and white ruffed lemur (Madagascar)

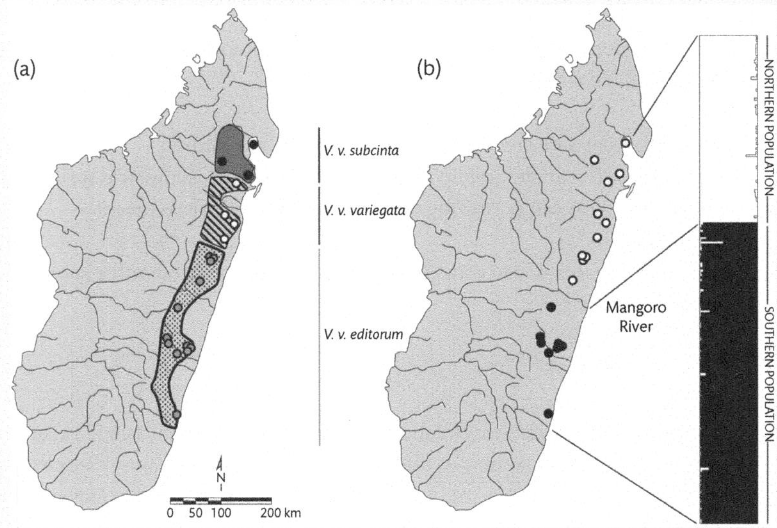

(after Baden et al. 2014, Fig. 1)

The European Endangered Species Program manages captive black and white ruffed lemurs as two isolated units (mixed *V. v. editorum* and *V. v. variegata*) and *V. v. subcinta* (based on Groves' [2001] delineations) and does not breed from hybrids between these. As their captive population of *V. v. subcinta* traces to only a handful of founders, it is unlikely to persist due to inbreeding problems (I. Porton, pers. comm.). Conversely, the Species Survival Program (SSP) of the Association of Zoos and Aquariums (centered in the USA) manages black and white ruffed lemurs as a single interbreeding population derived from founders belonging to all three PSC sub-species.

How do over-lumping and over-splitting arise?

Species delineations may be inappropriate if they create groups that are too inclusive (over-lumping), or many excessively small groups (over-splitting). How do these problems arise, and how can they be minimized?

Over-lumping

Excessively broad species delineations arise primarily from the use of characters (mainly morphological) with insufficient resolving power to delineate cryptic species

Many "species" have been shown to contain reproductively isolated segments that have subsequently been delineated as separate species (Bickford et al. 2007). Numerous species were delineated hundreds of years ago, based on limited information and sampling, both numerically and geographically, with insufficient resolving power. For example, combined molecular genetic, chromosomal, and morphological studies have shown that Australia is home to well over 100 locally distributed species of velvet worms (Onychophorans), rather than the seven widespread morphological species previously recognized (Briscoe & Tait 1995; Rowell et al. 1995, 2002; Reid 1996; Bull et al. 2013). Further, cryptic adaptively differentiated lineages of fluted gum (*Eucalyptus salubris*) exist in southwestern Australia and show no evidence of gene flow in areas of geographic overlap (Steane et al. 2014). Additionally, African elephants have recently been separated into savannah and forest species (Rohland et al. 2010).

In addition to cases with insufficient resolving power, excessive lumping may occur when strong adaptive differences drive reproductive isolation, despite gene flow in sympatry (Papadopulos et al. 2011). Alternatively, if neutral markers are used, recent adaptive divergence with reproductive isolation may result in no significant differentiation in neutral markers, and lumping of reproductively isolated populations. Further, populations that diverge in allopatry (Fig. 9.6), but later come into contact and form hybrid zones with introgression of alleles, may be misdiagnosed. Several rock-wallaby species in Australia exhibit hybrid zones and lack reciprocal monophyly for mtDNA and allozymes, but chromosomal, morphological, and crossing data indicate that they are distinct species (Box 9.2: Eldridge & Close 1992).

PSC and lineage based approaches may miss recently evolved chromosomal causes of outbreeding depression

Excessive lumping may also occur if there is recent allopolyploidy without corresponding morphological changes, as chromosomes are often not examined (Chapter 7). There is much enthusiasm for using genomics in species delineations, most of it using lineage based approaches (Fujita et al. 2012). However, even when based on genome sequences, GLSC and PSC based delineations will often fail to detect recently evolved chromosomal differences (autoploidy, translocations, inversions, and centric fusions) that result in outbreeding depression if such populations are crossed.

Use of multiple molecular loci or genome sequencing plus chromosomal assays, especially when combined with other characters, normally overcomes the issue of resolving power.

Over-splitting

Small population fragments of conservation concern are subject to large genetic drift effects, and thus may be inappropriately classed as different species, especially when mtDNA, highly mutable genetic markers, or genome sequence data are used together with the PSC and GLSC

In allopatric populations, especially those with small population sizes, genetic drift and mutation will lead to diagnosably different units that are not intrinsically reproductively isolated and may be ephemeral under natural patterns of population separation and re-connection. This is a major issue in the conservation of fragmented populations, because **the very measure that is used to delineate PSC and GLSC species (genetic differentiation among populations) is causally related to genetic problems within them** (Chapter 5). For example, the Illinois population of the greater prairie chicken was once connected to populations in Kansas, Nebraska, and Minnesota, but in historical times became a small and isolated genetic outlier (Fig. 9.3), yet it is not reproductively isolated from the other populations, and crossing yielded genetic rescue (Bouzat et al. 1998a, 1998b; Westemeier et al. 1998).

Greater prairie chicken (North America)

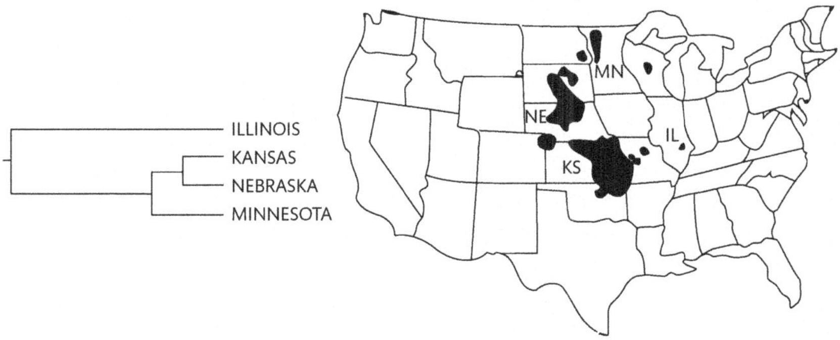

Fig. 9.3 The small Illinois (IL) population of greater prairie chickens is an outlier in the tree, compared to larger Nebraska (NE), Kansas (KS), and Minnesota (MN) populations, based on data on six microsatellite loci, but is not reproductively isolated from them (Bouzat et al. 1998a, Fig. 1). Map of the geographic distribution of the greater prairie chicken showing the location of studied populations. Estimated population sizes are Illinois < 50; Kansas > 100,000; Minnesota > 4,000; Nebraska > 100,000. The UPGMA phenogram (tree) was generated by Biosys-1 software, based on Nei unbiased genetic identity coefficients (Bouzat et al. 1998a, Fig. 3).

There are multiple problems with the use of mtDNA as the sole basis for delineating species

Small diverged/isolated populations are susceptible to being classified as different species according to the diagnostic version of PSC, especially when maternally inherited markers (mtDNA and cpDNA) are used in delineations. The problems inherent in using mtDNA as the sole data for species delineations are described in Box 9.4.

Box 9.4 What is mtDNA and why is its use in taxonomy problematic?

(Ballard & Rand 2005; Ballard & Whitlock 2004; Joseph & Omland 2009)

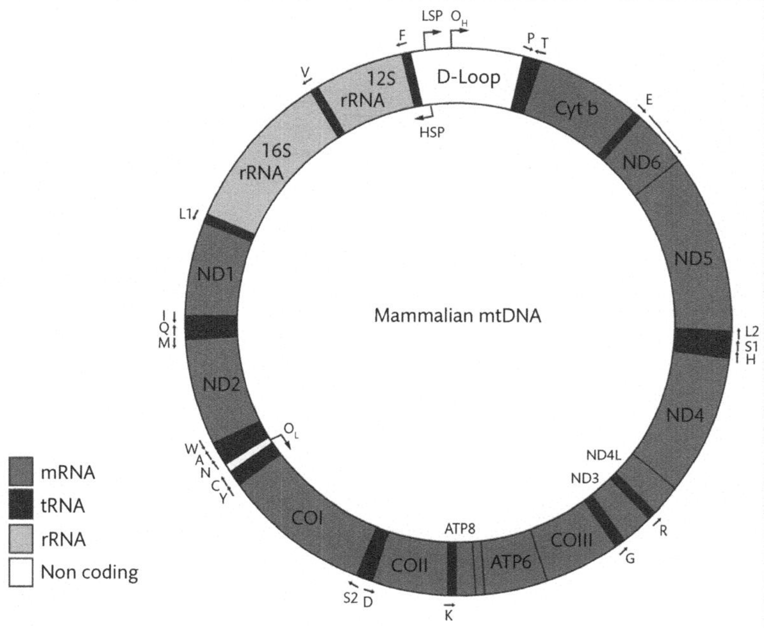

(after Park & Larsson 2011, Fig. 1)

Mitochondrial DNA (mtDNA) occurs as a circular DNA molecule of ~ 16,000 bp (as illustrated for mammals above) that originated as a captured bacterium within a eukaryotic cell. It codes for a small number of proteins involved in energy metabolism (shaded dark gray: Cyt b, COI, ND1, etc.). The mutation rate for mtDNA in vertebrates is ~ ten-times that for nuclear DNA (Ballard & Whitlock 2004).

It is maternally inherited and non-recombining in most species and has a lower N_e than nuclear loci (Barr et al. 2005; Frankham 2012). Further, it is subject to selective sweeps when a new favorable mutation in a single mtDNA base rises to fixation and sweeps away genetic diversity throughout the whole molecule. Consequently, genetic diversity in mtDNA is lost at a much greater rate than nuclear DNA (Frankham 2012). Phylogenetic trees based on mtDNA may differ from those determined for nuclear loci, and from the "true" species tree (Nichols 2001), as for example was observed in silvereye birds (*Zosterops*) in Australia (figures a and b). Further, primate phylogenies based on short mtDNA fragments are discordant with those from multiple nuclear loci and with morphological and other data (Perelman et al. 2011; Finstermeier et al. 2013).

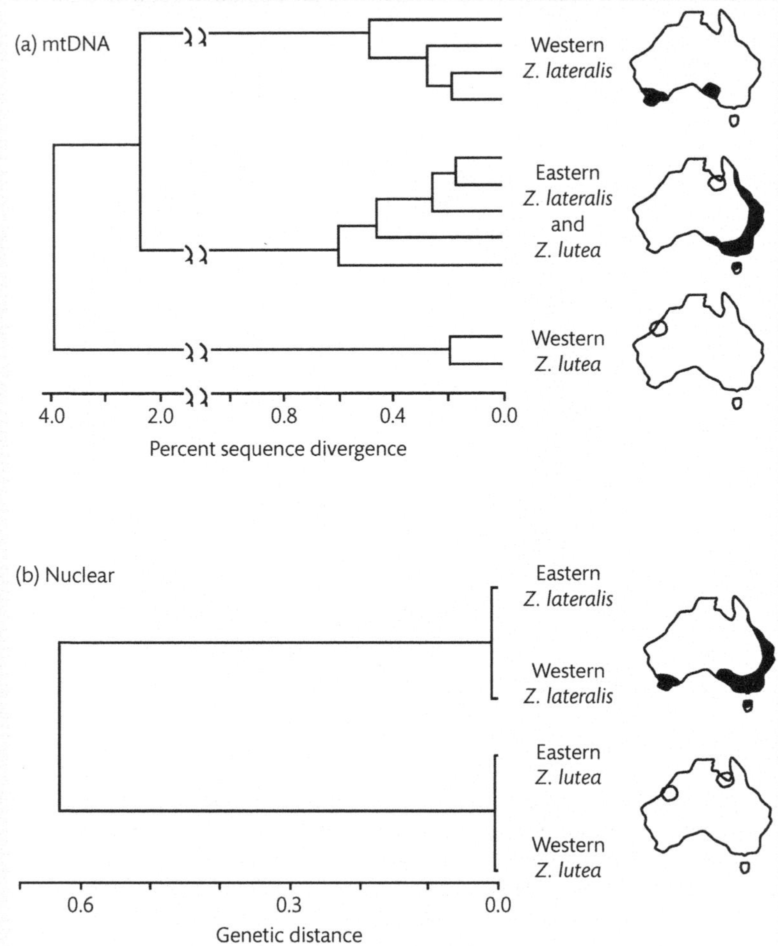

(Degnan 1993, Fig. 1)

As mtDNA diverges more rapidly than nuclear diploid loci (due to lower N_e and higher mutation rates), it has been widely used in taxonomy (Hebert et al. 2003a, 2003b; Rubinoff et al. 2006). However, **we do not recommend the sole use of mtDNA for taxonomic delineations due to its low precision, and the high potential for misleading inferences.**

Use of PSC or GLSC results in many problems

Theory and empirical studies show that small isolated populations may rapidly become diagnosably different, such that they are delineated as separate species under PSC and GLSC without exhibiting reproductive isolation

There are two serious additional problems with diagnosability under PSC and GLSC, namely (a) instability under technological change, and (b) over-splitting, especially in small populations. Technological advances, such as from morphology to allozymes, microsatellites, SNPs and whole genome sequencing have led to finer and finer diagnosable differences among populations using PSC and GLSC (Avise & Ball 1990; Carstens & Knowles 2007; Knowles & Carsten 2007; Winker 2010). Consequently, this has resulted in instability of species delineations, rather than improving certainty of delineations with increasing information.

Fixed gene differences, no shared alleles or reciprocal monophyly are required to delineate species under different implementations of the diagnostic PSC (Cracraft 1997; Groves 2004). How serious is over-splitting with the use of PSC or GLSC? To assess this, we ask "How long does it take for isolated populations to attain diagnosability under PSC?"

Most closed populations are predicted to attain diagnosability within about 2.5–4N_e generations for mtDNA and 10–16N_e generations for single nuclear loci, respectively under the Phylogenetic Species Concept

It takes ~ 4N_e generations from the time that two populations separate to reach a high probability of reciprocal monophyly for mtDNA and 16N_e generations for a nuclear locus (Fig. 9.4; Hudson & Coyne 2002). However, many pairs of populations are diagnosably different earlier than this. The corresponding durations are similar for the other two measures of diagnosability (Frankham et al. 2012).

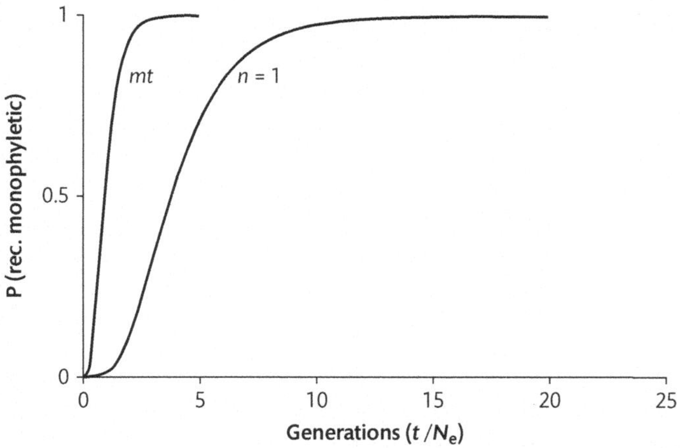

Fig. 9.4 Probability of reciprocal monophyly for mtDNA (mt) and a single nuclear DNA locus (n = 1) after increasing numbers of generations in isolated idealized populations (after Hudson & Coyne 2002). Generations are scaled in units of effective population size (t /N_e).

The number of generations to diagnosability will be less if multiple independent (unlinked) nuclear autosomal loci are genotyped and the criterion of at least one diagnosable locus is used (Frankham et al. 2012). Further, Carstens & Knowles (2007) have shown that a coalescent based analysis using DNA sequence data for multiple loci can detect distinct lineages long before reciprocal monophyly has been attained.

Frankham et al. (2012) estimated that reciprocal monophyly for nuclear autosomal loci is achieved for a threatened species of stable population size with only two equal sized populations, on average in < 52 generations for populations of critically endangered (CE) species, < 260 generations for populations of endangered (E) species, and < 1,040 generations for populations of vulnerable (V) species, and in only ~ 3, ~ 15, and ~ 58 generations, respectively for mtDNA. If such isolated populations inhabit similar environments, they will be diagnosably different long before their crosses show outbreeding depression (especially for CE and E species) because outbreeding depression has not evolved in populations isolated for 6,000 generations under these conditions (Chapter 7).

Threatened populations maintained for the number of generations required to achieve reciprocal monophyly have high probabilities of extinction from inbreeding, and are expected to exhibit large genetic rescue effects upon crossing (Frankham et al. 2012; Frankham 2015, 2016). These expectations have been verified in empirical studies.

In scenarios where a species has declined to small isolated populations, and there is no base population remaining, small populations are diagnosably different within relatively few generations and before they have evolved reproductive isolation

First, inbred strains of maize (typically developed using self-fertilization) are highly homozygous and diagnosably different within about seven generations (Liu et al. 2003), yet they typically show genetic rescue when crossed. Second, eight replicate populations of *Drosophila melanogaster* derived from the same wild source populations and maintained in isolation at N_e of 25 for 48–49 generations in the same environment were diagnosably different in 27 of 28 comparisons within ~ $2N_e$ generations, based on data for eight molecular loci (Frankham et al. 2012). These populations suffered 88% inbreeding depression in reproductive fitness in "wild" conditions and showed ~ sixfold genetic rescue effects (Woodworth et al. 2002).

Thus, over-splitting is highly likely with PSC and GLSC under conditions that frequently apply in threatened species with small isolated population fragments, as well as in small isolated populations of non-threatened species.

Which species definitions are appropriate for conservation purposes?

The species definition used in conservation biology needs to minimize harm and maximize potential benefits

The ideal species concept for conservation purposes would contribute positively to population and species persistence, by simultaneously (a) minimizing outbreeding depression, while (b) allowing maximum opportunities to cross small inbred populations with low genetic diversity to reverse inbreeding depression and restore genetic diversity (Frankham et al. 2012).

The consequences of different species delineations for six hypothetical populations are illustrated in Fig. 9.5. Too broad a delineation of species in case 1 leads to a high risk of outbreeding depression when populations a and b cross. Over-splitting in case 2, as a result of large genetic drift effects in small populations (as expected with PSC and GLSC), classifies the small a4 population as a distinct species, without any populations within its species being available to rescue it genetically, or reinforce it demographically. In case 3, use of reproductive isolation (defined to include any deleterious consequences of crossing on mating preference, pre- or post-zygotic reproductive fitness) to delineate species a versus species b both minimizes the risk of outbreeding depression and allows genetic rescue of small inbred populations within species.

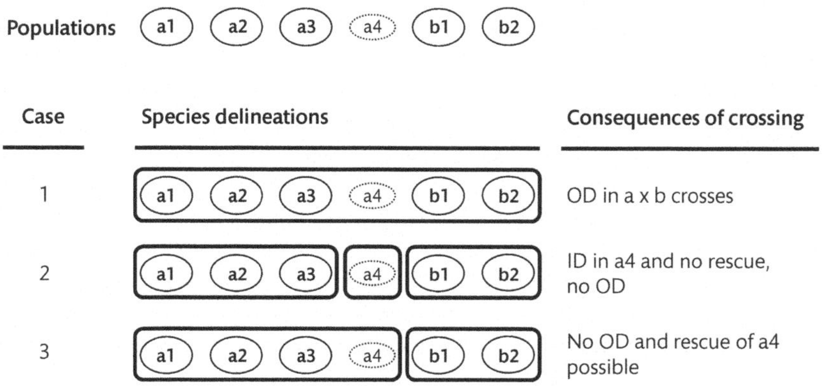

Fig. 9.5 Consequences of crossing populations following different species delineations in relation to outbreeding depression (OD), inbreeding depression (ID), and genetic rescue. Populations a and b are reproductively isolated (show OD on crossing), but populations within them reproduce freely. The small isolated a4 population has a small N_e, is inbred and has low genetic diversity, and shows elevated divergence from the other a populations (Frankham et al. 2012).

Use of species concepts based upon intrinsic reproductive isolation are recommended for conservation purposes

The preceding arguments lead us to recommend that substantial intrinsic reproductive isolation (pre-zygotic and/or post-zygotic) be used to define species of outbreeding sexual organisms for conservation purposes (Frankham et al. 2012). DFSC and exchangeability criteria satisfy these criteria, whilst BSC captures large components of it, especially when used in a "relaxed" form that accepts low rates of gene flow.

We do not recommend use of PSC and GLSC as they are prone to over-splitting and may miss chromosomally caused reproductive isolation. Such splitting, sometimes in an attempt to promote greater conservation of biodiversity, can actually prevent conservation actions necessary to preserve taxa with small population sizes, and thereby result in greater loss of biodiversity (Frankham et al. 2012; Zachos et al. 2013; Weeks et al. 2016). **We are opposed to the use of TSC**, because it does not allow managers to make knowledgeable decisions on whether to cross populations or not.

But aren't BSC and DFSC impractical to implement?

If experimental crossing of populations is impractical, reproductive isolation can be inferred from fixed chromosomal differences, and/or adaptive differentiations

It has been argued that it is difficult or impractical to determine whether populations are reproductively isolated, as experimental crossing through multiple generations is frequently impractical (Russello & Amato 2014). However, fixed chromosomal differences and/or adaptation to different environments predict the risk of outbreeding depression if populations are crossed (see Sexton et al. 2011; Frankham et al. 2012, 2014b; Frankham 2015, 2016). Notably, adaptive differentiation is a better predictor of intrinsic reproductive isolation than neutral markers across a diverse array of taxa (Nosil et al. 2002; Zigler et al. 2005; Stelkens & Seehausen 2009; Thorpe et al. 2010; Wang & Summers 2010).

Adaptive differentiation can be determined from common garden experiments, or inferred from long-term existence in different environments, or from differences in heritable attributes expected to reflect adaptation (Chapter 4). Increasingly, adaptive differences are being identified from genome sequencing, and it may soon be possible to predict reproductive isolation from changes in speciation loci (Grossman et al. 2013; Vitti et al. 2013; Hoban et al. 2016).

How much differentiation is required to classify populations as distinct species?

We recommend that the cut-offs for delineating species be based upon those for well-studied and well-recognized species defined under adaptive and intrinsic reproductive isolation criteria, as done using BSC, DFSC, or exchangeability

Since reproductively isolated species arise by relatively gradual genetic divergence as envisaged by Darwin (apart from more or less instantaneous origins of polyploids), there is a continuum from random mating to complete reproductive isolation (Hey & Pinho 2012). Species are at the higher end of the gradient of reproductive isolation, but requiring total reproductive isolation is too extreme, as many well-recognized species exhibit low residual levels of gene flow (Hausdorf 2011).

We recommend that the cut-offs be based upon those for well-studied and well-recognized species defined under adaptive and intrinsic reproductive isolation criteria. Tobias et al. (2010) have determined such cut-offs for morphological and song characteristics in birds (see also Brooks & Helgen 2010).

When do different species concepts matter in delineating species for conservation purposes?

For sympatric and parapatric distributions, all major species concepts lead to the same species delineations. However, for allopatric populations different species concepts often lead to different species delineations

Conclusions about appropriate species concepts for delineating species for conservation purposes are contingent upon whether populations have allopatric, parapatric, or sympatric distributions (Fig. 9.6).

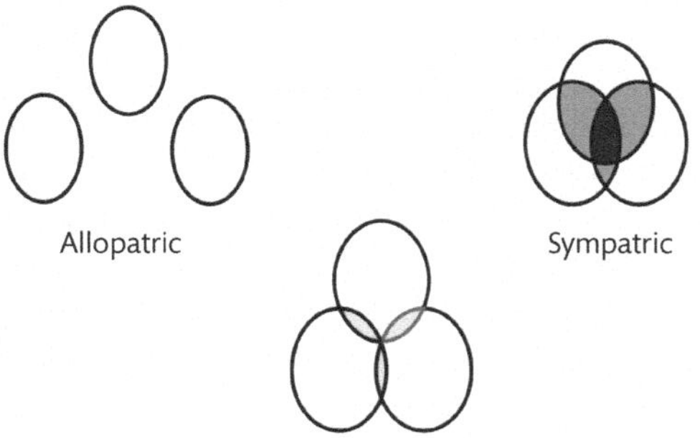

Fig. 9.6 Allopatric, sympatric, and parapatric distributions of populations.

Sympatric and parapatric populations

Lack of shared alleles at one or more autosomal loci (given sufficient sampling) is sufficient to establish intrinsic reproductive isolation between two sympatric or parapatric populations and for them to be classified as separate species under all major species concepts (even evidence of very limited gene flow should be acceptable) (Knowlton & Weigt 1997). For example, samples of velvet worms (Onychophora) from the same log in the Blue Mountains west of Sydney, Australia showed fixed gene differences at 70% of allozyme loci and were reclassified as two genera (D.A. Briscoe, pers. comm.; Briscoe & Tait 1995).

Velvet worm (Australia)

Allopatric populations

In the absence of crossing data, fixed differences may be due to isolation and genetic drift, or to intrinsic reproductive isolation. However, fixed chromosomal differences and adaptation to different environments predict reproductive isolation

Allopatric populations will be classified appropriately using the DFSC, exchangeability, and usually with BSC. However, allopatrically distributed populations delineated using diagnostic PSC or GLSC are often not intrinsically reproductively isolated.

So far, we have concentrated on outbreeding diploids; in what follows we consider asexual organisms. Frankham et al. (2012) discussed other breeding and mating systems.

What about asexual species?

Minimizing harm and maximizing potential conservation benefits, in combination with adaptive differentiation, provides a related means for delineating species in asexual taxa

Definitions of species based upon reproductive isolation do not apply to strictly asexual organisms, as they do not cross and do not exhibit outbreeding depression, inbreeding depression, or genetic rescue for reproductive fitness. However, Hausdorf (2011) has pointed out that "uniparental organisms are actually organized in units that resemble species of biparental organisms." Few asexual microbes lack gene flow, as they shift segments of DNA between individuals and species via transformation, transduction, and mobile genetic elements (transposons and integrons) (Gogarten & Townsend 2005; Stokes & Gillings 2011). There are benefits from enhancing clonal diversity in relation to adaptive evolutionary potential in small populations of asexual organisms with low genetic diversity. Thus, requiring adaptive differentiation and partial genetic isolation between asexual species, as for sexual ones, allows similar criteria to apply.

While species distinction in asexual organisms can be made only on relatively arbitrary levels of genetic and adaptive differentiation, as done in microbes (Gevers et al. 2005; Cohan & Perry 2007), this also applies to sexual species.

How do we decide whether taxonomic revision is required?

If the taxonomy of a species with an allopatric distribution was determined purely on morphology or mtDNA, on limited sampling, or on use of PSC, GLSC, or TSC, it likely requires revision

For sympatrically or parapatrically distributed populations, species delineations done under any of the species concepts should be reliable and repeatable, provided that sampling regimes and sample size have been adequate.

Taxonomic revisions are likely to be required for allopatric populations where PSC, GLSC, or TSC has been used, or delineations were based solely on mtDNA or morphology, and where sampling was limited. Conversely, if taxonomic delineations have been based on crossing results, or congruent data from multiple characters (morphology, chromosomes, several independent genetic loci, etc.) on geographically well-sampled distributions, and BSC, DFSC, or exchangeability has been used in the delineations, they should usually be adequate. Birds are less likely to require taxonomic revision, as their delineations are done using a well-organized system by Birdlife International, based upon use of BSC (del Hoyo et al. 2014).

How should a taxonomic re-evaluation be conducted if required?

There are no generally agreed scientifically robust methods for delineating species for conservation purposes with agreed species concepts, geographic sampling regimes, sample sizes, characters, records of habitat characteristics, and statistical analyses

While there are strict rules about naming new species, there are (to our knowledge) no minimum standards, widely recognized and used handbooks, or accredited institutions or workers for delineating them. This contrasts with the IUCN Red List categorization system (IUCN 2016) where criteria for listing are scientifically based and clearly defined, workers are now required to complete training programs for accreditation, and assessments are independently reviewed. Similarly, there are well-defined protocols for DNA forensics and a system exists for testing and accrediting laboratories (such as the Wildlife Forensics laboratory at the Australian Museum). **We strongly urge that standardized protocols for scientifically credible and repeatable taxonomic methods be devised for conservation purposes.**

The principles and issues we have discussed above provide a basis for species delineations for conservation purposes (Appendix 3). We endorse the moves for the use of multiple lines of evidence in species delineations, as in integrative taxonomy (Dayrat 2005; Schlick-Steiner et al. 2010; Fujita et al. 2012). However, we recommend that this should always include chromosomal data, and be based on reproductive isolation.

Conclusions

For conservation purposes, we recommend that taxonomic delineations be based on a wide array of information (chromosomes, multiple gene loci, morphology, life-history, ecology, etc.), and the use of defined species concepts based upon reproductive isolation in the broad sense (BSC, DFSC, and exchangeability). We advise strongly against use of PSC, GLSC, and TSC for populations with allopatric distributions.

Summary

1. A critical first step in conservation is to define species for conservation purposes, but species delineations have often been done inappropriately, or in a scientifically unsatisfactory manner.
2. The use of diverse species concepts often leads to widely different species delineations.
3. Species delineations that are too broad will often lead to outbreeding depression when populations cross, while those that split excessively may preclude genetic rescue of small inbred populations with low genetic diversity.

4. Minimum harm is done and maximum potential benefits in fitness and adaptive evolutionary potential accrue when reproductive isolation is used as the criterion to define distinct species: this minimizes the risk of outbreeding depression, while retaining the option for genetic rescue of populations suffering genetic erosion.
5. Use of the differential fitness or Biological Species Concepts, or exchangeability criteria, will typically yield classifications appropriate to conservation concerns.
6. Conversely, use of diagnostic Phylogenetic or General Lineage Species Concepts will often lead to inappropriate classifications (excessively splitting) for allopatric populations.
7. For sympatric or parapatric populations, distinct species are diagnosed by any genetically based distinctiveness that indicates lack of (or very limited) gene flow, and different species concepts typically yield concordant delineations.
8. For allopatric populations, crossing data are ideal, but if they are unavailable, reproductive isolation can be inferred from fixed chromosomal differences, and/or adaptive differentiation among populations, allowing delineation under BSC, DFSC, or exchangeability.
9. Reliable and scientifically sound species delineations require adequate geographic sampling, data on multiple characters (chromosomes, multiple independent genetic loci, morphology, life-history, ecology, etc.) and the use of DFSC, BSC, or exchangeability. When these conditions have not been satisfied, reassessment of taxonomy for conservation purposes is typically required, especially for species with allopatric distributions.

FURTHER READING

Fujita et al. (2012) Review of coalescent-based methods for identifying species trees from data on multiple loci.

Coyne & Orr (2004) *Speciation*: Outstanding textbook on speciation and its causes.

Crandall et al. (2000) A review and proposal for species delineation and units within species based on ecological and genetic exchangeability.

Frankham et al. (2012) Evaluated the implications of different species concepts for conserving biodiversity, and recommended concepts based on reproductive isolation.

Hausdorf (2011) Reviewed species concepts and proposed the differential fitness species concept.

Mace (2004) Review of the role of taxonomy in species conservation.

Rieseberg & Willis (2007) Authoritative review on plant speciation.

Wilkins (2009) *Species*: A book reviewing the diversity of species concepts.

SOFTWARE

*BEAST: Software for inferring phylogenetic trees and testing evolutionary hypotheses from molecular data (Heled & Drummond 2010). https://github.com/CompEvol/beast2

BPP: Coalescent-based phylogenetic inference software (Yang 2015). http://abacus.gene.ucl.ac.uk/software/

MrBayes 3: Free package for estimating trees by Bayesian methods (Ronquist & Huelsenback 2003). http://www.molecularevolution.org/software/phylogenetics/mrbayes

TreeBASE: Phylogenetic database of DNA sequences among species and phylogenies derived from them (Piel et al. 2009). www.treebase.org

Determining the number and location of genetically differentiated population fragments

CHAPTER 10

The number and geographic location of genetically differentiated populations must be identified to determine if fragmented populations require genetic management. Clustering of related genotypes to geographic locations (landscape genetic analyses) is used to determine the number of populations and their boundaries, with the simplest analyses relying on random mating within, but not across populations. Evidence of genetic differentiation among populations indicates either that they have drifted apart (and are likely inbred), or that the populations are adaptively differentiated. The current response when populations are genetically differentiated is usually to recommend separate management, but this is often ill-advised. A paradigm shift is needed where evidence of genetic differentiation among populations is followed by an assessment of whether populations are suffering genetic erosion, whether there are other populations to which they could be crossed, and whether the crosses would be beneficial or harmful.

TERMS

Assignment test, evolutionarily significant unit (ESU), gene genealogies, isolation by environment, landscape genetics

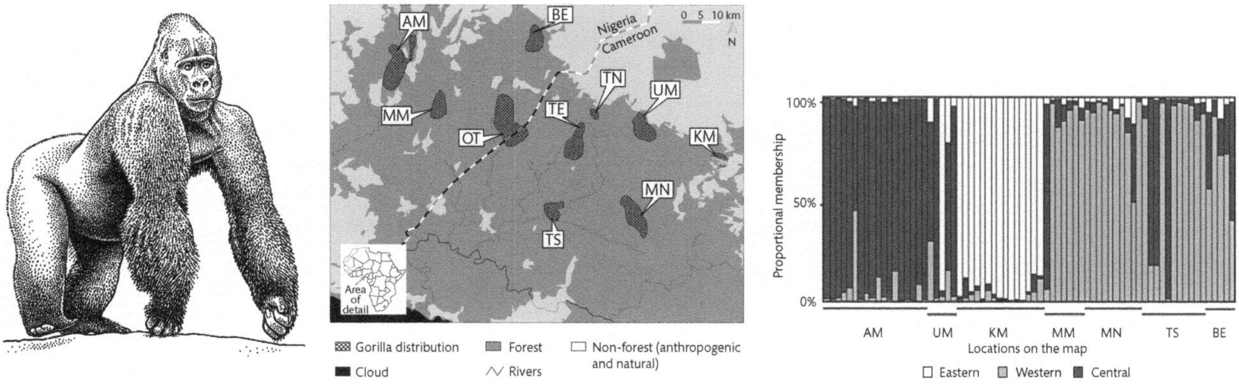

The small remaining population of Cross River gorillas is concentrated in 10 habitat fragments in Nigeria and Cameroon. A plot from the program STRUCTURE, based on 11 microsatellite loci, identified three major genetic clusters, and also evidence of recent gene flow between them. Each column in the STRUCTURE plot represents a single individual and the different shadings indicate the three genetic clusters. Individuals that are genetically different from the majority in their location but similar to those in another location are immigrants, and those with mixed origins likely reflect progeny of recent matings between individuals from different clusters (after Bergl & Vigilant 2007).

Genetic Management of Fragmented Animal and Plant Populations. R. Frankham, J. D. Ballou, K. Ralls, M. D. B. Eldridge, M. R. Dudash, C. B. Fenster, R. C. Lacy & P. Sunnucks. Oxford University Press (2017) © R. Frankham, J. D. Ballou, K. Ralls, M. D. B. Eldridge, M. R. Dudash, C. B. Fenster, R. C. Lacy & P. Sunnucks 2017. DOI 10.1093/oso/9780198783398.001.0001

What are our objectives?

We wish to determine the number and geographic locations of genetically differentiated population fragments, and to delineate genetic management options

A crucial first step in management is to define objectives. The first step in determining whether groups of individuals need to be managed separately, or as one unit, is to determine how many genetically distinct populations exist and where they are located.

The decision tree in Fig. 10.1 provides a perspective on where we are heading as we proceed through the genetic management phase of the book. The first question in the decision tree involves the issues covered in this chapter.

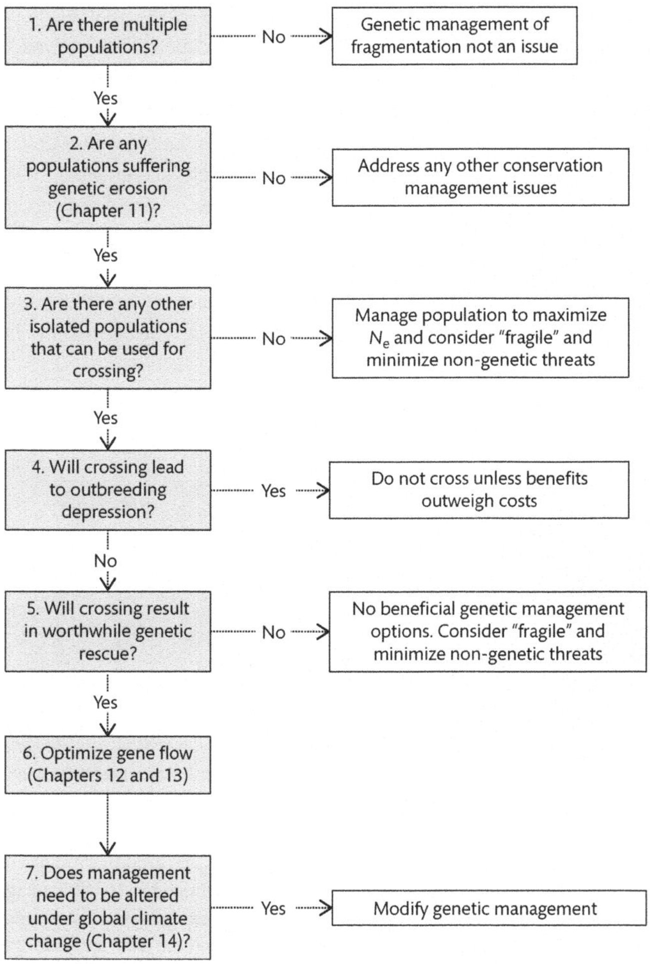

Fig. 10.1 Decision tree for genetic management of fragmented populations, designed to guide us through the genetic management chapters.

How do we identify the number of genetically differentiated populations?

The number of populations can be estimated from multilocus molecular data using clustering algorithms

Spatial structure, mating patterns, and progeny dispersal patterns usually provide clues as to whether multiple genetically differentiated populations are present. If groups of individuals occur further apart than the scale of mating and dispersal (if known), then they are likely to belong to different genetic populations. In the absence of resources to carry out genetic studies, this can guide genetic management.

However, such an approach is not definitive because low levels of gene flow may result in physically isolated population fragments acting effectively as a single genetic population. Alternatively, some species may be poorer dispersers than thought, such as habitat-specific birds, or insects, or plants whose pollinators or fruit dispersers have been extirpated (Aslan et al. 2013; Amos et al. 2014). Further, geographic isolation may vary over time, as for example with boom and bust cycles of small mammals inhabiting arid areas, or freshwater fish and other aquatic species affected by cycles of droughts and floods (Magoulick & Kobza 2003; Letnic & Dickman 2006).

What population genetic distribution patterns do we need to distinguish? Individuals may exist in a continuum with approximately random mating, or show isolation by distance, clines, or multiple genetically isolated populations (Fig. 10.2). In practice, such patterns are often modified by landscape features that may inhibit or facilitate gene flow such as rivers, lakes, mountains, soil changes, roads, and fences. Another common pattern is isolation by environment, where gene flow is greater among more similar environments (Sexton et al. 2014; Wang & Bradburd 2014).

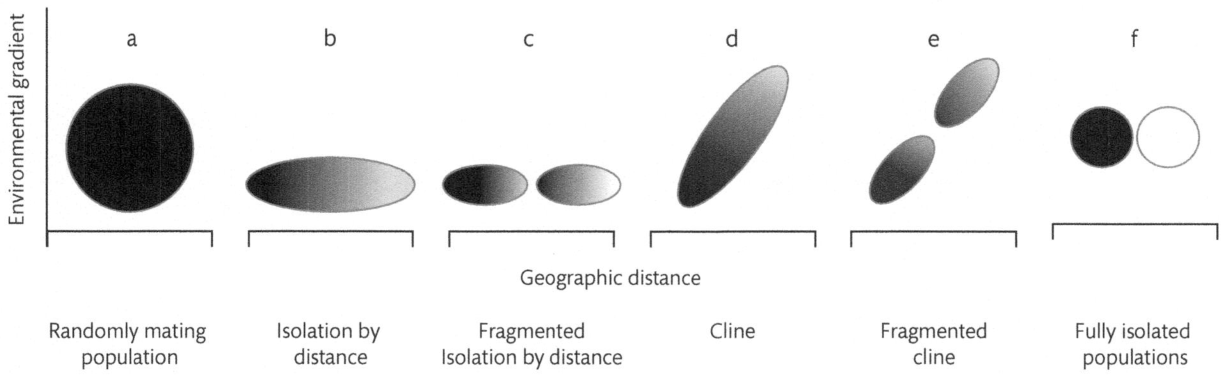

Fig. 10.2 Schematic view of population genetic structures we need to distinguish: (a) single random mating population, (b) isolation by distance where gene flow reduces with distance, (c) isolation by distance broken by recent habitat fragmentation, (d) cline where gene flow reduces over an environmental gradient (e.g. rainfall), often associated with adaptive differences, (e) cline broken by recent habitat fragmentation, and (f) geographically and genetically isolated populations. Individuals within the boundaries are connected by gene flow. The different intensities of shading indicate genetic differentiation along geographic and environmental gradients, with greater differences in intensities of shading indicating greater differentiation.

The question of whether or not there are multiple genetic populations can be answered by genotyping individuals and asking if there are genetic clusters or some other pattern. For example, the chapter frontispiece shows gorilla populations in three genetic clusters, with some dispersal and putative gene flow between them. Box 10.1 illustrates identification of the number of genetically distinct groups across Mexican wolves and several related taxa using a cluster analysis, along with the consequent genetic management actions and their outcomes.

Mexican wolf

Box 10.1 Identifying the number of distinct populations of Mexican wolf, whether there has been genetic introgression from other species, and subsequent management actions and their outcomes

(Hedrick et al. 1997; Fredrickson et al. 2007; Hedrick & Fredrickson 2010)

Three small inbred populations of presumed Mexican wolves, a sub-species of the gray wolf, were all that remained in the 1990s, and all were likely suffering inbreeding depression. How could the sub-species best be conserved? Did the three populations need to be managed as separate units, or could they be combined to alleviate problems of inbreeding, low genetic diversity, and inbreeding depression? Before genetic rescue could be contemplated, managers wanted to determine whether the three populations were "pure" Mexican wolves, or whether they had crossed with other canid species (introgression). Hedrick et al. (1997) used a clustering approach (multidimensional scaling) on allele frequencies for 20 microsatellite loci from three populations of Mexican wolves, plus coyotes (*Canis latrans*), dogs, and gray wolves from northern locations to determine whether the Mexican wolves belonged to a single genetic cluster that was distinct from the other canid taxa. Different coyote and northern gray wolf populations are indicated by state, province, or city.

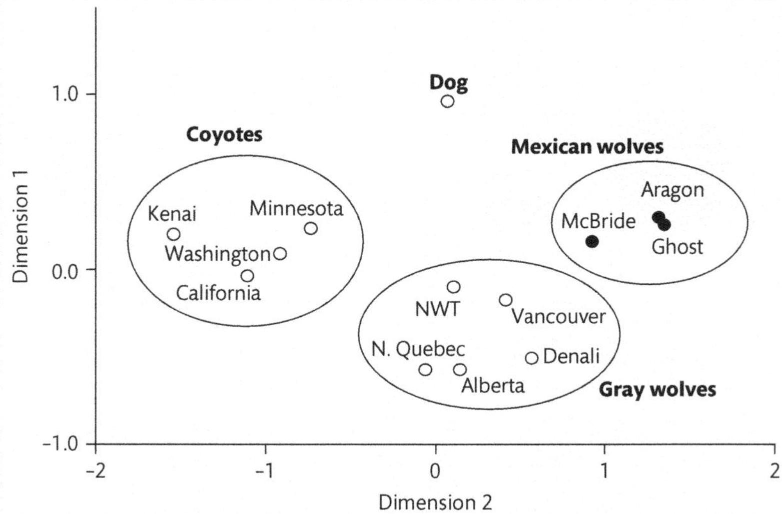

(after Frankham et al. 2010, Fig. 21.9)

Mexican wolves formed a cluster distinct from other canids (see also Cronin et al. [2015]), establishing that the Aragon and Ghost Ranch populations of Mexican wolves whose status was questionable, formed a single distinct cluster with the "pure" McBride population, rather than showing evidence of introgression from other canids. Had these two populations experienced such introgression, they would have been positioned between the McBride Mexican wolf and the introgressing canid taxon. Consequently, the three populations were combined, resulting in substantial genetic rescue effects for fitness and genetic diversity (Fredrickson et al. 2007; Hedrick & Fredrickson 2010).

The most widely used software packages for determining if there are single or multiple populations are based on random mating within but not among population segments

As we saw in Chapter 5, when distinct random mating populations are pooled they exhibit a deficiency in heterozygosity compared to Hardy–Weinberg equilibrium (Wahlund effect). Several of the most widely used software packages for determining the number of populations are based on tests for deficiencies of heterozygotes (Holderegger et al. 2010), including STRUCTURE (Falush et al. 2003, 2007), BAPS (Corander et al. 2008), TESS (Chen et al. 2007), and GENELAND (Guillot et al. 2005a).

STRUCTURE, the most commonly used package, estimates the number of distinct genetic populations from multilocus genotype data (typically microsatellites or SNPs) across individuals using clustering (Pritchard et al. 2000). It assumes that there are some number (k) of random mating genetic populations with different allele frequencies (unknown) and utilizes a probabilistic approach to assign individuals to populations and simultaneously estimate allele frequencies in them. This can be done without reference to where individuals were actually sampled. STRUCTURE and BAPS use Bayesian clustering and all use Markov chain Monte Carlo methods whose details are beyond the scope of this book (Lange 1997).

How well do these clustering software packages perform? They may lead to incorrect conclusions when (Excoffier & Heckel 2006; Chen et al. 2007):

- statistical power of the data is low
- there is isolation by distance or clines not accounted for
- sampling is inadequate
- the species has substantial selfing.

The number of populations may be underestimated if there is insufficient power in the analyses due to small sample size, limited geographic sampling, insufficient genetic loci, low genetic diversity, limited genetic differentiation among populations, or combinations of these. SPOTG software is available for determining power for such analyses (Hoban et al. 2013). The accuracy and precision of these analyses usually increases with the number of loci genotyped, the sample sizes, and the computational replication (Gilbert et al. 2012). More genetically distinctive populations will be easier to detect, so power to detect clusters will decrease with rising proportions of immigrant and admixed individuals. A simulation study showed that PARTITION correctly clustered populations as low as $F_{ST} > 0.09$ while STRUCTURE and BAPS yielded the correct clusters at $F_{ST} \geq 0.03$ (Latch et al. 2006). For perspective, European human populations have $F_{ST} \leq 0.02$, European and Chinese have $F_{ST} = 0.11$, and Europeans and Africans $F_{ST} = 0.15$ (Nelis et al. 2009).

The approaches considered so far assume that significant signals represent genetic clusters (Fig. 10.2 [a] versus [f]), rather than isolation by distance or clines (Fig. 10.2 [b] and [d]) (Guillot et al. 2009). Consequently, continuously patterned rather than discontinuous genetic variation needs to be routinely tested for, not least because sampling patchily along a continuum will suggest clusters that do not really exist. Puechmaille (2016) showed that STRUCTURE does not reliably recover the correct population structure when sampling is uneven, and developed sub-sampling and new estimators to alleviate the problem.

Isolation by distance can be detected by plotting F_{ST} against geographic distance between populations and carrying out regressions or related statistical analyses (Rousset 1997). It is often tested for using Mantel tests (Mantel 1967), a permutation procedure that assesses the correlation of any two matrices (e.g. genetic distance and geographic distance). Clustering software typically has a variety of options, with STRUCTURE, BAPS, and TESS having forms with and without admixture (gene flow). When admixture is present in the data, software without this option performs badly (Francois & Durand 2010). Notably, the forms with admixture can identify clines. In general, it will be advisable to run admixture models, which are more robust and flexible than ones without admixture, and to check inferred population structure using an independent method, such as principal component analysis (Patterson et al. 2006; McVean 2009). Continuous pattern methods such as spatial principal component analysis (sPCA), and approaches such as causal modeling can identify the relative importance of distance and barriers in driving genetic patterning (Jombart et al. 2008; Guillot et al. 2009; Pavlova et al. 2013; Amos et al. 2014).

The assumption of random mating within populations is often violated, especially in primarily selfing plants and invertebrates, meaning that use of STRUCTURE and related software is inappropriate. This can be overcome by using clustering analyses that do not assume random mating (Box 10.1). Alternatively, Gao et al. (2007) have developed a version of STRUCTURE (INSTRUCT) that does not assume random mating and allows simultaneous estimation of inbreeding and population structure by calculating expected genotypic frequencies based on inbreeding or selfing rates.

For example, INSTRUCT identified two genetically differentiated populations of *Oryza rufipogon* (the wild relative of domesticated rice) in Asia, and their selfing rates (Fig. 10.3).

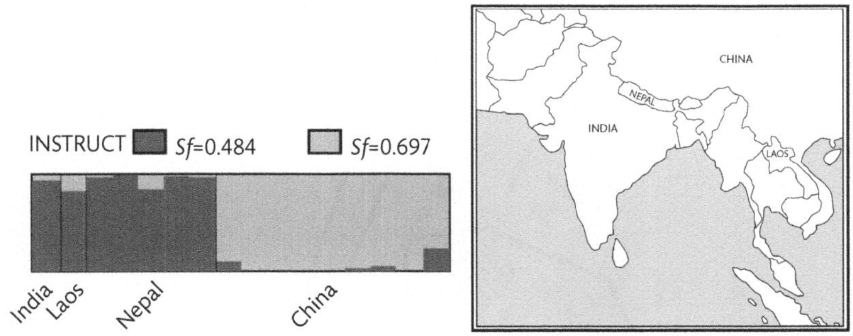

Fig. 10.3 Genetically differentiated populations of wild rice in Asia, identified using INSTRUCT software based on SNP data (Gao et al. 2007). The *Sf* values are the estimated selfing rates in the two population clusters.

Wild rice (Asia)

Isolated populations with low genetic diversity and strong genetic divergence from other populations should be relatively easily distinguished by the preceding methods, given multilocus genotypic data and adequate sampling.

How do we locate genetically differentiated populations in the landscape?

Analyses of patterns of genetic diversity across the landscape can locate differentiated populations geographically, delineate clines and boundaries to gene flow, and detect isolation by distance, metapopulations, and randomness

So far, we have ignored the geographic location of populations, but we must locate populations if we are to manage them. Geographic sub-structure within species can be defined by mapping patterns of genetic diversity onto geography (one element of **landscape genetics**) (Manel et al. 2003; Segelbacher et al. 2010; Balkenhol et al. 2015). Software such as GENELAND (Guillot et al. 2005a, 2005b) and TESS (Chen et al. 2007) achieve this using genotypic clustering on georeferenced individual multilocus genotypes. For example, GENELAND was used to identify and locate genetically differentiated populations of the endangered Florida scrub-jay (*Aphelocoma coerulescens*) and relate their presence to habitat and geographic features (Fig. 10.4).

Florida scrub-jay (USA)

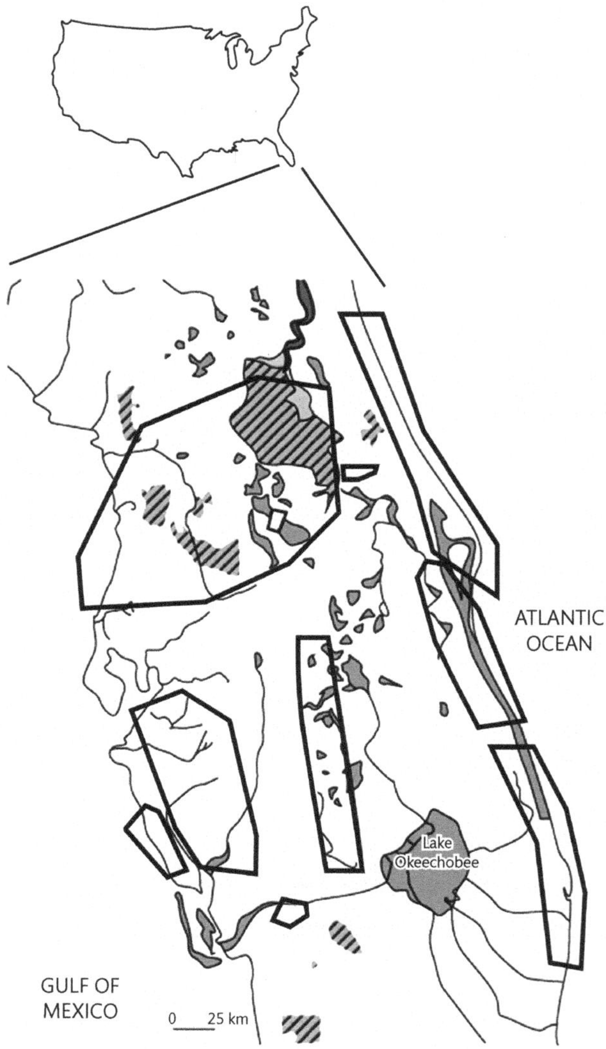

Fig. 10.4 Genetic distinctiveness mapped onto geographical regions using GENELAND for the non-migratory, endangered Florida scrub-jay, based on genotypes of > 1,000 birds for 20 microsatellite loci. The solid lines represent genetically differentiated populations and the crosshatched areas forest habitat suitable for the bird (Coulon et al. 2008). Coastal populations are genetically differentiated from each other and from the interior population, despite the short geographical distances separating them.

In comparisons of the ability of software to delineate genetically differentiated populations, GENELAND generally outperformed other Bayesian clustering methods and alternative methods (Safner et al. 2011; Blair et al. 2012). A version of GENELAND that allows use of both molecular and morphological data in clustering has performed well (Guillot et al. 2012). For example, vole (*Myodes rutilis* and *M. glareolus*) populations in Sweden have been successfully distinguished based on combined morphological and molecular (14 microsatellite loci) georeferenced data (Guillot et al. 2012).

It might be thought that we can use previously identified units within a species, such as sub-species and evolutionarily significant units (ESUs), as the basis for deciding whether populations should be managed separately or combined, but this is often not the case.

Are current units within species suitable for genetic management purposes?

Units such as sub-species and evolutionarily significant units (ESUs) have been defined to categorize differentiated units within species, but suffer from problems of multiple definitions and inconsistent delineations

Units within species are often misleading, as they suffer from similar shortcomings to those we identified for species, including a diversity of definitions (Chapter 9). For example there are several definitions for sub-species and ESUs (Table 10.1), and delineations based on different definitions are often incongruent (Neaves et al. 2012; Pavlova et al. 2013).

Sub-species

Designations of sub-species have been so haphazard that they cannot be relied upon for genetic management purposes

Sub-species is a widely used classification that has been employed with diverse criteria (Table 10.1; Ryder 1986; Haig et al. 2006). Worryingly, many sub-species reflect political boundaries (states and countries), rather than biological discontinuities, especially in less intensively studied biota. For example, koala sub-species in Australia reflected state boundaries (Houlden et al. 1999). Many sub-species have been delineated using the Biological Species framework, while others have been based on the Phylogenetic Species Concept (PSC) (Cracraft 1992). For example, the PSC has been used by Groves (2001) in taxonomic revisions of primates that include sub-species (e.g. for ruffed lemurs) and ungulates (Groves & Grubb 2011).

Table 10.1 Definitions used for units within species.

Unit	Definitions	References
Sub-species		
(a)	"an aggregate of local populations of a species, inhabiting a geographic subdivision of the range of that species, and differing taxonomically from other populations of the species"	Mayr (1963)
(b)	"share a unique geographic range or habitat, a group of phylogenetically concordant phenotypic characteristics, and a unique natural history relative to other subdivisions of the species"	O'Brien & Mayr (1991)
(c)	"Populations that show differentiation based upon data on multiple, independent, genetically based traits"	Avise & Ball (1990)
(d)	"Geographic segment of a species"	Groves (2001)
Evolutionarily significant units (ESUs)		
(a)	Distinct population segments based on concordance between data sets determined by different techniques (natural history information, morphometrics, range and distribution data, as well as protein electrophoresis, cytogenetic analysis, and nuclear and mitochondrial DNA data). Preceded by text referring to "actually represent significant adaptive variation."	Ryder (1986)
(b)	"Phylogenetically related populations characterized holistically on the basis of the simultaneous consideration of genetics, morphology, ecology and behavior."	Woodruff (1989)
(c)	"a population (or group of populations) that 1. is reproductively isolated from other conspecific population units, and 2. represents an important component in the evolutionary legacy of the species." "A population . . . that represents an important ecological adaptation for the species may be an ESU."	Waples (1991)
(d)	"Reciprocally monophyletic for mtDNA alleles and show significant divergences of allele frequencies at nuclear loci."	Moritz (1994)
(e)	Evolutionary significant units are groups that would respond to similar selective pressures differently in multivariate phenotypic space, as predicted by different variance-covariance matrices for multiple ecologically relevant quantitative traits, i.e. different evolutionary potentials.	Fenster & Dudash (1994)
(f)	"conservation units are delimited by characters that diagnose clusters of individuals or populations to the exclusion of other such clusters."	Vogler & DeSalle (1994)
Exchangeability		
	Different units are designated based on evidence of lack of genetic or ecological exchangeability, in recent and historical time frames.	Crandall et al. (2000)

ESUs

There are several definitions of ESUs and the practical designations are so varied that they often cannot be relied upon for genetic management purposes. However, some definitions are concordant with genetic management requirements

In a survey of the use of ESUs based on 98 studies, Crandall et al. (2000) found that authors had assigned the term ESU to all eight of their exchangeability categories (which cover the range from distinct species to single populations). Most of the definitions of ESUs in Table 10.1 approximately parallel the definitions of species given in Table 9.1. For example, the Ryder (1986) and Waples (1991) definitions are based largely on reproductive isolation, akin to applying the BSC (Mayr 1963) or DFSC (Hausdorf 2011) at a less stringent level.

Conversely, the Moritz (1994) ESU definition is based on reciprocal monophyly for mtDNA and significant divergence at nuclear loci and is thus akin to a within species PSC (Eldredge & Cracraft 1980; Cracraft 1983, 1997). Further, Vogler & DeSalle (1994) have recommended using PSC principles to define units within species. Fraser & Bernatchez (2001) recommend using an adaptive strategy to defining units within species, rather than a rigidly defined one (akin to TSC).

Which units within species are suitable for genetic management purposes?

We recommend use of partial reproductive isolation/outbreeding depression as the basis for delineating units within species for genetic management purposes

In evaluating units within species we use the same principles applied to species in Chapter 9, but at a less stringent level. Thus, we accept for conservation purposes sub-species delineation based on the BSC and related considerations (Avise & Ball 1990; O'Brien & Mayr 1991), but recommend against using those based on PSC and GLSC (such as Groves 2001; Groves & Grubb 2011) or those done with no specified basis (TSC).

The ESU concepts that include adaptive differentiation and/or partial reproductive isolation (Ryder 1986; Woodruff 1989; Waples 1991; Fenster & Dudash 1994) are similar to our designations and may be used, especially where funds are not available to do genetic clustering studies, as described earlier in the chapter. However, the Moritz (1994) and Vogler & DeSalle (1994) definitions do not consider adaptive differences, nor mention chromosomal differences, and are related to the PSC. Thus, we do not recommend them for determining whether or not allopatric populations require separate management.

The exchangeability framework (Crandall et al. 2000) is rather similar to our recommended delineations, as it involves genetic exchangeability (largely gene flow) and ecological exchangeability (adaptive differences) and short and long-term time frames. It was derived from different considerations (genetic diversity, rather than fitness) and does not mention chromosomes.

What is the current recommendation when genetically differentiated populations are identified?

Currently, evidence of population structure and limited gene flow often leads to the conclusion that populations should be managed as independent units, but separate management will often lead to extinction of small inbred population fragments

Evidence of population structure and limited gene flow for near neutral DNA markers predominantly leads to the questionable conclusion that the sub-populations should be managed as separate units, as there have been only ~ 30 genetic rescues done for conservation purposes, yet more than a million isolated populations of threatened species would likely benefit (Chapter 1). For example, separate management was recommended for the four isolated populations of the endangered eastern bristlebird (*Dasyornis brachypterus*) in Australia, despite negligible to slight genetic differentiation, and two remnants having reduced genetic diversity (Roberts et al. 2011).

However, separate management is a recipe for extinction of many fragmented populations, as there is expected to be an inverse relationship between genetic differentiation of recently isolated population fragments and their genetic diversity (Chapter 5), as has been observed in dwarf galaxia fish (*Galaxiella pusilla*) (Box 10.2) and several Australian mammals (Weeks et al. 2016). Consequently, **strongly differentiated population fragments will often have low genetic diversity and be suffering inbreeding depression.** Such inbred population fragments might be saved by augmentation of gene flow (Chapter 6), but current recommendations typically do not even consider this option (Chapter 1).

We seek a paradigm shift whereby evidence of genetic differentiation among populations triggers questions of whether any population segments are suffering genetic problems, and if they can be rescued by augmenting gene flow, rather than routinely recommending that segments be managed separately.

Box 10.2 High genetic differentiation among dwarf galaxia fish populations is associated with low within population genetic diversity

(Coleman et al. 2013)

Threatened dwarf galaxia fish from 51 populations (small dots) in many different rivers in southeastern Australia (see map) were genotyped at 10 microsatellite loci, and the association between F_{ST} and genetic diversity within populations investigated.

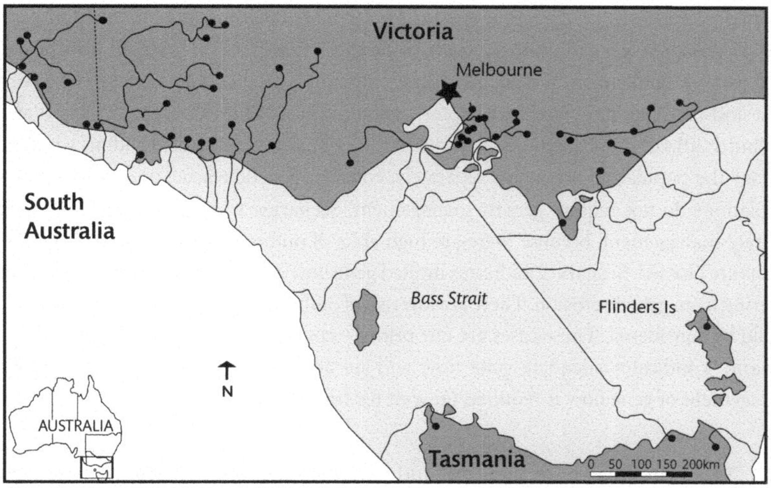

There were strong negative correlations between heterozygosity (H_e) of populations and their F_{ST} values with the other populations in both the eastern and western populations (left and right figures below), and also for allelic diversity (not shown), as predicted from theory (Chapter 5). Similar conclusions were also obtained for analyses using Jost's D, which scales 0–1, so they are not an artifact of F_{ST} not scaling 0–1 for microsatellites (Chapter 5). Thus, the most differentiated populations typically had the lowest genetic diversity, and were in general the ones in most need of augmentation of gene flow.

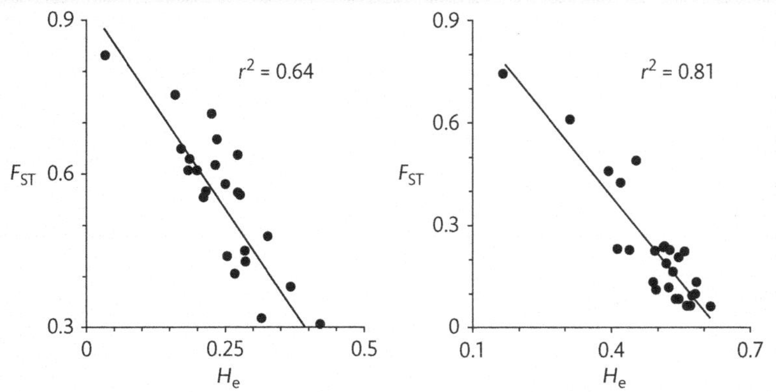

How should we proceed when there are genetically diverged populations?

Evidence for the existence of multiple populations may indicate either that some are suffering genetic erosion and would benefit from augmented gene flow, or that they are so differentiated (adaptively and/or chromosomally) that separate management is required. It is crucial that these alternatives be distinguished

Separate populations of the same species may exhibit differentiation for neutral genetic diversity, adaptive genetic diversity, both, or neither (Table 10.2). These can now be delineated using genomic data, and adaptive differentiation is predicted by isolation by environment and methods discussed in Chapter 4 (Funk et al. 2012; Sexton et al. 2014; Wang & Bradburd 2014). These distinctions provide the critical evidence for deciding whether to manage the population fragments separately, consider genetic rescue, or conclude that the populations do not require genetic management. Scenarios 1 and 3 in Table 10.2 require separate management, because there are high risks of outbreeding depression if the populations are crossed. Scenario 2 indicates limited gene flow and that some populations may be suffering from genetic erosion. There is a low risk of outbreeding depression so genetic rescue should be considered. These cases are our primary concern in this book (Chapters 11–13). Scenario 4 indicates adequate gene flow and no adaptive differentiation, so no genetic management of gene flow is required (at least for fragmentation).

Table 10.2 States of neutral and adaptive genetic differentiation among populations of the same species and their genetic management implications.

Scenario	Neutral	Adaptive	Implications	Proceed to chapter
1	Yes	Yes	High risk of OD[a] if crossed, manage separately	
2	Yes	No	Potentially suffering genetic erosion, probably low risk of OD, consider genetic rescue	11, 12, and 13
3	No	Yes	Likely adequate gene flow and selection, elevated risk of OD, so manage separately	
4	No	No	Adequate gene flow, check for other genetic problems	

[a] OD = outbreeding depression.

In the preceding material, we have referred to gene flow several times, but not addressed how to measure it.

How can we measure gene flow among population fragments?

Gene flow can be inferred from multilocus genotypic data on individuals from population fragments. Our primary interest is in recent gene flow, but most estimates represent either long-term (historical) gene flow, or inferences from short-term dispersal, and are inappropriate for management of gene flow

Gene flow rates are notoriously difficult to infer from dispersal measured by direct tracking of individuals, pollen, etc., because genetically relevant levels of gene flow may be below the usual detection limits of these methods (but see Fenster et al. 2003). Further, immigrants may not breed in their new population, or they may be at a selective advantage in inbred populations (Chapter 6). A variety of genetic methods have been used to infer gene flow or dispersal from multilocus genotypic data on individuals from multiple population fragments.

Many estimates of inferred gene flow do not provide the information we seek, namely gene flow on a recent time scale. They typically provide estimates of historical gene flow over very long durations (» 500 years) in populations assumed to be at equilibrium, or they estimate current dispersal rates, but not whether immigrants breed. We describe these methods and their limitations, as most available estimates of gene flow were obtained using them. A few methods (implemented in BAYESASS and MIGEST; Wilson & Rannala 2003; Wang 2014b) provide us with estimates of recent gene flow that are relevant to management of gene flow, as we describe later in this section.

Historical gene flow

Historical gene flow can be estimated from the degree of genetic differentiation among populations

Many available estimates of gene flow (e.g. from F_{ST}) represent long-term historical gene flow due to drift–gene flow equilibrium or mutation–drift–gene flow equilibrium. However, they may not represent current gene flow, especially if there has been recent habitat loss and fragmentation. They may be useful in the context of outbreeding depression, as very low rates of historical gene flow may be associated with an elevated risk of outbreeding depression (Chapter 7).

Several of the genetic methods are couched in terms of number of immigrants arriving per generation *Nm*, the product of the population size (*N*), and the rate of migration per generation (*m*) (migrants as a proportion of the population).

Nm from F_{ST}

Historical gene flow has frequently been inferred from F_{ST}, but this method has serious limitations

Many authors have estimated gene flow from F_{ST}, as it is related to population size and gene flow. For the island model (Chapter 5), the effective number of migrants per generation Nm is estimated by the following equation:

$$Nm = \frac{(1 - F_{ST})}{4F_{ST}} \quad 10.1$$

This estimate of Nm is the effective number of migrants (N_em). Using Pacific yew data (Example 5.1) where $F_{ST} = 0.078$, we estimate $Nm = (1 - 0.078)/(4 \times 0.078) = 2.96$. On average about three effective migrants per generation are entering Pacific yew populations. Related expressions have been derived for other migration models (see Neigel 1996).

While F_{ST} has been widely used to measure restrictions in gene flow, populations rarely adhere to the details of the island model, especially drift-gene flow equilibrium and equal migration rates between all pairs of populations in both directions, so this is, at best a crude estimate (Steinberg & Jordan 1998; Whitlock & McCauley 1999). Further, it is based on two alleles per locus and is not appropriate when using microsatellites, or other multiallelic markers (Chapter 5). Meirmans & Hedrick (2011) reviewed the consequences of this and suggested use of the following adjusted estimate of gene flow rates from multiallelic markers (see Chapter 5 for a description of F'_{ST}):

$$Nm = \frac{(1 - F'_{ST})}{4F_{ST}} \quad 10.2$$

However, this approach remains problematic in assuming that population differentiation is mostly determined by levels of gene flow, whereas population history and population size can sometimes be much more influential (Marko & Hart 2011). Use of F_{ST} and related methods is not recommended and they should only be used as a last resort.

Historical Nm may also be estimated from the frequency of private alleles (Slatkin 1985), but this method suffers from many of the shortcomings of the F_{ST} based method.

Coalescent (phylogenetic) estimates of gene flow

Coalescent based likelihood estimates of historical *Nm* involve more complex models with fewer assumptions, and make better use of the information in the data, especially in a landscape genetic context

Coalescent modeling involves moving backwards in time to estimate a common ancestral state (Kingman 1982; Rosenberg & Nordborg 2002; Wakeley 2007). Typically, such models do not need to assume equal population sizes, or constant symmetric gene flow, or mutation–drift–gene flow equilibrium. Thus, they are better than the F_{ST} based method, but they still estimate historical gene flow. The earlier coalescent methods were typically implemented using maximum likelihood estimation, as in the MIGRATE software (Beerli & Felsenstein 2001). Bittner & King (2003) used MIGRATE to estimate migration between the mainland and Lake Erie island population of garter snakes (*Thamnophis sirtalis*) in the USA (Fig. 10.5). The coalescent estimates of Nm were < 1 for both allozyme and microsatellite data, whereas estimates from F_{ST} were > 4, with the former estimates considered more credible. More recent coalescent methods estimate historical gene flow between populations based on DNA sequence data using Markov chain Monte Carlo simulation of gene genealogies (e.g. Hey & Nielsen 2007). For example, the program IMa can be used to estimate gene flow for recently diverged populations still exchanging migrants (Hey & Nielsen 2004; Hey 2010). IMa assumes an isolation-with-migration model, and uses multiple gene phylogenies (typically DNA sequences) to estimate effective population sizes (of the ancestor and the two daughter populations, A and B), levels of gene flow from A to B and B to A during their divergence, and time that divergence (restricted gene flow) began. However, these estimates of gene flow are not recommended for genetic management of fragmented populations, for reasons we have already specified.

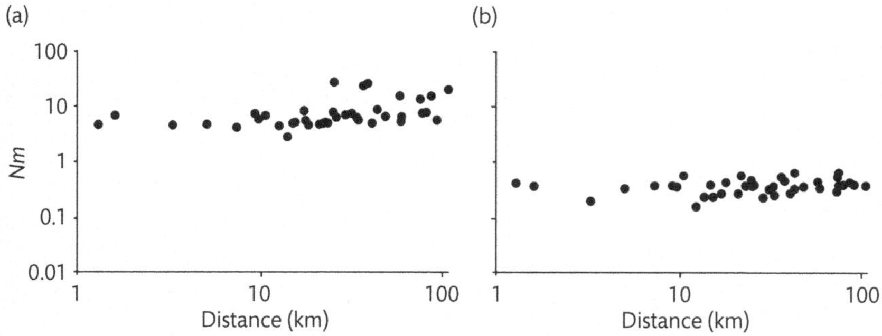

Fig. 10.5 F_{ST} based (a) and coalescent based (b) estimates of migration rates (*Nm*) between pairs of garter snake populations on the mainland and Lake Erie Island, plotted against geographic distance. Both were estimated from the same data on four microsatellite loci (after Bittner & King 2003).

Garter snake (North America)

Current dispersal rates from assignment tests

Multilocus genotypes can be used to assign individuals to different populations and to identify immigrants

Based on its genotype, an individual can be assigned to the population to which it has the greatest similarity (**assignment tests**) (Manel et al. 2005). If it is assigned to a population other than the one from which it was collected, it is likely to be an immigrant. This method provides an estimate of the actual number of immigrants (*Nm*), not the effective number (N_em). For example, if all individuals in geographic areas 1 and 2 have genotypes $A_1A_1B_1B_1C_1C_1D_1D_1$ and $A_2A_2B_2B_2C_2C_2D_2D_2$, respectively, then an individual in region 2 with the former genotype must be an immigrant. An identical principle applies when populations differ in frequencies at several loci, but the computations are more complex and assignments are expressed as probabilities (Rannala & Mountain 1997).

For example, Eldridge et al. (2001) compared the ability of various methods to identify known-source migrants in a reintroduced population of black-footed rock-wallabies, based on data for 11 microsatellite loci. Bayesian clustering (STRUCTURE), interpopulation differences in allele frequency distributions (Paetkau et al. 1995), and Bayesian methods (Rannala & Mountain 1997) were more consistent and accurate than distance methods. Using the better performing methods, they determined that the (unknown) source of immigrants that re-established the Gardner's outcrop population was the eastern end of Mt. Caroline, the nearest population (Fig. 10.6).

Black-footed rock-wallaby (Australia)

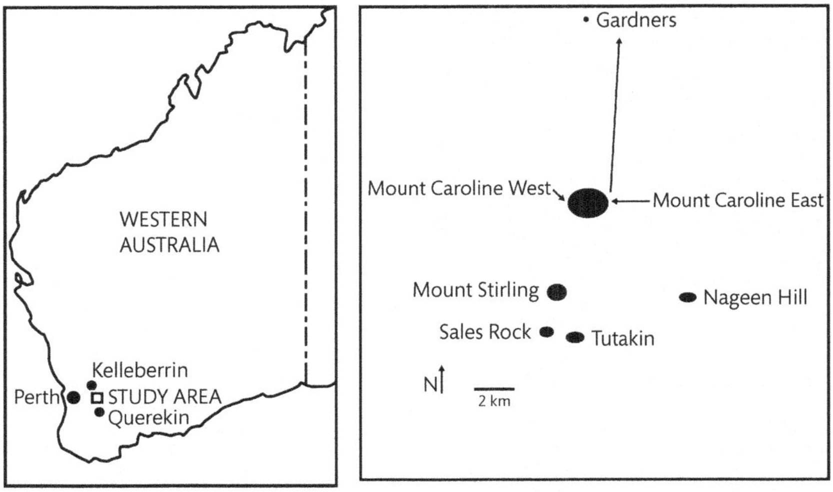

Fig. 10.6 Source population of dispersing rock-wallabies assigned from multilocus microsatellite genotypes (after Eldridge et al. 2001). According to Bayesian clustering the probability that the Gardners animals derived from the Mt. Caroline East population was 0.967, whilst the probabilities for the other four populations ranged from 0.012 to 0.005.

However, immigrants do not necessarily breed successfully and contribute to gene flow (Hansson et al. 2004; Peery et al. 2010), potentially decoupling migration and gene flow. Consequently, we prefer estimates of recent gene flow for genetic management of fragmented populations.

Estimating recent gene flow

We recommend the application of recent gene flow in genetic management of fragmented populations, as our focus is on mitigating the impacts of recent human-caused fragmentation

By contrast to the preceding methods, BAYESASS and MIGEST software implement estimation of recent gene flow, our primary concern (Wilson & Rannala 2003; Wang 2014b). Based on multilocus genotypes, BAYESASS searches for individuals in the population sample that have migrant ancestry in the two previous generations using assignment methods. Individuals likely to have resulted from F_1 or F_2 immigrant x resident matings should have a genetic composition half way between the two parental populations, while the two backcrosses should have compositions ¼ and ¾ ways between them, etc., as can be seen in the STRUCTURE plot for gorillas on the chapter frontispiece. This method does not assume equilibrium between drift and gene flow, nor symmetrical gene flow, and can be used on populations that are not in Hardy–Weinberg equilibrium. For example, Yuan et al. (2012) found low recent gene flow among recently fragmented island populations of the herb *Hedyotis chrysotricha*. This method does not provide good estimates when there is only limited differentiation among the populations, or when the assumptions of the model are violated (Faubet et al. 2007; Meirmans 2014).

MIGEST (Wang 2014b) software estimates recent gene flow by assigning parentage from multilocus genotypes using likelihood methods. This software requires sampling and data sufficient for parentage assignment, but then often provides superior estimates to BAYESASS, based on computer simulations. Using MIGEST on genotypic data for nine polymorphic microsatellite loci, Wang (2014b) estimated that gene flow among proboscis bat (*Rhynchonycteris naso*) colonies in Brazil was higher in females than males, an unusual observation for a mammal, but consistent with behavioral observations and other genetic analyses.

In Chapter 11, we ask whether any populations are suffering genetic erosion (and consider how to decide this), and whether they would benefit from augmented gene flow.

Summary

1. If fragmented populations are to be appropriately managed, genetically differentiated populations within species need to be identified and their geographic locations determined.
2. The number of populations and their boundaries are typically inferred from clusters of related genotypes linked to geographic locations (landscape genetic analyses), with the simplest analyses relying on random mating within, but not across populations.
3. Sub-species and ESUs are usually not suitable for use as genetic management units.
4. The current inference when populations are genetically differentiated is usually to recommend separate genetic management, **but this is often ill-advised and a paradigm shift is needed.**
5. If populations are genetically differentiated, we should determine whether any are suffering genetic erosion, and if so, are there other populations to which they could be crossed, and whether the crosses be beneficial or result in outbreeding depression?
6. Estimates of recent gene flow are important for deciding appropriate genetic management for fragmented populations: they can be obtained from multilocus genotypic data.

FURTHER READING

Balkenhol et al. (2015) *Landscape Genetics*: Book on landscape genetics of animals and plants with contributions from many of the leaders in the field.

Coleman et al. (2013) Show that genetic differentiation among populations of dwarf galaxia fish is associated with low genetic diversity.

Eldridge et al. (2001) Compares different methods for detecting immigrants based on an example with known immigrants, and then applied the better performing methods to determining the (unknown) source of immigrants.

Falush et al. (2007) Describes means for estimating the number of populations using multilocus genotype data using Bayesian clustering with the STRUCTURE software.

Guillot et al. (2005a) Describes means to estimate the number of populations and to locate them in the landscape, based on multilocus genotypes for georeferenced samples using GENELAND software.

Wang (2014b) Describes a method for estimating recent gene flow by assigning parentage from multilocus genotypes using likelihood methods.

SOFTWARE

BAYESASS: Bayesian approach to estimating recent gene flow rates using multilocus genotypes (Wison & Rannala, 2003). http://rannala.org

GENELAND: a free program for landscape genetics, akin to a spatial version of STRUCTURE for identifying population units and delineating their landscape locations (Guillot et al. 2005a). http://www2.imm.dtu.dk/~gigu/Geneland/

INSTRUCT: a version of STRUCTURE modified to account for inbreeding by simultaneously estimating inbreeding (selfing) and levels of genetic differentiation (Gao et al. 2007). http://cbsuapps.tc.cornell.edu/InStruct.aspx

MIGEST: software designed to estimate recent gene flow by assigning parentage from multilocus genotypes, using likelihood methods (Wang 2014b). http://www.zsl.org/science/software/migest

POWSIM: software for determining the power of tests for genetic differentiation among populations (Ryman & Palm 2006). http://www.zoologi.su.se/~ryman

SPAGeDi (Spatial Pattern Analysis of Genetic Diversity): computer package for characterizing the spatial genetic structure of mapped individuals or populations, using genotypic data of any ploidy level (Hardy & Vekemans 2002). http://ebe.ulb.ac.be/ebe/SPAGeDi.html

STRUCTURE 3.2: most frequently used software to delineate number of populations. Uses Bayesian clustering to identify population units (Falush et al. 2007). http://pritch.bsd.uchicago.edu/structure.html

Are there populations suffering genetic erosion that would benefit from augmented gene flow?

CHAPTER 11

Having identified small geographically and genetically isolated populations, we need to determine whether they are suffering genetic erosion, and if so, whether there are any other populations to which they could be crossed. We should next ask whether crossing is expected to be harmful or beneficial, and if beneficial, whether the benefits would be large enough to justify a genetic rescue attempt. Here we address these questions based on the principles established in the preceding chapters.

TERMS

DNA fingerprint, monomorphic

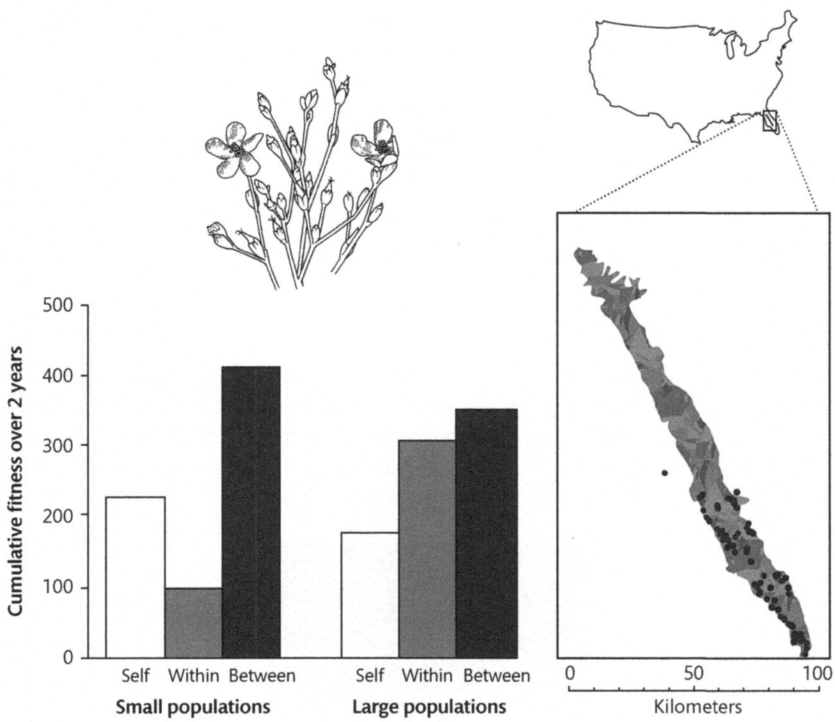

Small isolated populations of the outcrossing highlands scrub hypericum on the xeric soils of Lake Wales Ridge (shaded) in central Florida exhibit low cumulative fitness (within population crosses), no inbreeding depression upon selfing, but large genetic rescue effects from between population crosses (left portion of graph). Conversely, large populations have higher fitness following within population crosses, exhibit greater inbreeding depression when selfed, and smaller genetic rescue effects following between population crosses (right portion of graph) (Dolan et al. 1999; Oakley & Winn 2012).

Genetic Management of Fragmented Animal and Plant Populations. R. Frankham, J. D. Ballou, K. Ralls, M. D. B. Eldridge, M. R. Dudash, C. B. Fenster, R. C. Lacy & P. Sunnucks. Oxford University Press (2017) © R. Frankham, J. D. Ballou, K. Ralls, M. D. B. Eldridge, M. R. Dudash, C. B. Fenster, R. C. Lacy & P. Sunnucks 2017. DOI 10.1093/oso/9780198783398.001.0001

What are our objectives?

In this chapter we seek to determine if populations are suffering genetic erosion, and whether any would benefit from augmented gene flow.

Box 11.1 illustrates the considerations and evidence used in evaluating the case for genetic rescue of the endangered Florida panther.

Florida panther (USA)

Box 11.1 On what basis was the decision made to attempt genetic rescue of the Florida panther?

(Roelke et al. 1993)

By the early 1990s, the endangered Florida panther was restricted to a small relict population of ~ 20–25 individuals in southern Florida. Prior to European settlement, they ranged across the entire southeastern USA, and other sub-species were spread throughout North and South America.

The following are the considerations and evidence that went into the decision to attempt genetic rescue of this (unhybridized) population.

Was the Florida panther suffering genetic erosion?

The Florida panther population had low levels of genetic diversity, as measured using allozyme, DNA fingerprint, and mtDNA data, compared to earlier museum specimens, other puma populations, and felids generally (table). Subsequently, this was confirmed with evidence from microsatellites (Culver et al. 2008).

Taxon	Allozymes H_e (%) (range)	DNA fingerprints H_e (%)	Microsatellites H_e (%)
Florida (unhybridized)	1.8	10.4	14.7
Florida (museum)			31.1
Western USA	4.3 (2.0–6.7)	46.9	34.8
Other felids	3–8	–	
Domestic cat		44.0	

The panthers had attributes indicative of inbreeding depression, including kinked tails (photograph), cardiac defects, a high prevalence of infectious disease, and very poor semen quality (pair of photographs), compared to other panthers and felids. Further, about half of the males had at least one undescended testis (likely an inherited defect) and the incidence had increased with time (Mansfield & Land 2002).

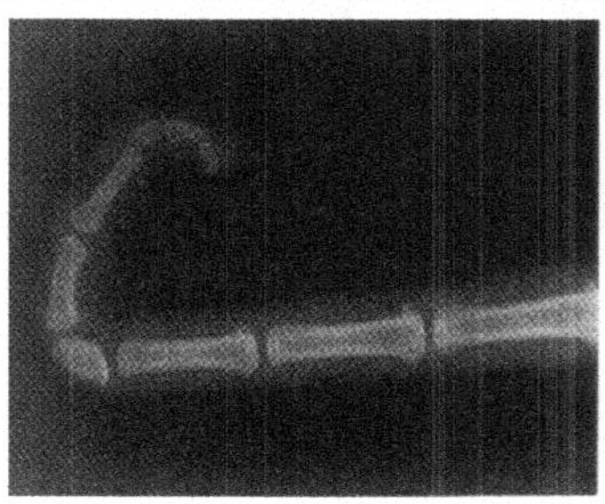

(After Roelke et al. 1993)

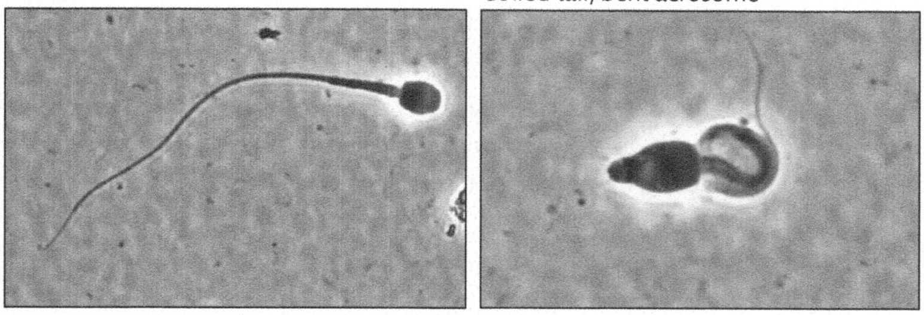

(Supplied by J. Howard & B. Pokazhhenthi, Smithsonian Institution)

Were there other populations that could be used to rescue the Florida panthers?

Yes, panthers/cougars/mountain lions are widely distributed throughout the USA. In particular, the Florida and Texas populations had been connected in historical times (Hedrick 1995).

What was the risk of outbreeding depression in a cross of Florida panthers × Texas cougars?

Outbreeding depression was a potential problem, because Florida panthers and Texas cougars were designated as distinct sub-species at that time. However, the main distinguishing feature of the Florida panther, its kinked tail, was likely an effect of inbreeding depression. Subsequent molecular genetic analyses showed that there are not genetic sub-units within the North American puma population, so Florida panthers are now included with all other North American populations in the *Puma concolor cougar* sub-species (Culver et al. 2000).

There remained a concern about adaptation to somewhat different habitats in Florida and Texas, but the species is wide-ranging and inhabits environments throughout North America. Concerns about possible chromosomal divergence do not seem to have been raised, but all felid species have similar karyotypes, implying that fixed chromosomal differences are unlikely (O'Brien et al. 2006).

What was the final decision?

Following extensive consultations and genetic modeling studies (Hedrick 1995), the Florida population was augmented with eight wild-caught Texas puma females. Box 6.1 details the subsequent beneficial consequences of this augmentation.

Before addressing the above objectives, we first consider existing guidelines for genetic rescue attempts.

Are there guidelines for genetic rescues?

We present guidelines for genetic rescue attempts

The recent guidelines in Table 11.1 reflect advances in predicting the risk of outbreeding depression (Frankham et al. 2011), a meta-analysis on genetic rescue (Frankham 2015), and other advances made during the writing of this book, and thus are the ones we endorse.

Table 11.1 Guidelines for management of genetic rescues (after Frankham 2015).

When should we contemplate genetic rescue?

1. When there is a (recipient) population that is inbred and/or has low genetic diversity for fitness (or evolutionary potential), especially when it is known or suspected to be suffering from inbreeding depression for fitness.
2. When there is another genetically isolated (donor) population(s) of the same species to which it can be crossed to reverse inbreeding and loss of genetic diversity.
3. When the risk of outbreeding depression in crosses between the donor and recipient populations is low through to the F_3, as determined, for example by using the decision tree of Frankham et al. (2011).
4. When the potential benefits of genetic rescue are sufficiently large to justify the financial costs of translocation and any risk of outbreeding depression, as determined by a cost–benefit analysis. The benefits are expected to depend upon:
   a. The magnitudes of $\Delta F_{zygotic}$ and $\Delta F_{maternal}$
   b. The mode of reproduction in the species (sexually reproducing > asexual)
   c. The mating system in the species (self-incompatible > other naturally outbreeders > mixed mating > selfers)
   d. The ploidy in the species (diploid > polyploid > haplodiploid > haploid)
   e. The intended environment (stressful/wild > benign/captive)
   f. Inbreeding level in immigrants (outbred > inbred)
   g. Demographic history of and genetic diversity in the base population from which the recipient and donor populations were derived (numerically large [N_e] population with high genetic diversity > numerically small with low genetic diversity).
5. How many immigrants should be used? Any are better than none when the risk of outbreeding depression is low. At the upper end there is the risk of genetically swamping the recipient population, so immigrant alleles should generally be $\leq$ 50% of the crossed population (see Chapters 12 and 13).
6. Will more than one augmentation of gene flow be required? The need for additional rounds of crossing will depend upon the proportion of immigrants (low > high), their inbreeding level (high > low), and the N_e in the crossed population (low > high).
7. Should the program be monitored? Yes, essentially as suggested by Hedrick & Fredrickson (2010).

Earlier and more restrictive guidelines were proposed for genetic rescue by Edmands (2007) and Hedrick & Fredrickson (2010). Edmands (2007) recommended that genetic rescue only proceed when there is clear evidence of inbreeding depression in the isolated populations and where the effects of crosses have been evaluated over two generations (wherever possible). The Hedrick & Fredrickson (2010) guidelines emphasize evidence of fitness loss in the target population and beneficial effects of crossing from an experiment in captivity before proceeding.

We favor the Frankham (2015) guidelines, based on:

1. Urgency, as conservation biology is a crisis discipline where inaction often leads to further decline and extinction of threatened populations (Soulé 1985)
2. Ubiquity of inbreeding depression in adequately studied naturally outbreeding diploid and polyploid species (Chapter 3)
3. Low power of most experiments on threatened populations to detect inbreeding or genetic rescue effects, even if they exist (e.g. Lacy 1997; Kalinowski & Hedrick 1999). Conversely, meta-analyses usually provide a more reliable overview of likely effects (Ralls & Ballou 1983; Crnokrak & Roff 1999; Frankham 2015, 2016). Experiments in captivity typically underestimate the effects of inbreeding and crossing in the wild, and are not feasible for the many species that have not been successfully bred in captivity (e.g. Northern hairy-nosed wombats and shoebill storks [*Balaeniceps rex*])
4. Experiments are not feasible within reasonable time frames in large long-lived species such as elephants, or long-lived trees where > 100 years would be required.

We disagree with the Edmands (2007) recommendations for all of the reasons above. Edmands (2007) also recommended that the population(s) chosen for crossing be as similar genetically and adaptively as possible. We agree with the need for adaptive similarity (Frankham et al. 2011), but not for genetic similarity for near neutral markers, because the benefits of crossing increase with differences in allele frequencies between target and donor populations when there is no outbreeding depression (Falconer & Mackay 1996).

We disagree with the Hedrick & Fredrickson (2010) emphasis on evidence of fitness decline in the target population for reasons 1–4, and their requirement for evidence from crossing experiments in captivity for reasons 3 and 4. However, we endorse Hedrick & Fredrickson's (2010) advice to consider the overall situation and other management options, including doing nothing (see Chapter 12), plus the desirability of having translocation protocols in place and a detailed monitoring plan (see Chapters 6 and 7). Clearly, it is desirable for management to continue over time, if needed.

Ottewell et al. (2016) proposed more general guidelines for genetic assessment, conservation prioritization, and decision making, with brief consideration of genetic rescue. Their guidelines, developed for endangered plants, are based on high versus low values for three genetic parameters, F_{ST} as a measure of population differentiation, heterozygosity for neutral markers to characterize genetic diversity, and F_{IS} to indicate inbreeding. These are used to define eight management strategies (Fig. 11.1). Their system has several weaknesses. Strategies 5–8 involve differentiated populations where they warned about the risk of outbreeding depression, but provided little guidance about detecting it. They did not clearly distinguish cases of differentiation due to drift versus adaptation (e.g. strategy 5), a fundamental distinction in our treatment. Genetic rescue by augmenting gene flow is considered an option in categories 4, 7, and 8, with the risk of outbreeding depression needing to be excluded for the latter two.

Ottewell et al. (2016) use $F_{ST} > 0.15$ to indicate high differentiation, but they do not address the problems with this measure. First, it estimates historical equilibrium gene flow that may not reflect recent fragmentation, so a direct measure of recent gene flow would be preferable (Chapter 10). Second, F_{ST} does not scale 0–1 when there are more than 2 alleles at each locus, as is typical with microsatellites. Use of G'_{ST} would provide more comparable measures of differentiation (Chapter 5), while population mean kinship is a more desirable measure for genetic management purposes than any form of F_{ST} (Chapter 13). Additionally, F_{IS} encompasses only inbreeding in the most recent generation due to non-random mating, while inbreeding often accumulates over many generations in small closed populations: more inclusive estimators of inbreeding are given in the next section.

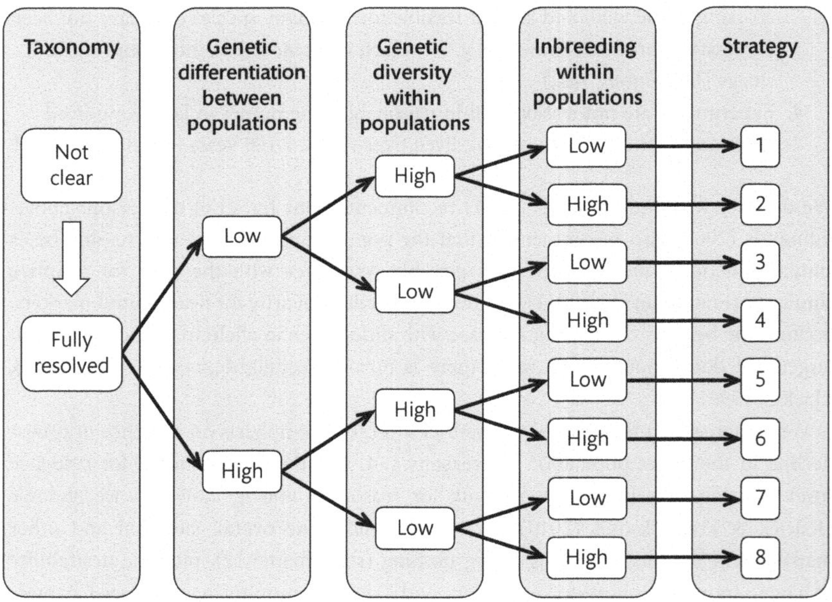

Fig. 11.1 Ottewell et al. (2016) process for determining population level genetic management strategies based on genetic differentiation (F_{ST}), genetic diversity (heterozygosity), and inbreeding (F_{IS}).

The remainder of the chapter is concerned with implementation of genetic management of (small) isolated inbred populations by answering each of the questions posed in Fig. 11.2.

What questions do we need to answer?

The decision tree in Fig. 11.2 specifies the questions that should be considered and lays out our path through these issues.

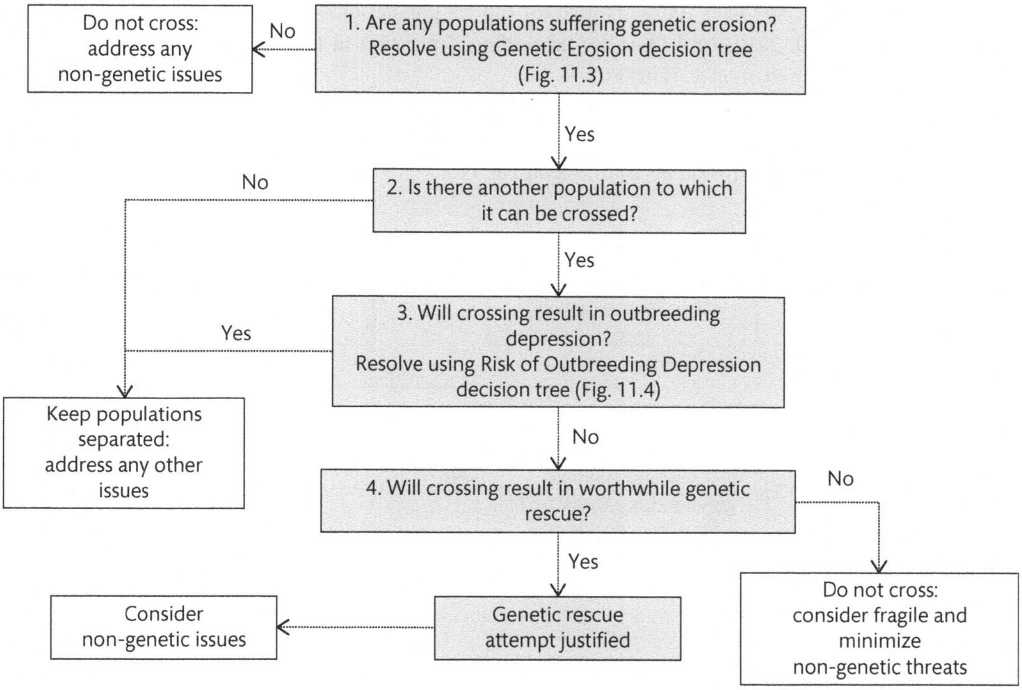

Fig. 11.2 Decision tree for determining whether there are populations suffering genetic erosion that would justify genetic rescue attempts.

We now proceed to address questions 1–4 in turn below.

Are any populations suffering genetic erosion?

To determine whether a population is likely to be suffering from problematic levels of genetic erosion in fitness and/or evolutionary potential, we need evidence for one or more of the following:

- lower fitness than in a non-inbred control (Chapter 3)
- lower molecular genetic diversity than in a related non-inbred population (within the same species or a related one) (Chapter 4)
- elevated pedigree inbreeding coefficient (Chapter 2)
- mean $F > 10\%$ estimated from population size (and inferred N_e) and number of generations of isolation (Chapter 3).

What are problematic levels of these indicators? If a physically isolated outbreeding diploid population has suffered at least a 10% loss of fitness (ideally in total or composite fitness, but any fitness component is acceptable), a known or inferred 10% inbreeding coefficient, or 10% loss of genetic diversity, there is a high probability that it is suffering genetic erosion. The thresholds of at least 10% follow those for tolerable losses of genetic diversity over 100 years used for captive populations of threatened species (Frankham et al. 2010), and those for inbreeding depression in the wild under the guidelines of Frankham et al. (2014a).

In many cases, we will not have direct information on inbreeding depression and will have to assess the state of the population from more indirect information, using the Genetic Erosion decision tree in Fig. 11.3.

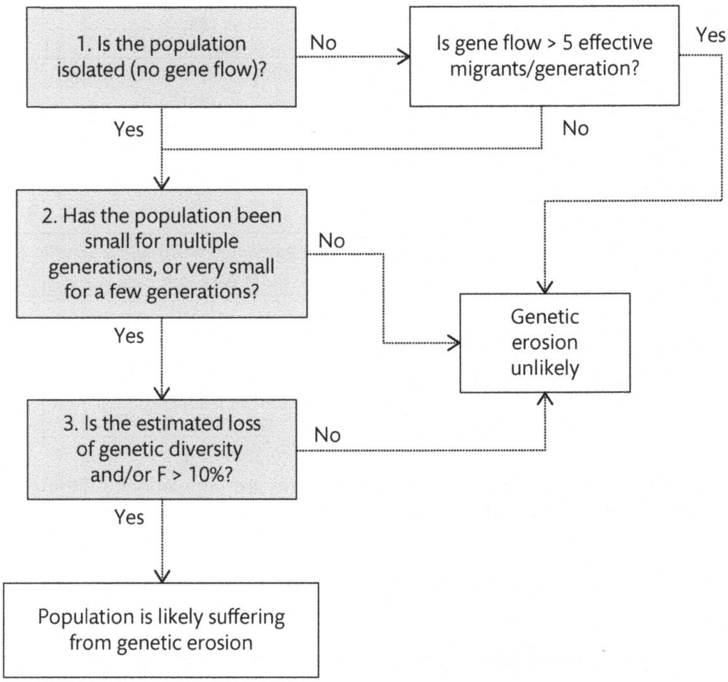

Fig. 11.3 Genetic Erosion decision tree to determine, in the absence of documented inbreeding depression, whether a population of a sexually reproducing outbreeding diploid species is likely to be suffering from meaningful genetic erosion. Small populations (N_e) and multiple generations (t) are defined as combinations of N_e and t that result in $F \geq 10\%$ (see later), e.g. $N_e = 100$ for 21 generations, or $N_e = 10$ for 3 generations.

There are four main approaches for obtaining estimates of F when there are no pedigrees for individuals within populations. First, kinship and individual inbreeding coefficients can be estimated from multilocus genotype data (e.g. SNPs, microsatellites, or allozymes) based on identity by descent of alleles in the different individuals within the population, using software such as COANCESTRY (Wang 2011). This method is being used to estimate kinship and inbreeding in California condors (*Gymnogyps californianus*), based on many SNPs in the genomes of the founders of the population (Garner et al. 2016). Improved genomic methods that estimate F back to remote common ancestors (equivalent to deep pedigrees) from the size of retained homozygous haplotypes have recently been developed and validated in birds and mammals (Keller et al. 2011; Knief et al. 2015). These are recommended, but they require data on many SNP markers in each individual. These and other genomic methods will typically have greater precision than those from the following methods (Kardos et al. 2015; Wang 2016).

Second, population mean inbreeding coefficient can be estimated from eqn 11.1 using heterozygosities for "neutral" markers such as SNPs, microsatellites, or allozymes in the target populations (H_{Inbred}) and in an outbred population (H_{Outbred}) (Frankham et al. 2010):

$$F = 1 - \frac{H_{\text{Inbred}}}{H_{\text{Outbred}}} \qquad 11.1$$

Example 11.1 illustrates estimation of the inbreeding coefficient for the Barrow Island black-footed rock-wallaby population using this equation.

Example 11.1 Estimating the inbreeding coefficient in the Barrow Island rock-wallaby population from microsatellite data (Eldridge et al. 1999)

The Barrow Island population of black-footed rock-wallabies has an average microsatellite heterozygosity of 0.05, while two mainland populations have heterozygosities of 0.62 and 0.56 (average 0.59) (Eldridge et al. 1999). Consequently, the estimated inbreeding coefficient (F) for the Barrow Island population is:

$$F = 1 - \frac{H_{\text{Inbred}}}{H_{\text{Outbred}}} = 1 - \frac{0.05}{0.59} = 0.915$$

Thus, the Island population is highly inbred. Further, the frequency of lactating females in the Barrow Island rock-wallabies is only 56%, compared to 92% in mainland rock-wallabies, suggestive of inbreeding depression.

In natural outbreeders, it matters little whether observed or expected heterozygosity is used in the equation, as they are typically very similar, but in inbreeding species, the use of observed heterozygosity captures the inbreeding effects of both drift and non-random mating.

Black-footed rock-wallaby (Australia)

Third, population mean F can be estimated from the population size and generations of isolation (t), using eqn 3.2. While the N_e is generally unknown, as a first approximation it can be obtained from the mean of the adult population sizes, and N_e/N ratios from related species (Chapter 2). Box 11.2 illustrates the use of this approach to estimate the inbreeding coefficient of isolated populations of the greater bamboo lemur (*Prolemur simus*) in Madagascar.

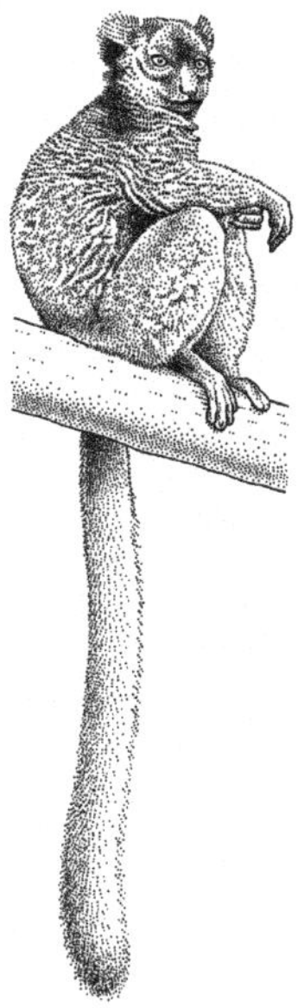

Greater bamboo lemur (Madagascar)

Box 11.2 Estimating the mean inbreeding coefficient for isolated populations of greater bamboo lemurs

(Data from Tony King, pers. comm.)

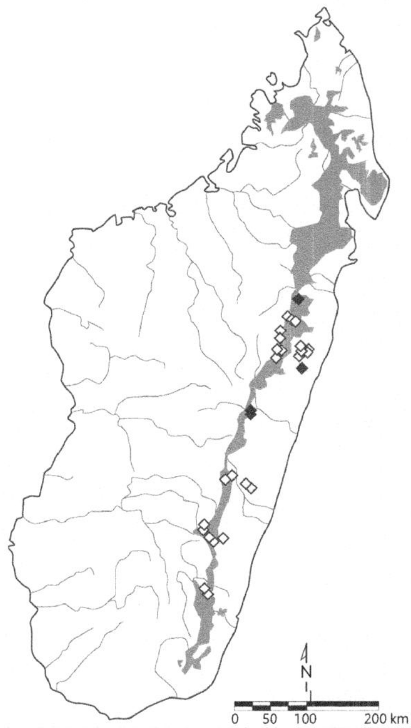

(Map after Rakotonirina et al. 2011)

Small isolated populations of Critically Endangered greater bamboo lemurs, mostly consisting of ~ 5–30 individuals, occur in deforested lowland agricultural landscapes (map: closed diamonds represent recent lemur sightings and open diamonds earlier ones). Tony King asked RF if the lemur populations might be suffering genetic erosion and needed genetic rescue.

A completely isolated population of size 30 individuals would likely have an effective size of ~ 3–10 (Chapter 2). The "isolated" sites have probably been isolated for less than 100 years. With a generation time of 14 years (the mid-point of female breeding duration from ~ 3 to ~ 25 years), there have been ~ 7 (100/14) generations of isolation.

Given seven generations of isolation at an N_e of 10, the inbreeding coefficient in our target population will be (eqn 3.2):

$$F = 1 - \left(1 - \frac{1}{2N_e}\right)^t = 1 - \left(1 - \frac{1}{2 \times 10}\right)^7 \sim 0.3$$

If N_e is only 3, then $F \sim 0.72$. These two estimates should bracket the true value of F, and both indicate that augmentation of gene flow should be seriously considered.

Fourth, simulation modeling (e.g. using VORTEX: Lacy & Pollak 2014) can be used to predict the decline in heterozygosity and the inbreeding coefficient from known trajectories in adult population sizes. The simulated decline of heterozygosity in the banteng (*Bos javanicus*) population in northern Australia was similar to that estimated from microsatellite analyses (Bradshaw et al. 2007).

Is there another population to which it can be crossed?

Candidate populations for use in genetic rescue will typically belong to the same species and have been genetically isolated (and typically geographically separate) from the recipient population for a substantial number of generations, with the amount of genetic differentiation determined by the combinations of effective population sizes and the number of generations. Total genetic isolation is not necessarily required, but one of the populations should not be a recent derivative of the other.

If there is no population to which the inbred population can be crossed, then management of the population is restricted to non-genetic procedures designed to minimize the risk of extinction (fire management, disease sanitation, captive propagation, etc.), as applied in the Wollemi pine (NSW National Parks and Wildlife Service 1998).

If there is another population to which it could be crossed, then we proceed to evaluate the risk of outbreeding depression in the cross.

Will crossing result in outbreeding depression?

The risk that crossing between populations will be harmful to the progeny in the F_1 or later generations depends on whether or not there are fixed chromosomal differences, or adaptive differences between populations, and the number of generations the populations have been isolated

We used these insights to develop a decision tree with five questions to predict the risk of outbreeding depression (Fig. 11.4). Questions 2–5 are based upon considerations above, while question 1 addresses whether the taxonomy is resolved (Chapter 9). This tree correctly identified the risk of outbreeding depression in between population crosses for almost all cases examined (Chapters 6 and 7).

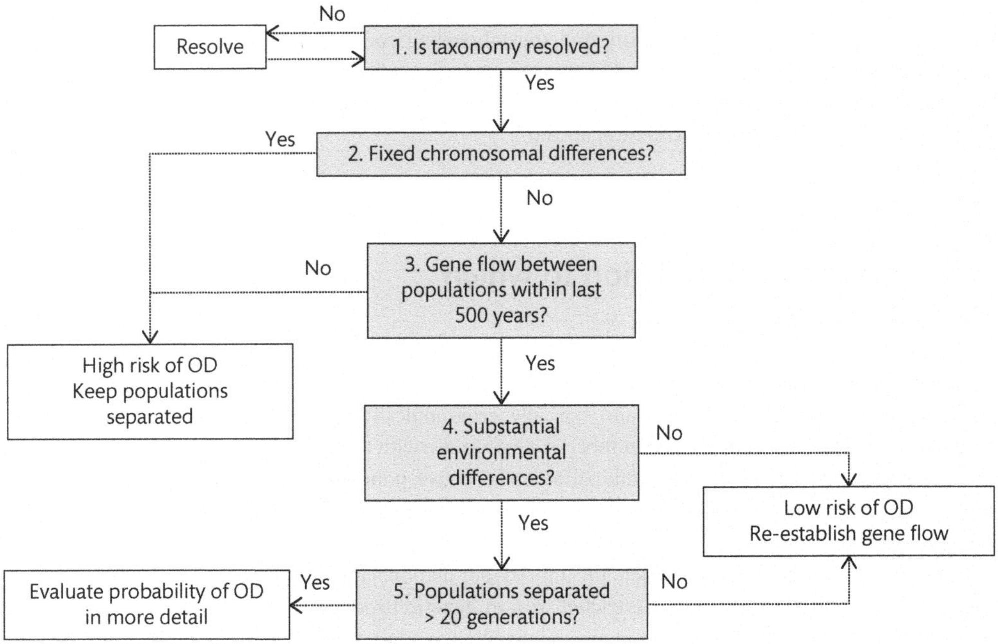

Fig. 11.4 Risk of Outbreeding Depression (OD) decision tree for determining whether the crossing of populations is likely to be harmful (Frankham et al. 2011).

The question about gene flow within the last 500 years seeks to avoid outbreeding depression due to crosses between cryptic species (morphologically similar, but reproductively isolated)

The risk of outbreeding depression in population crosses increases with the duration of isolation between them (Coyne & Orr 2004; Bolnick & Near 2005). We used a 500 year time span in an attempt to simplify this, in a similar manner to the use of years in the Vulnerable category of the IUCN Red List categorization system (IUCN 2016). While the time frame can be defined in generations, 500 years captures much of the human associated habitat fragmentation that is our primary concern (see Preface).

If populations have been isolated for 500 years or more, we recommend managing them as separate populations, or proceeding on a limited, experimental basis. This recommendation errs on the side of caution as it is often possible to cross populations that have been separated more than 500 years in the same environment, with no adverse effects (Table 7.3). Thus, we would not rule out a genetic rescue based solely on the time two populations have been separated. If genetic rescue seems like a good management option in all other respects, we recommend evaluating the risks in more detail.

Populations with reliable taxonomy, no fixed chromosomal differences, gene flow within the last 500 years, and that inhabit similar environments have low risks of outbreeding depression following crossing

There are many examples of populations with these attributes and a low risk of outbreeding depression (Frankham et al. 2011; Frankham 2015). For example, golden lion tamarin populations in Brazil all belong to a single species, have these attributes, and no outbreeding depression was observed when individuals from different populations were crossed. Further, isolated diploid populations of the threatened button wrinklewort daisy from the Canberra region in southeastern Australia have these attributes and exhibited genetic rescue effects when populations were crossed (Pickup & Young 2008; Pickup et al. 2013).

Conversely, populations with high risk of outbreeding depression when crossed exhibit the opposite states (different, rather than similar) for one or more of the above characteristics, and we recommend that they be managed separately (Table 7.1).

We recommend that populations with > 20 generations of adaptation to different environments (but no other risk factors) be subject to a detailed assessment of risk, based upon the variables identified in Chapter 7 (Frankham et al. 2011).

Button wrinklewort daisy (Australia)

How do we determine the duration of isolation between populations?

Historical or molecular genetic data can be used to estimate the length of time populations have been isolated

Duration of isolation between populations can be determined from historical evidence on the timing of fragmentation (if available). For example, habitat fragmentation due to the impacts of European settlement in Australia began no earlier than 1788 (when the first European settlers arrived). This may be distinguishable from fragmentation caused from aboriginal activities beginning ~ 50,000 years ago by using information in diaries of early explorers (Gammage 2011). Alternatively, divergence time can be estimated from molecular genetic data, using coalescent approaches (Box 11.3).

Brown trout (Europe)

Box 11.3 Estimating isolation times for Danish brown trout populations due to dams, based on microsatellite data

(Hansen et al. 2014)

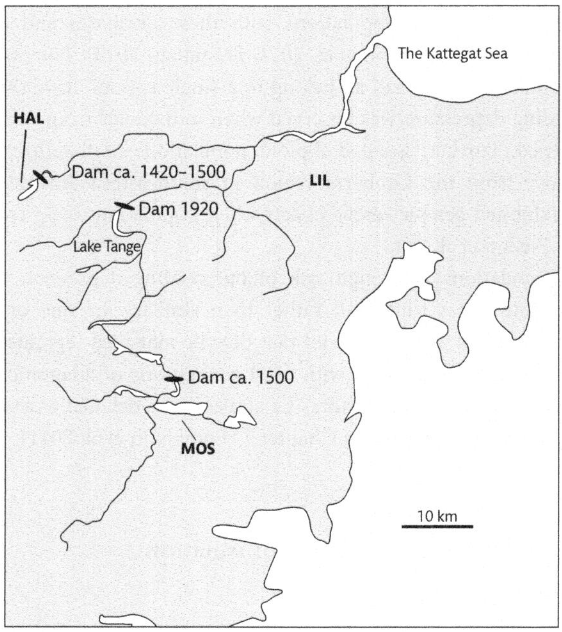

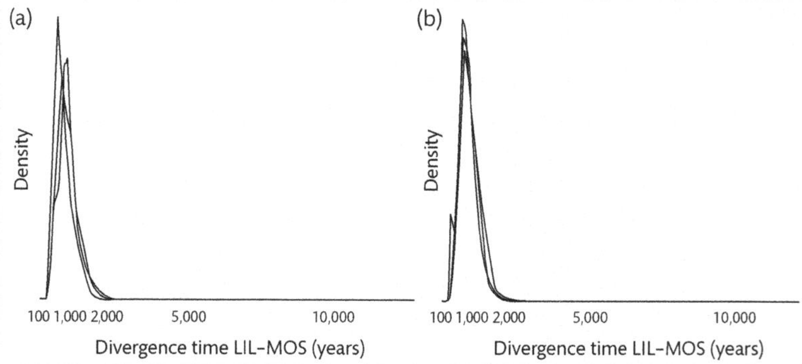

Hansen et al. (2014) estimated divergence times between brown trout populations above (HAL and MOS) and below (LIL) old dams in the Gudena River system, Denmark (map) using (a) approximate Bayesian computation and (b) a coalescent based isolation with gene flow model applied to data from 40 microsatellite loci. Both methods suggested divergence times ~ 600–800 years ago, concordant with the historical record of dam building. Three replicate estimates of divergence times with different inputs (N_e, mutation rates, and priors) gave essentially indistinguishable estimates (figures [a] and [b]).

What constitutes meaningfully different environments for two populations of a species?

Similar versus different environments may be defined using expert opinion, ecological niche modeling, or by comparing environmental similarities and differences associated with and without outbreeding depression in related groups

Expert scientific opinion can be used in the current context, as it appears to be widely used to distinguish similar versus different environments in planning for translocations and reintroductions (Fiedler & Laven 1996; Bremner-Harrison & Cypher 2007). Determinations should rely on the range of variation of key features of the two environments, such as the non-overlap of environmental features to which a species is either adapted or sensitive, and whether the species is a narrow specialist (such as some butterflies and plants) or a generalist (such as some wide-ranging birds and large carnivores).

Ecological niche modeling (also termed multidimensional environmental scaling) is used to associate species or populations with habitat envelopes (such as precipitation, elevation, mean annual temperature, and vegetation) (Osborne & Seddon 2012), using software such as BIOCLIM and GARP (Peterson et al. 2002; Hirzel & Le Lay 2008; Thorn et al. 2009). For example, Sattler et al. (2007) found that the widely distributed common pipistrelle bat in Europe actually consisted of two cryptic species, *Pipistrellus pipistrellus* and *P. pygmaeus* (rare), that had partially different niches based on ecological niche factor analysis, consistent with genetic differentiation and differences in echolocation signals. Further, multidimensional environmental scaling was used by Rutter & Fenster (2007) to predict the performance of *Arabidopsis thaliana* accessions in a novel environment using a global climate data set. Most current multidimensional environmental scaling concentrates on physical environmental features, but we recommend adding biotic environmental features where possible.

We are not aware of comparisons of environmental similarities and differences associated with presence versus absence of outbreeding depression in taxonomically related groups being used in this context, but a similar philosophy is used in taxonomy (Chapter 9).

How should clines be considered?

For populations that previously exhibited clines and are now fragmented, we recommend augmentation of gene flow between nearby populations (with similar environments), in preference to distant ones that may inhabit different environments

In general there is an increased risk of outbreeding depression when populations from increasingly distant geographic locations are crossed, especially when there are clines and adaptive differences. Consequently, if species that previously exhibited continuous clines, but now exist in isolated fragments (Fig. 10.2e) need to be augmented, gene flow should come from nearby populations rather than distant ones. For example, the red-cockaded woodpecker previously had an essentially continuous distribution, but is now fragmented and shows isolation by distance in molecular markers, and clines in wing and tail lengths associated with habitat temperature (Chapter 5 frontispiece and Box 5.1: Mengel & Jackson 1977). Gene flow was re-established, but only between near-neighbor populations (US Fish and Wildlife Service 2003).

How should we handle chromosomal polymorphisms?

We do not preclude augmentation if two populations with otherwise identical karyotypes share a chromosomal arrangement, but one is polymorphic and the other monomorphic

Some populations are polymorphic for chromosomal arrangement, while other populations within the species share only one of the arrangements (White 1973; Hoffmann et al. 2004). For example, some common shrew (*Sorex araneus*) populations in Europe are polymorphic for centric fusions and others monomorphic (Zima et al. 1996). Similarly, one population of the plant *Calycadenia ciliosa* (Asteraceae) in the western USA was polymorphic for a pericentric inversion, but others were monomorphic (Carr & Carr 1983). As the long-term existence of chromosomal polymorphism within a population implies that karyotypic heterozygotes are neutral or advantageous (as with many polymorphic inversions in *Drosophila*; Krimbas & Powell 2000), crossing a polymorphic population to a monomorphic one should not lead to outbreeding depression, and may be beneficial.

Many plant species have variable numbers of non-pairing B chromosomes consisting mainly of heterochromatin with few functional loci (Jones & Houben 2003). These typically appear to be selfish elements that are dispensable (Burt & Trivers 2006), so they should be of little importance in relation to the risks of significant outbreeding depression.

Diploid and polyploid populations in close proximity do not represent a polymorphism, and crossing them should be avoided.

How do we proceed if some information is missing?

If some information required to answer a question in the decision tree is missing, especially for chromosomes (Severns & Liston 2008), we recommend that managers pass to the next question and complete a "preliminary" assessment. If this yields a low probability of outbreeding depression, it is advisable to obtain the missing information, or only to proceed to augment gene flow on an experimental basis. Obtaining chromosomal information is particularly important in groups with varying ploidies (e.g. plants) and/or high rates of chromosomal evolution (e.g. primates and rock-wallabies) (Levin & Wilson 1976; Murphy et al. 2005).

What happens if we make an incorrect prediction about outbreeding depression?

Even if there are occasional crosses of inbred populations that exhibit modest outbreeding depression, this should only be a temporary problem, as natural selection will almost certainly improve fitness in future generations (Chapter 7), and evolutionary potential will be enhanced.

If the risk of outbreeding depression is low, we need to establish whether the benefits of augmenting gene flow will be sufficient to justify the effort and costs of implementing a genetic rescue attempt.

Will crossing result in worthwhile genetic rescue?

On the basis that genetic rescue for fitness is recovery from inbreeding depression (Chapter 6), the potential benefits of gene flow on fitness depend upon:

- inbreeding depression in the target population (high > low)
- the extent to which augmentation of gene flow reduces inbreeding (ΔF)
- whether the environment is wild/stressful or captive/benign (stressful > benign)
- whether the donor population is inbred or outbred (outbred > inbred)
- the mating system (naturally outbreeding > inbreeding).

Since inbreeding depression is approximately linearly (or log linearly) related to the inbreeding coefficient (Chapter 3), change in the inbreeding coefficient (ΔF) between the inbred population and that in the population cross in the F_2 generation can be used to predict the gain in fitness in crosses between an inbred population and another population. For example, if a population exhibiting 60% inbreeding depression in total fitness has its inbreeding coefficient halved by crossing to another population (in F_2 and beyond), it is expected to show a 30% increase in fitness in the augmented population compared to the source population, i.e. have a fitness 70/40 = 1.75 times that in the inbred population.

If the extent of inbreeding depression is unknown, the proportionate change in total fitness (ΔW) due to a genetic rescue attempt can be estimated following Ralls et al. (1988) as:

$$\Delta W = 1 - e^{-\Delta FB} \qquad 11.2$$

where B is the number of haploid lethal equivalents for total fitness. Values of B for total fitness in the wild will often be 6–10 (Chapter 3 text and Table 3.2). Example 11.2 illustrates the use of this equation for a number of scenarios.

Example 11.2 Predicting the magnitude of genetic rescue effects

If the assessment is carried out in a wild environment, the relevant *B* value will typically be 6–12. For a reduction in inbreeding coefficient due to crossing of 0.2 and $B = 6$, we obtain using eqn 11.2:

$$\Delta W = 1 - e^{-0.2 \times 6} = 1 - 0.3 = 0.7$$

Thus, we predict 70% improvement in total fitness.

If the fitness assessment is done in captivity, *B* values for total fitness in outbreeding plant species are 2.77 and 1.55 for the two data sets in Table 8.2 (mean 2.16). Using $B = 2$, we obtain:

$$\Delta W = 1 - e^{-0.2 \times 2} = 1 - 0.67 = 0.33$$

Thus, we predict a 33% improvement from crossing for total fitness in captivity.

How do we determine the best cross(es) to make for genetic rescue purposes?

If there are several potential populations to cross with an inbred target population, then the best choice in terms of fitness is to make a cross with low risk of outbreeding depression, and high potential benefits following crossing

The highest projected genetic rescue effects typically involve the most genetically differentiated population pairing (Falconer & Mackay 1996), but this often involves a higher risk of outbreeding depression. Box 11.4 illustrates the risk–benefit trade-off between these issues regarding potential genetic rescue of the Sonoran pronghorns (*Antilocapra americana sonoriensis*). There may be other logistical considerations that modify the choice, especially in large species, e.g. geographical distance, political boundaries, cost of translocating individuals, cuttings, or gametes, flowering synchronization in plants, etc.

Box 11.4 Potential donor populations for genetic augmentation of the Sonoran pronghorn

(Stephen et al. 2005)

The endangered Sonoran pronghorn exists in only two isolated populations in Arizona (< 100 individuals) and Mexico (~ 300 individuals), with reduced microsatellite genetic diversity, compared to populations of four sub-species across Arizona, New Mexico, Texas, Wyoming, Oregon, and Canada. A tree derived from microsatellite data indicated isolation by distance, without deep sequence divisions, and did not support the sub-specific designations.

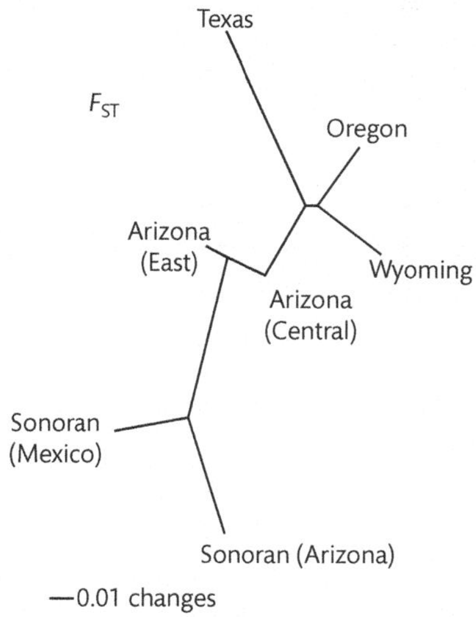

The authors suggested augmentation of gene flow between the two Sonoran pronghorn populations. This will require translocation, because this section of the US–Mexican border is fenced. While the other Arizona populations are more genetically differentiated and have higher genetic diversity, crossing with them involves a risk that translocated individuals, or their non-desert adapted progeny will not survive in the desert (the habitat of the Sonoran pronghorns).

Sonoran pronghorn (Mexico and USA)

An alternative approach is to augment gene flow from a combination of several donor populations. This has the most immediate and sustained benefits (see next section for computation of resulting F).

Estimating inbreeding coefficients in populations derived from several populations

If a population is formed by crossing several different populations (n) with diverse inbreeding levels in dissimilar proportions, the inbreeding coefficient after random mating is established in the F_2 is (Margan et al. 1998):

$$F_{\text{pooled}} = \sum_{i=1,}^{n} f_i^{\,2} F_i \qquad 11.3$$

where f_i is the proportion of genetic material contributed by the ith population. Example 11.3 illustrates the use of this equation. Subsequent to the F_2, the inbreeding coefficient will increase at a rate that depends inversely on the effective population size.

If equally inbred populations are crossed, the equilibrium inbreeding levels in the F_2 will decrease with the number of populations entering the mix: with two fully inbred populations ($F = 1$) mixed in equal proportions, the equilibrium pooled F after crossing is ½, while if four completely inbred populations are crossed the F of the combined populations is ¼.

Example 11.3 Computation of inbreeding coefficients for populations formed by pooling several populations with different *F* values in unequal proportions

If we pool three populations with F values of 0.1, 0.3, and 0.5 in the proportions 0.5, 0.4, and 0.1, the F of the pooled populations after random mating equilibrium is established is given by eqn 11.3 as:

$$F_{\text{pooled}} = \sum_{i=1,}^{n} f_i^{\,2} F_i = (0.5^2 \times 0.1) + (0.4^2 \times 0.3) + (0.1^2 \times 0.5) = 0.078$$

Thus, the pooled population has a zygotic F of 7.8% in the F_2.

How do we predict the benefits of genetic rescue for evolutionary potential?

Evolutionary potential is expected to be improved by augmenting gene flow to a degree that depends on the proportionate increase in genetic diversity

The benefits of crossing populations on evolutionary rescue are expected to depend on changes in the heritability and the selection differential between the recipient and the F_2 crossed population (Chapter 6).

The benefits of gene flow on heritability should depend approximately upon the proportionate increase in genetic diversity for fitness, provided quantitative genetic effects are additive (Falconer & Mackay 1996). This proportion should typically be similar to that for neutral markers in the F_2 compared to the parent populations (Chapter 6). If there are heterozygosity data, then the improvement in evolutionary potential should be given by $(\text{heterozygosity}_{\text{F2 cross}}/\text{heterozygosity}_{\text{Recipient}})-1$, whereas applying inbreeding coefficients in random mating populations, the improvement should be $[(1-F_{\text{F2 cross}})/(1-F_{\text{Recipient}})]-1$. For example, crossing the small inbred Mt. Jasmin population of the jellyfish tree to the Bernica population in the Seychelles was expected to increase the microsatellite heterozygosity from 0.26 to 0.59 (Finger et al 2011), yielding a 127% expected increase in evolutionary potential.

In addition, fitness improvements due to population crossing will often increase the selection differential, as illustrated in Box 4.3, but the required information will often not be available to predict the effect.

How do we cope with simultaneous harmful and beneficial effects of crossing?

Where beneficial and harmful effects of crossing are both expected, the situation requires a risk–benefit analysis

In some crosses, beneficial and harmful impacts occur simultaneously. For example, both impacts were found by Lacy in crosses of beach and old-field *Peromyscus* subspecies (Box 7.2) and by Fenster & Galloway (2000) in long-distance crosses of partridge peas. Since we are not able to make quantitative estimates of the risk and magnitude of outbreeding depression, we cannot do quantitative risk–benefit analyses, but can evaluate relative risk in a range of relevant circumstances.

If the risk of outbreeding depression is modest, but the expected benefits of augmented gene flow are large, then genetic rescue attempts are desirable. Conversely, if there is a significant risk of outbreeding depression and only small expected benefits, augmentation of gene flow is unlikely to be warranted.

In Chapters 12 and 13 we address the management of rates of gene flow, based on different amounts and types of information.

Summary

1. The decision on whether an isolated population fragment is suffering genetic erosion and would benefit meaningfully from crossing to another population involves first determining whether it is inbred and/or has suffered loss of genetic diversity, and is suffering from known or suspected inbreeding depression.
2. If a target population is suffering genetic erosion, we should ask if there are any donor populations within the species to which it could be crossed.
3. Next we ask, what is the risk of outbreeding depression in crosses between the two populations? This will be low for populations with the same karyotype, that are adapted to similar environments, and that were isolated within the last 500 years.
4. If an interpopulation cross has a low risk of outbreeding depression, the potential fitness benefits depend upon the mating system (outbreeders > inbreeders), extent of inbreeding depression in the target population (large > small), ΔF between the F_2 cross and the inbred parent population (large > small), the environment (wild > captive), and inbreeding level in the immigrants (outbred > inbred).
5. The improvement in evolutionary potential after augmentation of gene flow depends on the ratio of heterozygosities in the F_2 cross and the inbred parents.

FURTHER READING

Frankham et al. (2011) Reviews causes of outbreeding depression, presents a decision tree for assessing its risk of occurrence, and provides evidence that it worked.

Frankham (2015) Meta-analysis (a) substantiating the effectiveness of the Frankham et al. (2011) screen against outbreeding depression, (b) documenting the consistency and magnitude of genetic rescue effects when inbred populations are crossed to another population, and (c) delineating the variables determining the magnitude of rescue effects. Also provides updated guidelines for genetic rescue.

Hansen et al. (2014) Estimates divergence time between wild brown trout populations from molecular data, and validates the methods for populations isolated by dams of known antiquity.

Oakley & Winn (2012) Document genetic rescue by crossing of small inbred populations to other populations in an endangered Florida plant species.

Wang (2011), Keller et al. (2011), and Knief et al. (2015) Described methods for estimating levels of inbreeding from SNP data that are applicable to species in the wild.

SOFTWARE

COANCESTRY: software to estimate kinship and inbreeding coefficients from multi-locus genotype data (Wang 2011). http://www.zsl.org/science/software/coancestry

VORTEX: population viability analysis software that tracks the predicted heterozygosity and inbreeding coefficients of individuals (Lacy & Pollak 2014). www.vortex10.org/

Managing gene flow among isolated population fragments. I. Limited information

CHAPTER 12

When the decision is made to augment gene flow into an isolated population, managers must decide on how to augment gene flow, when to start, from where to take the individuals or gametes to be added, how many, which individuals, how often, when to cease, etc. Even without detailed genetic data, sound management strategies for augmenting gene flow can be instituted by considering population genetics theory, and/or computer simulations. When detailed data are lacking, moving some individuals into isolated inbred population fragments is better than moving none, as long as the risk of outbreeding depression is low.

TERMS

Corridor, translocation, genetic swamping, maximum avoidance of inbreeding

Translocating a black rhinoceros in South Africa using a helicopter.

Genetic Management of Fragmented Animal and Plant Populations. R. Frankham, J. D. Ballou, K. Ralls, M. D. B. Eldridge, M. R. Dudash, C. B. Fenster, R. C. Lacy & P. Sunnucks. Oxford University Press (2017) © R. Frankham, J. D. Ballou, K. Ralls, M. D. B. Eldridge, M. R. Dudash, C. B. Fenster, R. C. Lacy & P. Sunnucks 2017. DOI 10.1093/oso/9780198783398.001.0001

Why do we need to manage gene flow?

Levels of gene flow required to provide genetic rescues need to be determined such that genetic benefits are obtained at reasonable costs and genetic swamping is avoided

Gene flow can range from a single immigrant, to a majority of immigrants (genetic swamping). We need to define reasonable and practical levels of gene flow to rescue fitness and genetic diversity, and to avoid genetic swamping. We need to relate these to cost and frequency of augmentation. The current number of immigrants affects how soon another round of augmentation is needed. Should we have a large augmentation now or a small augmentation every generation, or something in between? These proposals have to be integrated with non-genetic issues, such as the organism's behavior, and which sex should be used to augment gene flow. How does all this relate to the amount of genetic and demographic information we have, or can reasonably obtain?

In the previous two chapters, we defined means for determining whether populations are suffering genetic threats to their viability, and whether they might benefit from gene flow, as indicated by our decision trees for genetic management of fragmented populations (Figs 10.1, 11.2, 11.3, and 11.4). This chapter and the next describe methods to manage gene flow between populations, depending on how much genetic information is available, by using population genetic theory, computer simulations, and molecular genetic analyses. In this chapter, we present basic strategies that can be used in the absence of detailed information on the history or genetic structure of a population's fragments. In the next chapter, we discuss strategies based on more detailed genetic information. Box 12.1 illustrates how decisions on gene flow were made for Florida panthers, as a prelude to detailed treatment of the issues in this chapter.

Florida panther (USA)

Box 12.1 Use of analytical theory to evaluate impacts of gene flow into inbred Florida panthers from Texas cougars

(Hedrick 1995)

The management of the Florida panther genetic rescue was based on expert opinion, aided by analytical work by Hedrick (1995) that used basic equations from evolutionary genetics to examine the impact of gene flow into the inbred Florida panthers. As we saw in Box 11.1, the small isolated Florida panther population was suffering from inbreeding depression and reduced genetic diversity. Hedrick investigated the consequences of gene flow (m) of 20% of the population deriving from immigrants (Texas cougars) in the initial generation and 2.5% thereafter for 10 generations (these values having been derived from expert opinion). We present some sample calculations to illustrate the principles, and readers are referred to Hedrick (1995) for full details. Hedrick assumed that Florida panthers and Texas cougars were fixed for different alleles, which we designate as A_1 and A_2, respectively. Ignoring drift and selection, the effects of one generation of gene flow on the change in allele frequencies at a locus is derived in the following figure:

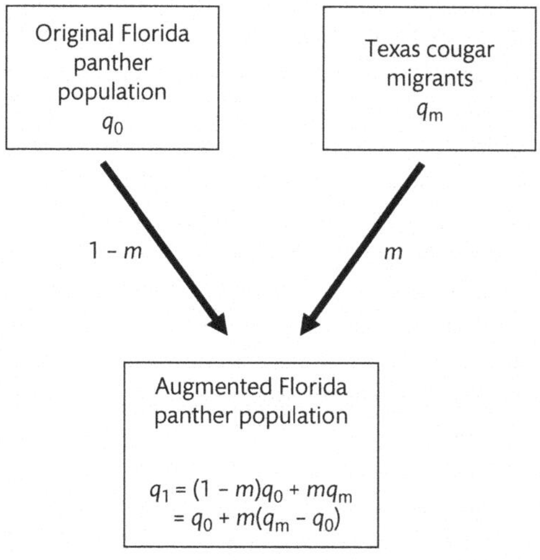

where m = proportion of immigrant alleles, q_0 = initial frequency of allele A_2 in Florida panthers (q_0 = 0), q_m = initial frequency of allele A_2 in Texas cougar immigrants (q_m = 1.0), q_1 = the frequency of A_2 in Florida panthers after one generation of gene flow, and Δq = the change in its frequency. After one generation of immigration with m = 0.2, the A_2 allele would then have a frequency q_1 of:

$$q_1 = 0.8 \times 0 + 0.2(1.0 - 0) = 0.2.$$

After a generation of random mating, heterozygosity at the locus would be 0.32 (having been 0 prior to gene flow).

Under this same scenario, but with t generations of migration at levels of m_i in the ith generation, the frequency of allele A_1 (p_t) is given by

$$p_t = p_0 \prod_{i=1}^{t} (1 - m_i)$$

where Π is the product. Thus, with the proposed scenario of 20% migration in the first generation and 2.5% for the remaining 10 generations, the frequency of the A_1 allele would drop from 1 to $0.8 \times 0.975^{10} = 0.62$ and the donor immigrant allele (A_2) increase correspondingly from 0 to 0.38.

How can we augment gene flow?

Gene flow can be increased by linking habitat fragments, or moving genetic material between them

Gene flow may be augmented by habitat management, or by direct translocation of individuals or gametes between fragments. A long-term solution to inadequate gene flow among fragmented populations is to connect populations by improving the habitat matrix between them or by establishing habitat corridors through which individual animals can move or plants can colonize (Franklin and Lindenmayer 2009; Beier et al. 2011). Changes in management practices such as modifying fences to allow passage of some wildlife species, building hedgerows to facilitate habitat connectivity, or reducing grazing, pesticide, and rodenticide use may improve gene flow. Corridors among habitat fragments (frequently recommended for non-genetic reasons) can re-establish gene flow among isolated populations (Christie & Knowles 2015). For example, 21 ribbons of habitat have been established between isolated habitat fragments for the golden lion tamarin in Brazil (Ana Maria de Godoy Teixeira, pers. comm.). The most ambitious proposals of this kind aim to provide continental-scale connectivity, particularly north to south along latitudinal temperature gradients, as for "The Wildlands Project" in North America, and related ones in Australia and elsewhere (Davis 1992; Soulé & Terborgh 1999; Pulsford et al. 2003). The time frame for achieving these visions could be within many decades, but they face multiple political, social, and financial challenges. Nonetheless, such systems are essential if we are to conserve biodiversity in the long term, especially in the face of global climate change (Chapter 14).

Where humans have built barriers to gene flow, such as roads or dams and weirs, these can be removed or their effects mitigated. Fish ladders are often used to assist fish to migrate past dams, but may be ineffective in some instances (Williams et al. 2012). Consequently, dam removal is increasingly used to restore connectivity in rivers. For example, 72 dams in 19 states in the USA were removed in 2014, restoring more than 730 miles of streams for the benefit of fish and other aquatic wildlife (American Rivers 2016). The effects of roads in preventing gene flow have been alleviated with wildlife underpasses or overpasses for terrestrial animals, as in Banff National Park in Canada (Fig. 12.1; Beckman et al. 2010, Sawaya et al. 2014). For arboreal species, canopy bridges, glider poles, and vegetated medians improved connectivity over a freeway for marsupial gliders in southeastern Australia (Soanes et al. 2013).

When restoring gene flow through habitat management is not feasible or practical, human intervention is required to move individuals or gametes, as for example translocating black rhinoceroses by helicopter (chapter frontispiece). The relative merits of habitat management versus human translocation of individuals or gametes to restore gene flow are considered during comprehensive planning for each species, as in the One Plan approach (Byers et al. 2013). Sometimes both strategies can be combined, initiating habitat management to improve connectivity and genetic rescue. This strategy increases the probability that isolated populations will survive until corridors provide sufficient connectivity. The remainder of this chapter assumes that movement of individuals or gametes has been deemed necessary to augment gene flow and offers guidance for implementing this option. The first step is to define the objective of genetic management.

Fig. 12.1 Highway overpass built as a wildlife corridor over the Trans-Canada Highway in Canada's Banff National Park to facilitate access to food, shelter, and gene flow (http://conservationcorridor.org/2012/10/banff-national-park/). Photo courtesy of Adam Ford, ARC Solutions (http://arc-solutions.org).

What is the objective of genetic management for isolated population fragments?

The overall objective of genetic management for isolated population fragments is to reduce the risk of population and species extinctions by decreasing inbreeding and augmenting genetic diversity within population fragments

We can achieve this objective by augmenting gene flow to increase the probability of successful genetic rescue. Specifically we want to recover:

- reproductive fitness
- adaptive evolutionary potential.

The former mainly involves reducing inbreeding and inbreeding depression (see Chapters 3 and 6), while the latter involves enhancing genetic diversity to recover the ability to evolve in response to environmental change (Chapters 4 and 6). To successfully reach this main objective, we must answer the following questions about the details of augmenting gene flow:

- Which population(s) needs genetic augmentation?
- When should we begin genetic augmentation?
- From which population(s)?
- Augmentation from one or several populations?

- Should we use inbred or outbred donor populations? Captive and/or wild?
- How many individuals or gametes to move?
- Which sex and age to move?
- Which individuals?
- Will genetic rescue be required again in the future?
- When to stop?

Our ability to answer these questions depends on the level of information available on the population(s) being managed.

In this chapter and the next, we examine different situations where the amount of information varies from very little to detailed. Thus, we seek to encompass genetic management situations from those in programs with very limited funds and little basic information, to well-funded programs with access to the most up-to-date molecular genetic and genomic techniques where more precise management of gene flow is possible.

Genetic management with different amounts of information

The type and intensity of genetic management that can be implemented ultimately depends on the amount of information that is available or can be obtained for any particular case

Figure 12.2 illustrates the types of genetic management that can be instituted in fragmented populations with differing amounts of information.

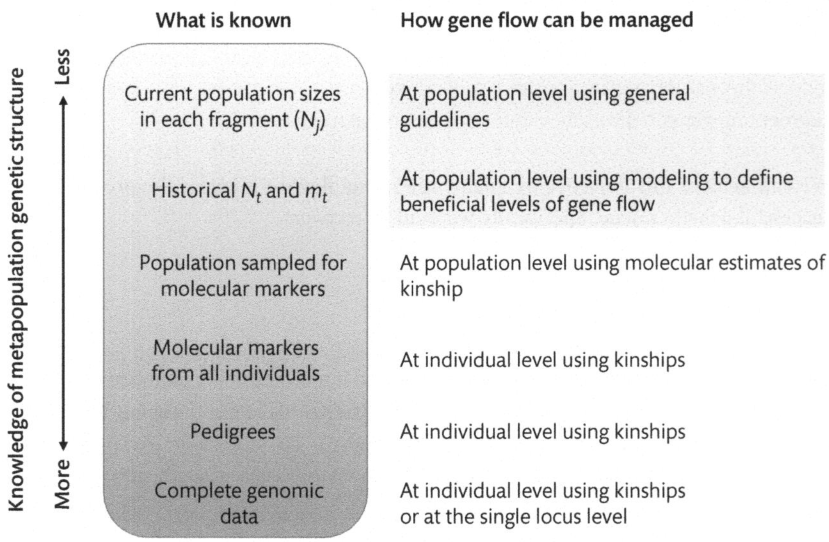

Fig. 12.2 Types of gene flow management that are possible with different levels of information on current or historical population size, migration rates, population structure, and kinships (the tinted shading on the right indicates topics addressed in this chapter).

The following sections illustrate options for genetic management of fragmented populations when only basic information is available on fragments (e.g. only current population sizes in the fragments (N_i) are known; Fig. 12.2).

What can be done with little or no genetic information?

Effective genetic management can be implemented even if little basic genetic information is available

From Chapter 6, we know that only a modest number of contributing immigrants each generation is needed to cause meaningful improvements in fitness and genetic diversity, even as small as a single contributing immigrant. We emphasize that **any gene flow into isolated inbred populations is better than none** (assuming the risk of outbreeding depression is low). Otherwise, when a small population is inbred, failing to increase gene flow is a risky strategy likely to lead to eventual extinction of the population. Consequently, **the default should not be inaction, but to augment gene flow** to minimize extinction risk (provided the risk of outbreeding depression is low).

When little information is available on the demographic history or genetic structure (top of the continuum in Fig. 12.2), we can still provide general guidelines to implement when considering the issues in Box 12.1.

Using computer simulations to guide management of gene flow

Computer models that simulate the consequences of gene flow can be used to assess the effectiveness of genetic management strategies

Population viability analysis (PVA) is a systems modeling approach for predicting the fate of a population due to the combined effects of all its systematic and stochastic threats (Shaffer 1981; Beissinger & McCullough 2002; Frankham et al. 2014a). VORTEX software (Lacy & Pollak 2014) includes projections of genetic changes over time, and allows assessment of the long-term impacts of genetic management on inbreeding, genetic diversity, population divergence, and population viability (see Appendix 2 for more details). Multiple population fragments can be modeled, and users can evaluate the effects of gene flow strategies between them. Example 12.1 illustrates the use of VORTEX to help guide genetic management of the Allegheny woodrat (*Neotoma magister*) in a fragmented metapopulation spread across eight sites in Indiana, USA.

Example 12.1 Modeling genetic management of fragmented populations of Allegheny woodrat to identify effective strategies

(T. J. Smyser, pers. comm.)

Allegheny woodrats in southern Indiana are restricted to eight small fragments, with only a few close enough to allow natural dispersal. A VORTEX PVA of this metapopulation showed that if the fragments were all fully interconnected, the single large population would quickly grow to fill the habitat carrying capacity (top line in figure), would be demographically resilient, and would retain 91% of its initial heterozygosity for 100 years (mean $F = 0.09$ [table]). However, without intervention the woodrats become highly inbred due to the fragmentation (mean increase in $F = 0.37$, in addition to any inbreeding that existed at the outset), most sub-populations are likely to crash to extinction, and the metapopulation is highly unstable (bottom line in figure). Local inbreeding depression in the unmanaged metapopulation must be the cause of its downward trajectory, because runs of the model that included no inbreeding depression in the metapopulation (middle line) had demographic performance similar to that of the panmictic metapopulation scenario (despite inbreeding rising to $F = 0.34$).

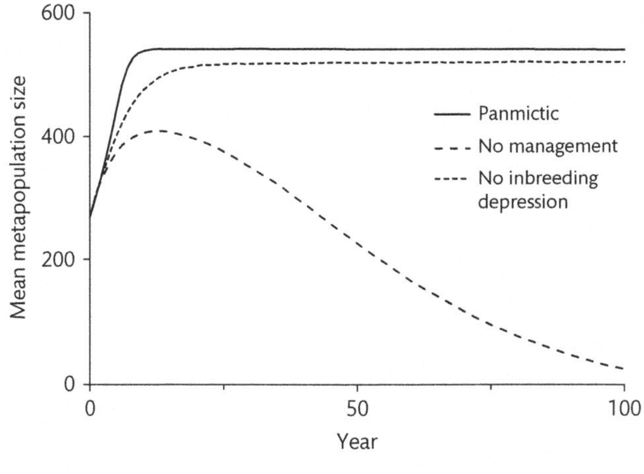

Different levels of genetic management were tested by moving random individuals between fragments at rates of one per population per generation ($Nm = 1$) up to $Nm = 8$. Increasing levels of managed translocations reduce inbreeding, prevent sub-population extinctions, and result in better population growth (next graph and table).

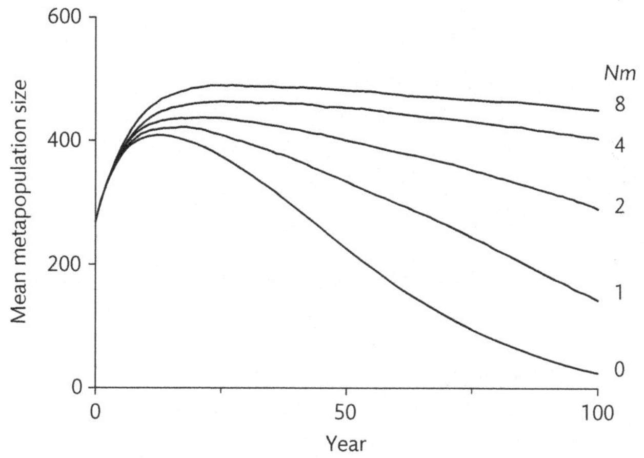

Managed translocations	Mean probability of sub-population extinction	Final metapopulation N	Mean inbreeding (F)
$Nm = 0$	0.81	26	0.37
$Nm = 1$	0.18	144	0.24
$Nm = 2$	0.02	288	0.18
$Nm = 4$	0.00	404	0.14
$Nm = 8$	0.00	450	0.11
Panmixia	0.00	540	0.09

When is gene flow needed?

This question was addressed in Chapter 11: population fragments with known or inferred inbreeding coefficients of at least 10% or a 10% reduction in genetic diversity are candidates for attempted genetic rescue. This level of inbreeding ($F = 0.10$) has the potential to reduce total fitness in the wild by almost 45%, assuming the population contains six haploid lethal equivalents (O'Grady et al. 2006; Table 3.2), and thus likely warrants genetic augmentation. Thus, the Allegheny woodrat requires slightly more than eight immigrants per generation into each population fragment to keep the overall inbreeding level below our recommended level (Example 12.1 table).

From where?

Crosses between donor and recipient populations that are candidates for genetic rescue should have a low risk of outbreeding depression. Donor and recipient populations should be substantially isolated from each other, but ideally have been connected to one another in their historical past

If genetic rescue is to succeed, the donor and recipient populations must be genetically differentiated by having been isolated for many generations, with the number of generations being dependent on the effective population sizes (Falconer & Mackay 1996). In fish, there is a positive correlation between the benefits of crossing between populations and the genetic distance between the populations, but there is an increasing likelihood of outbreeding depression as populations become more genetically differentiated (McClelland & Naish 2007). The safest approach is to use populations that were previously connected and isolated within the last 500 years by human activities (Chapter 10).

Care needs to be taken to avoid closely related populations, such as those with source–sink relationships, or those in a metapopulation where one population has been recently derived from the other, because the gene flow would provide little benefit. If there is sufficient information, the donor population with least kinship with the recipient population is ideal (Chapters 5 and 13), or the one with the most differentiation in allele frequencies for near neutral genetic markers.

Inbred versus outbred donor populations

Inbred or outbred donor populations can both result in genetic rescue, but crossing with an outbred population is preferable, because it reduces F by more in the F_2 and later generations, and so requires less frequent augmentations

The donor population can be outbred or inbred provided it differs genetically from the recipient population. However, the inbreeding coefficient in the F_2 cross will be lower if outbred immigrants are used (Table 6.5), and empirical results confirmed the expected benefit on fitness (Table 6.3). For example, beneficial effects of crossing were found in Mexican wolves when three inbred populations were combined, and in Florida panthers when outbred immigrant animals from Texas were used (Fredrickson et al. 2007; Johnson et al. 2010). Beneficial effects from crossing to inbred and outbred donor populations have also been reported in plants (Dudash et al. 1997; Finger et al. 2011; Frankham 2015).

Augmenting from one or several populations

When several acceptable populations are available for augmenting gene flow, the best long-term solution is to use as many populations as possible

When several populations exist, individuals for augmenting gene flow may come from a single population, from multiple populations, or from crosses among individuals from different populations. As we saw in the previous chapter, the decline in the inbreeding coefficient in the F_2 as a consequence of augmented gene flow will be greater if more populations are used and the subsequent rise in inbreeding with each subsequent generation will be slower. Further, genetic diversity and fitness are expected to be higher when more populations are used to augment gene flow, and using multiple sources makes more different alleles available as material for future adaptive evolution. For example, when two versus four equally inbred *Drosophila* populations (N_e = 25 for 50 generations) were crossed and allowed to mate randomly for 8–10 generations, crosses of four populations (combined in equal proportion) had lower inbreeding (0.16 vs 0.33), 25% higher allozyme heterozygosity, 23% higher "wild" fitness, and 18% greater evolutionary potential than the crosses of two populations (Margan et al. 1998).

Alternatives to crossing multiple populations at the same time are to use different populations as sources for successive augmentations of gene flow, as for example by progressively crossing between populations in a rotational scheme.

Rotational movement of individuals between fragments

A regular rotational pattern of moving individuals between populations over generations can maintain genetic diversity in fragmented animal and plant populations

If there are several small isolated populations, one option is to implement a management strategy of regular exchange of individuals between pairs of populations. This breeding strategy, illustrated in Box 12.2 for eight population fragments, utilizes a maximum avoidance of inbreeding scheme (in this example moving only males). In the first generation in Box 12.2, males are moved one population in a clockwise direction. In the second generation, males are moved two populations in a clockwise direction. In this example, inbreeding cannot be avoided after the third generation (Princée 1995).

Box 12.2 Low intensity genetic management for fragments using maximum avoidance of inbreeding

(Princée 1995)

Application of maximum avoidance of inbreeding (MAI) across fragments is illustrated with eight fragments in the succeeding figures. MAI strategies utilize a circular mating system to attempt to equalize reproduction among all individuals and delay inbreeding in a population for as long as possible. Boxes represent fragments and the numbers in the boxes show the lineages contributing to that box. Each fragment starts in generation 0 with its own lineage. Arrows indicate transfer of males from natal fragments to host fragments. In generation 1, males are moved one fragment in a clockwise direction so that, for example, at the start of generation 2, fragment 2 has lineages [21], fragment 3 has [32], etc.

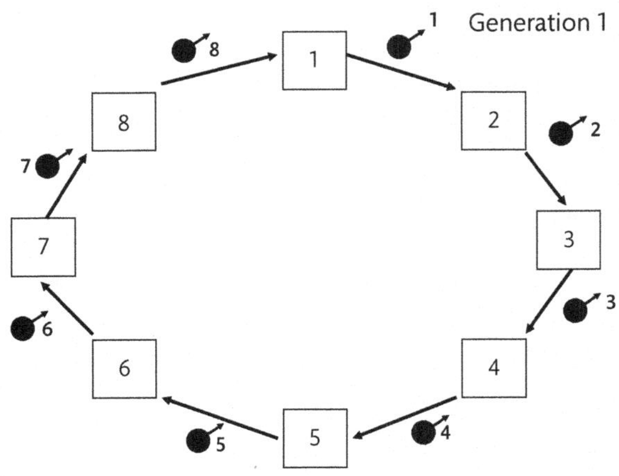

(after Princée 1995)

In generation 2 (below), male offspring are moved two fragments in a clockwise direction, as moving males by one fragment would result in inbreeding.

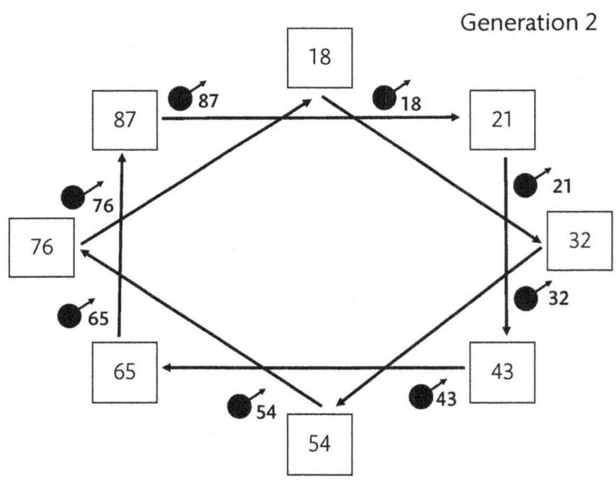

(after Princée 1995)

This results in fragments having a mix of lineages, fragment 1 with lineages [1876], fragment 2 with [2187], etc. In generation 3, males are moved four fragments in a clockwise direction, 1 to 5, 2 to 6, 3 to 7, and 4 to 8, again avoiding inbreeding.

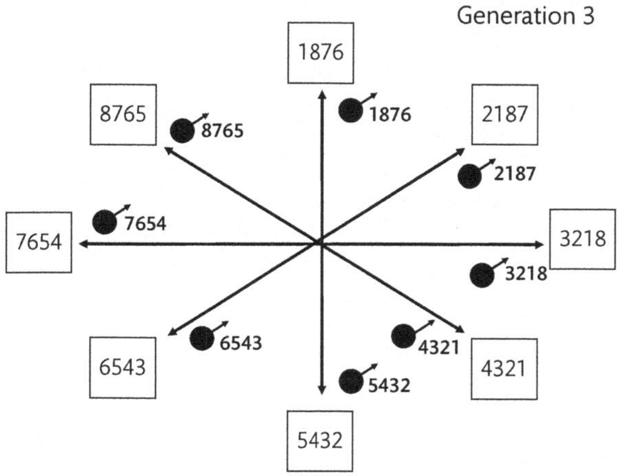

However, inbreeding can no longer be avoided after the third generation as all fragments in generation 4 have identical lineages [12345678]. In further generations, the pattern of exchange reverts to that of generation 1 and the cycle is repeated.

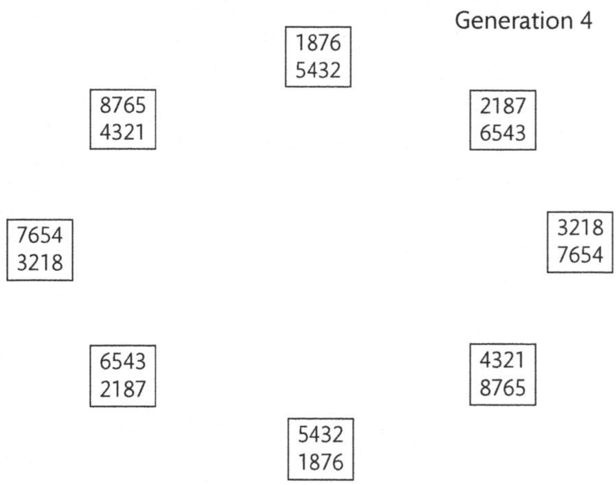

Detailed guidelines for applying this system to different numbers of fragments are provided by Princée (1995). Population fragments with unequal population sizes can be managed using an approximation of this strategy, or an advanced version where the number of individuals moved from each population is weighted according to population sizes (Wang 2004b). These schemes are similar to those used by indigenous peoples, including those of South America, to avoid inbreeding (Levi-Strauss 1973).

Partula snail (Tahiti)

We not aware of Princée (1995)'s group management scheme being used for populations in the wild, but it has been applied with captive populations of *Partula* snails. The procedures now being used are an advance on the scheme in Box 12.2 because they account for differences in group size and prior gene flow in a manner analogous to mean kinship strategies for individuals (Ballou & Lacy 1995). Using information on previous transfers of individuals between groups, splitting and merging of groups, observations on births and deaths within groups, and the breeding system within groups (e.g. sexual, selfing, or clonal), it is possible to estimate levels of genetic kinships among populations and manage to minimize it (Wang 2004b), as will be discussed in detail in the next chapter.

How many?

When the recipient population is inbred, any immigrants are better than none. The optimal level of gene flow depends on the goals of genetic management

Only a modest number of **contributing immigrants** is needed to cause meaningful improvements in fitness and genetic diversity; even single contributing immigrants typically produce worthwhile benefits if the recipient population is small and highly inbred (Spielman & Frankham 1992; Vilà et al. 2003; see Frankham 2015). A classic rule has been that a single effective migrant per generation is adequate for managing fragmented populations (Mills & Allendorf 1996). However, one effective migrant per generation is usually inadequate to avoid genetic problems and is not recommended as a general guide (Wang 2004a; Fernández et al. 2008; Sánchez-Molano et al. 2013).

The optimal number of migrants depends on whether the goal is to prevent fixation of alleles in fragments, to avoid damaging levels of inbreeding from accumulating, to reverse substantial existing inbreeding, or to provide enough variation to allow for adaptation to changing environments. If fewer immigrants than optimal are used, the duration of benefits is shorter.

Avoiding fixation of alleles

One migrant per generation has been recommended as adequate gene flow to avoid fixation of different alleles across fragments, but it is insufficient if the population structure is not ideal

Theory indicates that approximately one effective migrant per generation is sufficient to keep different alleles from becoming fixed in population fragments (Wright 1951). However, real populations do not have idealized structures, such that many more than one migrant is required per generation (up to 10; Lacy 1987; Mills & Allendorf 1996; Vucetich & Waite 2000) to achieve one genetically effective migrant (Wang 2004a). Since composite N_e/N ratios average ~ 0.13, and are diverse across species with different life histories (Frankham 1995; Palstra & Ruzzante 2008; Palstra & Fraser 2012; Waples et al. 2013), the average species needs $1/0.13 = 7.7$ adult immigrants per generation (assuming immigrants have equal reproductive fitness to residents—unlikely).

Additionally, most fragmented populations do not have island structures, so Wright's theory, on which the one effective migrant per generation rule depends, is often not applicable to real-world populations. In particular, metapopulations with cycles of extinction and recolonization or source–sink structures suffer much more serious genetic deterioration than theoretical island populations and require much higher effective levels of migration to achieve genetically healthy populations (Gilpin 1991; Pannell & Charlesworth 1999; Wang & Caballero 1999; Whitlock 2004). Further, prevention of fixation is rarely, if ever, the goal of genetic management, and more migrants are needed to meet other genetic management goals for small populations. To define the actual levels of immigration for any particular species, computer modeling using actual life history and demographic attributes is recommended, or otherwise to convert from N_e to N using N/N_e ratios for species with similar life histories (Waples et al. 2013).

Preventing damaging inbreeding

More than one effective migrant per generation is needed to prevent inbreeding from accumulating to damaging levels

Even under ideal conditions, a single effective immigrant per generation is not sufficient to prevent substantial inbreeding from accumulating over time, as it results in an equilibrium inbreeding coefficient = 0.2 due to population differentiation (Mills & Allendorf 1996). Thus, one effective migrant per generation would not keep the population below the recommended $F = 0.1$ level that should trigger further genetic augmentation.

The approximate relationship between the equilibrium F and immigration rates in an idealized island model is $F = 1/(4Nm + 1)$ (see Chapter 10; Wright 1969), in which Nm is the number of migrants for each sub-population of size N each generation. Thus, with $Nm = 1$, equilibrium $F = 0.20$. We can determine the effective number of migrants to keep F below 0.10 by rearranging this equation, as done in eqn 10.1. By substituting $F = 0.1$ we estimate that we need $Nm > 9/4$. Using the average N_e/N ratio from above to convert this from N_e to N yields at least 17.3 immigrants per generation.

Since real populations do not have ideal structures with random gene flow among an infinite number of fragments, **we recommend that typically 5 effective migrants per generation** (and at least 38 actual immigrants for an average species, based on an average N_e/N ratio of 0.13) are needed for each sub-population to prevent potentially damaging accumulation of inbreeding. Such numbers of migrants per generation might be easily achieved for large plant populations (e.g. with $N > 1{,}000$), but represent a more substantial proportion of the population and can be much more challenging to achieve for small population fragments (e.g. $N < 100$).

The above refers to gene flow every generation, but how much gene flow is needed to immediately reverse existing inbreeding?

Reversing existing inbreeding

Genetic rescue of populations already suffering from substantial inbreeding can require many immigrants

The above calculations apply to the equilibrium conditions that are achieved after many generations of gene flow (100s) among fragments. However, if a population is currently threatened by inbreeding that accumulated during past generations of isolation, we might wish to determine the number of immigrants introduced in just one or a few generations that are needed to quickly restore genetic health and population viability. If we know the approximate level of inbreeding in a population, we can use eqn 11.3 to estimate the number of unrelated migrants needed to reduce that inbreeding with a single generation of augmentation. From eqn 11.3:

$$F_{\text{Pooled}} = f_{\text{I}}^2 F_{\text{I}} + f_{\text{O}}^2 F_{\text{O}} \qquad 12.1$$

where F_{Pooled} is the inbreeding coefficient in the pooled population after migration and random breeding (typically in the F_2), f_{I} is the proportion of the pooled population derived from the inbred population, and f_{O} is the proportion from the unrelated migrants ($f_{\text{I}} + f_{\text{O}} = 1$). F_{I} and F_{O} are the inbreeding levels in the inbred and outbred (unrelated) populations, so $F_{\text{O}} = 0$. Then the proportion of the pooled population derived from unrelated migrants is:

$$f_{\text{O}} = 1 - \sqrt{F_{\text{Pooled}}/F_{\text{I}}} \qquad 12.2$$

This relationship is shown in Fig. 12.3 when the F_{Pooled} is set at 0.1 as the goal for genetic rescue. For example, if a population fragment has $N = 50$ breeding adults and an F of 0.2, it takes 15 non-inbred immigrants to reduce its F to 0.1 (Example 12.2).

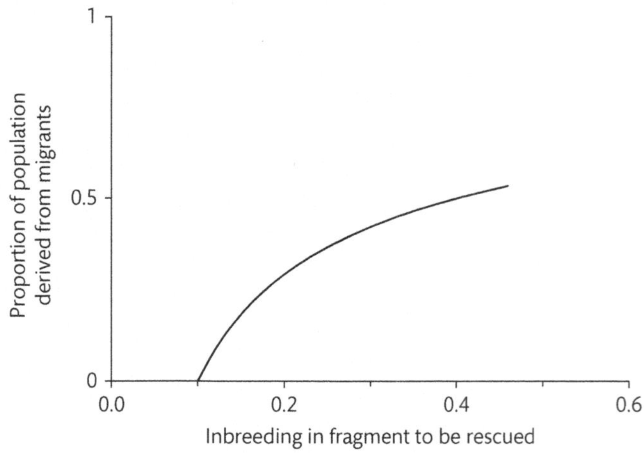

Fig. 12.3 The proportion of non-inbred, unrelated migrants needed to bring levels of inbreeding down to $F_{\text{Pooled}} = 0.10$ in fragments with inbreeding levels from 0.1 and higher.

Example 12.2 Calculating the number of migrants needed to reduce inbreeding

A fragment of size 50 but with an inbreeding level of $F_I = 0.2$ should be augmented with non-inbred, unrelated migrants. How many migrants are needed to reduce the level of inbreeding to $F_{Pooled} = 0.1$? Using eqn 12.2

$$f_O = 1 - \sqrt{{}^{0.1}/_{0.2}} = 0.29$$

Thus, 29% of the pooled population needs to derive from migrants, or about 29% of 50 (~ 15 immigrants that breed).

Numbers of immigrants will need to be higher if the source population is partly inbred or related to the recipient population. These calculations also highlight the value of beginning a genetic rescue before an isolated population has become so inbred that a substantial augmentation with unrelated immigrants will be needed to bring it back to genetic health.

Providing variation for adaptation

The migration rate necessary to maintain adequate variation for adaptive evolution might be approximately the same as the rate needed to avoid possibly damaging inbreeding, but this will depend on the rate of environmental change

The rate of response to selection is proportional to additive genetic variation, and approximately proportional to heterozygosity (see Chapters 4 and 6, especially eqns 4.5 and 6.1). Thus, a 10% increase in F will translate to about a 10% slower rate of adaptation to new selective pressures. We have not defined a goal for evolutionary rescue, but as a starting point, if we assume that more than a 10% decline in the rate of adaptation could place a population at risk, then the gene flow required to sustain adaptive potential will be approximately the same as was calculated above for the migration necessary to avoid damaging inbreeding. However, under global climate change many species will need to adapt more rapidly than ever before (see Chapter 14). Thus, the numbers of migrants will probably need to be greater and source populations need to have genetic variation conferring adaptations to the new conditions (Chapter 14).

Will genetic rescue be required again in the future?

If populations remain small and isolated after genetic rescue they will again become inbred, and need additional rounds of rescue in the future

Unless connectivity (natural gene flow) can be restored, genetic rescue may have to be undertaken multiple times in the future to ensure the survival of genetically isolated population fragments. For example, the Illinois population of greater prairie chickens was genetically rescued, but is again becoming inbred (Bouzat et al. 2009). Similarly, natural rescue of the Isle Royale gray wolves occurred through immigration of a single male from the mainland. This reduced the inbreeding from 0.81 to 0.09 (or 0.33 to ~ 0.03 using alternative estimates) within a generation, but the population rapidly became inbred again, and after another 2.5 generations F had risen to 0.22, and the population has subsequently collapsed (Adams et al. 2011; Hedrick et al. 2014; Mlot 2016).

The frequency of translocations between donor and inbred recipient populations depends on the gene flow rate, the effective size of the donor population, and whether the donor population is outbred or inbred. Lower immigration rates, smaller N_e, and inbred donors mean that more frequent translocations are needed

Given the number of variables, we recommend that computer modeling be used to predict the number of generations until augmentation of gene flow is needed. However, we failed to find an example where this had been done. Using our prior guidelines, we recommend another augmentation when $F \geq 0.10$ and that gene augmentation should continue until population F goes below 0.10.

Other things being equal, the use of several donor populations should extend the time before another genetic rescue attempt is required, but may limit future options for genetic rescue. Use of different donor populations in successive genetic rescue attempts is desirable, if possible, especially if the donor populations are inbred. Benefits of crossing for both fitness and evolutionary potential are likely to incrementally decline with successive genetic rescue attempts using the same donor populations, as the genetic divergence between the two populations will typically become less and less (Hedrick & Fredrickson 2010). An alternative approach is to introduce a few immigrants every generation to provide continual connectivity between populations (see Chapters 6 and 13). Similar genetic rescue benefits can be obtained by different combinations of low gene flow rates per generation and frequent translocations versus high immigration rates and less frequent translocations across generations.

In general, levels of inbreeding or genetic diversity should be monitored in populations with augmented gene flow, via genetic markers, pedigrees, or modeling (Chapter 13). The benefits might be less or more if there is selection against or in favor of individuals carrying the donor alleles. Selection favoring immigrant alleles is more likely if the donor individuals are outbred, and the recipients are inbred (Chapter 6).

Which individuals?

The best individuals to transfer from the donor to the recipient populations would be minimally inbred, unrelated to other individuals being transferred, taken from across the donor population's range, disease free, and of the sex and age to maximize the likelihood of successful genetic augmentation

The approach one chooses to decide which individuals to transfer from the donor population to the recipient depends on the depth of prior genetic knowledge and the known basic biology and ecology of the species involved. If there is no genetic information on individuals in the donor population, then individuals can be chosen at random and from across the range of the population to minimize the chance of selecting close relatives and to sample across the genetic structure of the population, if there is any. This approach is likely to achieve much of the potential benefit obtainable from a more refined analysis.

Sex of donors

Either sex (or both) can be used to rescue inbred populations, but one sex may be preferred for genetic, ecological, demographic, or behavioral reasons

Augmenting an inbred population with either females or males, or both can reverse inbreeding and loss of genetic diversity. However, there may be fundamental reasons to prefer one sex, including which sex will be more readily accepted into the social environment, and logistical considerations such as the availability of individuals of a given sex in the source population. But the decisions over sex ratio of migrants are complex (Havird et al. 2016). As a starting position, introduction of both sexes may typically be sensible, since most species have sex-specific genetic and epigenetic variation. But it may be easier to introduce one sex than the other. Transferring females from a large, genetically diverse population may alleviate inherited problems in mtDNA (in animals and plants) or cpDNA problems in plants (with maternal inheritance). While introduction of very divergent mitochondrial DNA (typically via females in animals) may be inappropriate in its own right, or drive fitness-harming mitochondrial-nuclear incompatibilities (Gemmell et al. 2004, Havird et al. 2016), such issues should be minimized by our requirement for populations to be adapted to similar environments.

Age and life stage of donors

Sex and age of the transferred individuals are also relevant to ecology, demography, and potential behavioral or incompatibility problems. For example, young animals have greater lifetime reproductive potential and may be better accepted by resident animals. Aggressive encounters between resident and introduced males may be a problem in many animal species. Dispersal of introduced animals from the release site and attempts to return to the capture site can be a severe problem, and the probability of such homing behavior may differ by age, life stage, or sex (Rathbun et al. 2000). Considering the age of first reproduction in long-lived plants and self-incompatibility in plants can further enhance successful donor integrations. Guidelines that consider other aspects of reintroductions and translocations are provided by IUCN/SSC (2013).

Genetic swamping

Genetic swamping is loss of alleles or genotypes (especially locally adapted ones) through augmentation, using high rates of gene flow

While low levels of gene flow into inbred populations are often beneficial, very high levels of gene flow can lead to dilution of local genetic composition to the point where its unique attributes are lost (Lenormand 2002). This may be important if locally adapted alleles and genotypes are lost. For example, exotic rainbow trout introduced into the habitats of native cutthroat trout in the western USA led to hybrid swarms that eventually overwhelmed the local species (Kovach et al. 2015). Related problems have also been reported in canids, ducks, and plants (Rhymer & Simberloff 1996; Randi 2008; Todesco et al. 2016). The issue also occurs within species. For example, release of large numbers of hatchery adapted fish typically has harmful effects on the fitness of wild fish (Ryman et al. 1995; Lamaze et al. 2012).

Genetic swamping should not be an issue in the current context for two reasons. First, we require that donor and recipient population fragments be adapted to similar environments, so differential adaptation is minimal. Second, cost constraints in genetic management of fragmented populations of threatened species will typically lead to management with minimum acceptable rates of gene flow, rather than high rates of gene flow that could result in genetic swamping. However, in species where the cost of augmenting gene flow is low (especially plants, invertebrates, and fish), care needs to be taken to err towards the minimum required gene flow for genetic rescue (as defined in this chapter and the next).

More precise genetic management with greater conservation benefits is possible when we have more information on the degree to which populations and/or individuals are related, as detailed in Chapter 13.

Summary

1. Sound management strategies for gene flow into populations suffering genetic erosion can be instituted without detailed genetic data, although additional information and/or modeling can provide information for more precise and effective management.
2. The overall objective of genetic management is to reduce the risk of population extinction by decreasing inbreeding and enhancing genetic diversity.
3. When limited information is available, the default should be to initiate genetic augmentation based on general principles rather than doing nothing. If no management action is taken, the genetic condition of the isolated population will continue to deteriorate, increasing its risk of extinction.
4. Any appropriate gene flow into an isolated inbred population is better than none. More than one effective migrant per generation is needed to prevent accumulation of genetic problems. We recommend at least five per generation to avoid accumulation of damaging levels of inbreeding and increased vulnerability due to reduced adaptive potential.
5. Higher levels of gene flow will be required to reverse substantial existing inbreeding. Thus, waiting for inbreeding to become severe before restoring gene flow is ill-advised because it will require more aggressive action later.

6. The best individuals to transfer from the donor to the recipient populations would be minimally inbred, unrelated to other individuals being transferred, taken from across the donor population's range, disease free, and be of the sex and age to minimize behavioral problems and maximize the chances of incorporating the translocated individuals' genetic diversity into the recipient population.
7. When several acceptable source populations are available to augment gene flow, use of individuals from as many populations as possible is recommended, either all at once, or over several sequential augmentations.

FURTHER READING

Ewen et al. (2012) *Reintroduction Biology*: an edited book on reintroduction, see especially Chapters 11, 12, and 13 on genetic issues.

IUCN/SSC (2013) *Guidelines for Reintroductions and Other Conservation Translocations*: provides useful guidelines for translocations.

Princée (1995) Proposes a group management scheme for populations without pedigrees.

Todesco et al. (2016) Recent review of genetic swamping.

Wang (2004b) Proposes an improved method for managing genetic variation in group breeding populations without individual pedigrees.

SOFTWARE

NEWGARDEN: a program to model the growth and population genetics of plant populations developing from different founding and life history conditions (Pelikan and Rogstad 2013). https://math.uc.edu/~pelikan/NEWGARDEN/

VORTEX: population viability analysis software that tracks the predicted heterozygosity and inbreeding coefficients of individuals, and thus can be used in management of gene flow when there is limited genetic information (Lacy & Pollak 2014). www.vortex10.org/

CHAPTER 13

Managing gene flow among isolated population fragments. II. Management based on kinship

TERMS

Coancestry, kinship coefficient, mean kinship, kinship matrix

With more detailed genetic information, more precise genetic management of fragmented populations can be achieved, leading to improved retention of genetic diversity and lower inbreeding. Using mean kinship within and between populations (estimated from modeling, pedigrees, genetic markers, or genomes), and moving individuals among fragments with the lowest between fragment mean kinships provides the best means for gene flow management. Populations should then be monitored to confirm that movement of individuals has resulted in the desired levels of gene flow, and that genetic diversity has been enhanced.

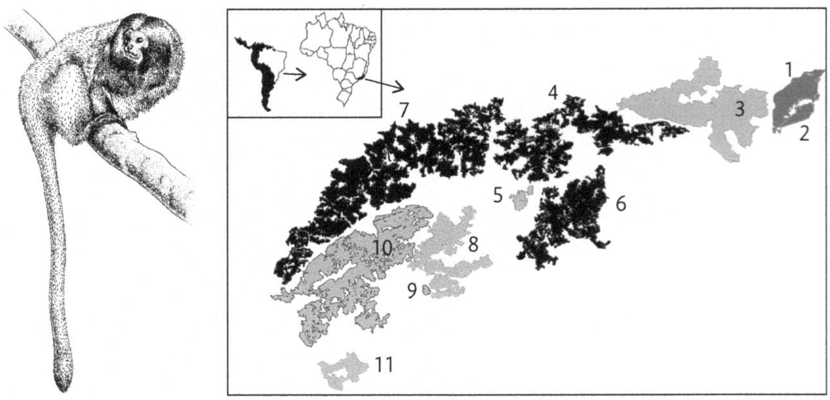

The golden lion tamarin's fragmented populations in Brazil. Fragments 4, 6, and 7 (shaded black) were the original wild populations. The lightly shaded fragments (3, 5, 8, 9, 10, and 11) were founded and managed with gene flow from the captive population. The mid-shaded fragments in the upper right (1 and 2) were formed by translocating individuals from several extremely small vulnerable wild fragments. (Image provided by the Associação Mico-Leão-Dourado)

Genetic Management of Fragmented Animal and Plant Populations. R. Frankham, J. D. Ballou, K. Ralls, M. D. B. Eldridge, M. R. Dudash, C. B. Fenster, R. C. Lacy & P. Sunnucks. Oxford University Press (2017) © R. Frankham, J. D. Ballou, K. Ralls, M. D. B. Eldridge, M. R. Dudash, C. B. Fenster, R. C. Lacy & P. Sunnucks 2017. DOI 10.1093/oso/9780198783398.001.0001

How does detailed genetic information aid management of gene flow?

Data on the genetic diversity and structure of population fragments, ideally complemented with their demographic history, allow for more precise genetic management of gene flow, leading to higher genetic diversity and lower inbreeding than possible with management based on less information

Why should we go to the extra expense of obtaining detailed genetic or genomic information on fragmented populations? What are the management benefits that can be obtained from detailed genetic information? What genetic parameter should be managed? How should fragmented populations be managed at the population level, and at the individual level?

This chapter addresses management of gene flow between populations with increasing levels of genetic information (Fig. 13.1). We address the issue of what genetic parameter to use and show that the widely used F_{ST} is unsuitable for management purposes. We present ways of using pedigrees, and/or molecular genetic and genomic analyses to manage gene flow more precisely. Box 13.1 provides a case study of genetic management of population fragments of the golden lion tamarin, including wild, translocated, captive, and reintroduced populations.

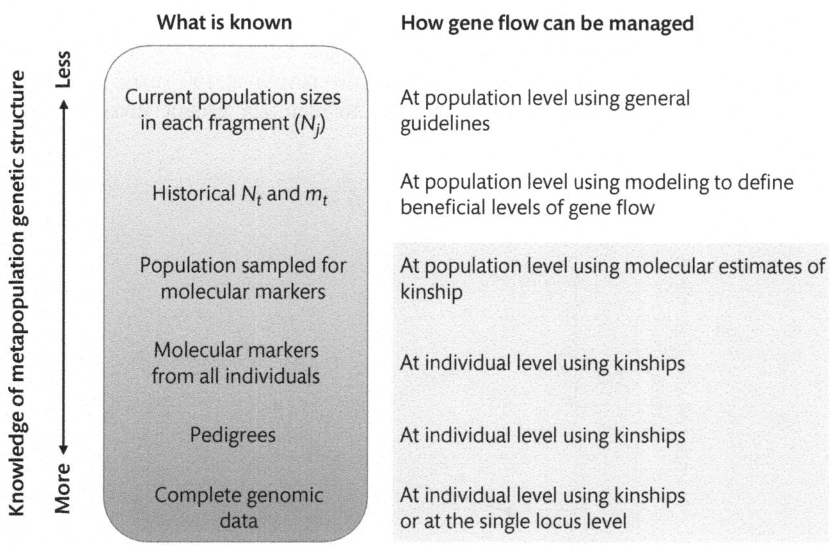

Fig. 13.1 Types of gene flow management possible with different levels of information (the tinted shading on the right indicates topics addressed in this chapter).

Box 13.1 Genetic management of wild, captive, and reintroduced populations of golden lion tamarins

(Mickelberg 2011; Dietz 2016, pers. comm.)

Golden lion tamarins are small, arboreal, monogamous primates from Brazil (see chapter frontispiece). This species declined in abundance and became endangered following habitat reduction in the Atlantic Rainforest to less than 2% of its original area, along with fragmentation. The Golden Lion Tamarin Conservation Program was developed as a collaboration including the Smithsonian National Zoological Park, the Golden Lion Tamarin Conservation Association, and the Brazilian Government. The program integrates captive breeding, reintroductions, translocations, and studies on the ecology of wild tamarins, habitat restoration, community education, and outreach (Kleiman & Rylands 2002).

The wild population has recovered from < 150 to ~ 3,200 since 1970 as a result of this program (figure). The individuals are now dispersed among 11 isolated or nearly isolated population fragments (see frontispiece map). The largest habitat fragment (7) contains ~ 1,100 tamarins. The second-largest (10) contains ~ 500 tamarins, all descendants of reintroduced zoo-born tamarins. Translocations of animals from the most severely threatened populations were used to establish a new protected population (fragments 1 and 2), now numbering about 130 in total. Census information is used to design a program of regular translocations among fragments to minimize inbreeding and maximize the effective size of the entire population. Further, habitat corridors have been planted to reconnect the largest forest fragments. Once reconnected, the resulting block of forest should meet conservation targets of near-zero probability of extinction risk and 98% retention of genetic diversity over 100 years. At that point, translocations will be necessary only to maintain genetic diversity in small unconnected fragments.

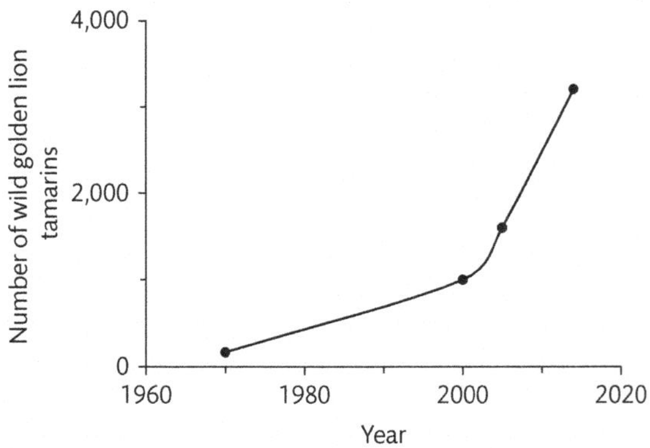

Genetic management of reintroduction

The reintroduction program was initiated in 1983 to re-establish populations in areas of their former range. A total of 153 tamarins (from 43 captive breeding institutions) were released between 1984 and 2001 and the reintroduced population has flourished, currently numbering over 750. Individuals were primarily selected for reintroduction based on the goal of transferring as much captive population genetic diversity as possible to the wild (see Fig. 13.3). Use of mean kinship to select animals for reintroduction resulted in transfer of most of the captive population's founder genetic diversity to the wild population (Mickelberg 2011), as shown in the following figure. Most of the captive founders are now represented in the reintroduced population, and no further introductions are planned.

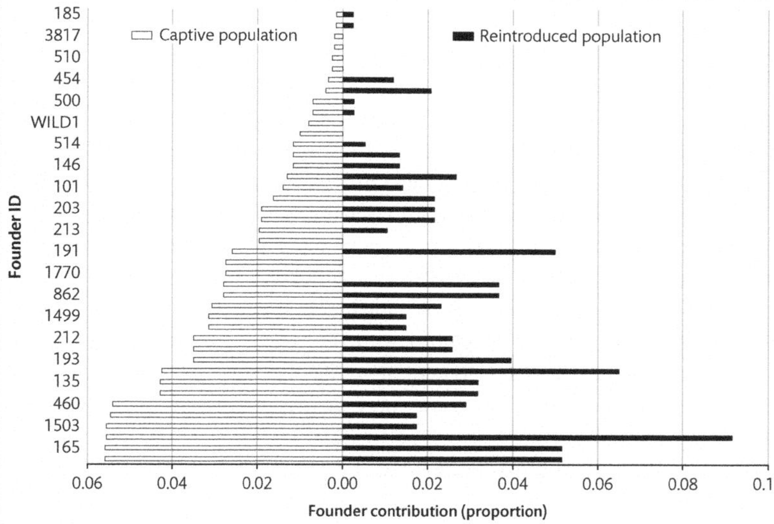

Before proceeding, we must first define our objectives.

What are our objectives?

Our objectives are to maximize retention of genetic diversity within and across populations and to minimize inbreeding (in natural outbreeders), thereby maximizing the persistence of population fragments and species

We are seeking to augment gene flow between population fragments such that we maximize heterozygosity, conserve allelic diversity, and minimize inbreeding. Fortunately, these objectives can largely be achieved simultaneously.

Before we can address genetic management, we must first determine what genetic parameter we need to manage.

What population genetic parameter should we manage?

While many measures of population differentiation exist, mean kinship is superior for managing genetic diversity within populations and fragments, as well as determining the direction and extent of gene flow between fragments

Which of the metrics of population differentiation presented in Chapter 5 are best for management of gene flow? F_{ST} (and related measures) are widely used to describe population structure, and Ottewell et al. (2016) recommended that it be used in plant conservation management along with heterozygosity and F_{IS}. However, F_{ST} and related parameters are not optimal for managing fragmented populations, because their values are extremely sensitive to levels of genetic diversity in the populations, rather than reliably reflecting the relatedness of individuals among the populations (Hedrick 1999; Jost 2008; Box 13.2). Populations with the highest F_{ST} to a target managed population will usually have low genetic diversity themselves (Chapters 5 and 10): a more sensible choice would be to use more genetically diverse populations as the source for genetic rescue.

Minimizing mean kinship has been shown by computer simulations and analytical studies to be better than alternative means for managing pedigreed populations to maximize retention of genetic diversity (Ballou & Lacy 1995), and its benefits have been verified experimentally in *Drosophila* (Montgomery et al. 1997). It is also highly effective in conserving allelic diversity (Fernández et al. 2004) and minimizing inbreeding (Ballou & Lacy 1995). It is the recommended procedure for managing threatened species in captivity, and is also used in conservation of domesticated livestock breeds (Falconer & Mackay 1996; Ballou et al. 2010; Frankham et al. 2010). While the minimizing kinship strategy was devised for pedigreed populations, kinships can also be inferred from genetic marker information for individuals and populations, as we will detail later (Russello & Amato 2004; Fernández & Toro 2006; Dasmahapatra et al. 2008).

How well does managing using kinship perform in practice, compared to F_{ST}? Box 13.2 provides evidence that illustrates minimizing mean kinship is superior to other measures for managing puma populations, especially when compared with F_{ST}.

Box 13.2 Comparing *mk* and F_{ST}: Which puma population would be best for genetically rescuing Florida panthers?

Is managing by mean kinship superior to managing using F_{ST}? We used Culver et al.'s (2000) data on allele frequencies at 10 microsatellite loci in six regional populations of puma (*Puma concolor*) across the Americas to identify the optimal source population for restoring genetic diversity to the depleted Florida panther population.

Levels of genetic diversity in the remnant Florida (FL) population and Culver's six regional populations that might be considered as possible sources of immigrants are:

Region	# Alleles/locus	Heterozygosity
FL (Florida)	1.2	0.04
NA (North America)	6.4	0.55
CA (Central America)	5.4	0.67
NSA (Northern S America)	8.7	0.80
ESA (Eastern S America)	8.3	0.76
CSA (Central S America)	6.7	0.76
SSA (Southern S America)	5.8	0.67

Thus, the Florida population is very low in heterozygosity and alleles/locus, reflecting its history of small size and inbreeding and was clearly in need of genetic rescue.

Reduction of inbreeding will be optimized by choosing the source population with the lowest mean kinship to the inbred population, because the inbreeding coefficient produced by crossing individuals at random between two populations is equal to the mean kinship between the populations. The following matrix contains the pairwise differences between populations measured as $mk_{between}$ (below diagonal, high values indicate low divergence) and F_{ST} (above diagonal, high values indicate high divergence). The final row of the table shows the heterozygosity of the FL population prior to genetic rescue and the heterozygosities that would result if 20% of the FL population were to be replaced by immigrants from each of the other populations.

	FL	NA	CA	NSA	ESA	CSA	SSA
FL		**0.245**	0.338	0.327	0.270	0.332	0.350
NA	0.452		0.087	0.113	0.089	0.133	0.163
CA	0.234	0.275		0.075	0.076	0.092	0.116
NSA	0.156	0.161	0.144		0.089	0.133	0.163
ESA	0.278	0.217	0.164	0.159		0.035	0.085
CSA	0.184	**0.149**	0.142	0.158	0.184		0.044
SSA	0.212	0.152	0.156	0.126	0.151	0.219	
Resulting H	0.043	0.225	0.299	**0.330**	0.289	0.319	0.307

Thus, to restore genetic diversity to the Florida population, *mk* identifies NSA as the best source. NSA is the most genetically diverse population (first table in Box) as well as being most divergent from FL (*mk* = 0.156, second table in Box: shaded). A population with 20% of its genes from NSA and the rest from FL results in higher heterozygosity than any of the alternative augmentations, with an increase from 0.043 to 0.330 (bold). By contrast, F_{ST} identifies SSA (F_{ST} = 0.35, shaded) as the most divergent population and thus the best source of immigrants, but SSA has lower heterozygosity than do the other South American populations (part of the reason why it has a higher F_{ST} to FL). Introducing animals from SSA to FL would not increase heterozygosity as much as introducing them from NSA.

We also considered a second hypothetical case: which puma population would be the best for genetically rescuing the NA population, if such a genetic augmentation were desired? *Mk* identifies CSA as the best source population (bold). Conversely, F_{ST} pinpoints FL as the source for improving diversity in NA (bold), even though FL is highly inbred and in need of rescue itself, and contains no alleles not also found in NA. Indeed, F_{ST} erroneously indicates that the highly inbred FL population would be the best choice for genetic supplementation of any of the five other populations.

Other metrics of divergence discussed in Chapter 5 partly correct the deficiencies of F_{ST} as a measure of divergence and often lead to better choices for source populations than does F_{ST}. For example, G'_{ST} (not shown) points to NSA as the best source for FL, but selects SSA as a source for supplementation of NA, and still selects FL as the preferred source for many of the other populations. Neither F_{ST}, G'_{ST}, nor any of the other tested metrics (Nei's [1987] D_N—standardized genetic distance, *I*—normalized identity of genes, D_{ST}—numerator for G_{ST}, and D_M—absolute genetic differentiation) consistently perform as well as *mk* in identifying source populations that would most effectively reverse inbreeding and increase gene diversity.

This is a hypothetical example as only genetic diversity has been considered, whereas our approach involves a prior screen against gene flow between differentially adapted population fragments that may affect practical management of these populations.

So what is kinship and how do we manage fragmented populations using it?

How do we manage using mean kinship?

Genetic management that minimizes mean kinship retains maximum genetic diversity within the population

To preserve genetic diversity in captive populations, breeding programs prioritize the breeding of individuals with low mean kinships to the current population (Ballou & Lacy 1995). Similarly, when managing wild or captive populations, an immigrant from the source population with the lowest mean kinship to the genetically depleted population will minimize inbreeding of the offspring. If the kinships among populations are recalculated with each transfer of migrants (e.g. each year), then a new preferred source population might be identified if prior immigrants successfully reproduced and increased the kinship between the recipient and previous source. Thus, dynamic management with constantly updated kinships results in maximal restoration and retention of gene diversity over generations, and is the preferred approach.

We now more precisely define kinship within and between populations. The kinship (or coancestry) (k_{ij}) between individuals *i* and *j* is the probability that two alleles at a locus taken at random, one from each individual, are identical by descent (Wright 1969; Frankham et al. 2010). The mean kinship of an individual *i* (mk_i) is the mean of all kinships between that individual and all living individuals in the population, including itself:

$$mk_i = \sum_{j=1}^{N} k_{ij} \Big/ N \qquad 13.1$$

where k_{ij} = kinship between individuals *i* and *j*, and N = the number of individuals in the population.

The mean kinship of the population is the average of all pairwise kinships between individuals in the population or, equivalently, the average of all individuals' mean kinships:

$$\overline{mk} = \frac{\sum_{i=1}^{N} mk_i}{N} \qquad 13.2$$

Example 13.1 illustrates estimation of kinship and mean kinship from pedigrees for a hypothetical example with two populations.

Example 13.1 Calculating kinships in two hypothetical population fragments

Consider an original group of six unrelated individuals (top row of circles) that are paired and bred and their descendants (bottom row of individuals) then placed into two fragments (A and B). Individuals B_2 and B_3 are inbred ($F = 0.25$, because their parents are a brother/sister pair).

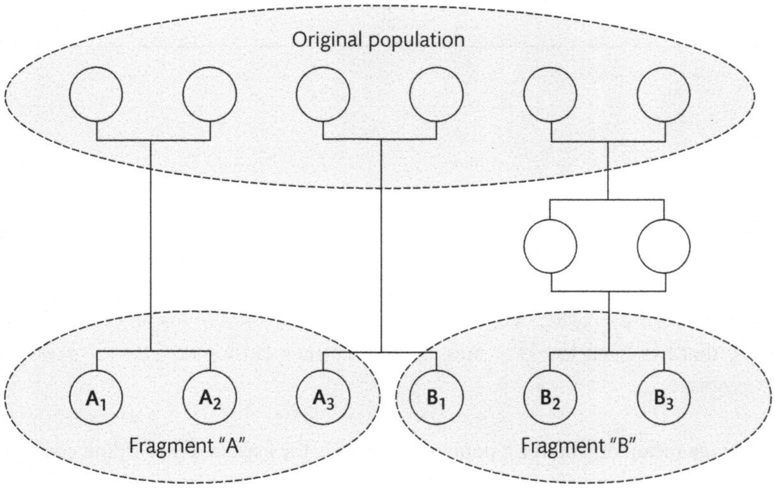

Kinship matrix

The kinship matrix shows the kinships, based on the pedigree above, of all possible pairwise combinations (see Ballou [1983] and Frankham et al. [2010] for methods of calculating kinships from pedigrees). Note that the inbred individuals have self kinships that are greater than those for non-inbred individuals (0.625 vs 0.50). Within fragment kinships are shown in light shading.

Individuals

Individuals	A_1	A_2	A_3	B_1	B_2	B_3
A_1	0.5	0.25	0	0	0	0
A_2	0.25	0.5	0	0	0	0
A_3	0	0	0.5	0.25	0	0
B_1	0	0	0.25	0.5	0	0
B_2	0	0	0	0	0.625	0.375
B_3	0	0	0	0	0.375	0.625

Calculating within population mean kinships for population A (mk_{AA}) and B (mk_{BB}), based on eqn 13.2 we obtain:

$$mk_{AA} = \frac{0.5 + 0.5 + 0.5 + 0.25 + 0.25}{9} = 0.22$$

$$mk_{BB} = \frac{0.5 + 0.625 + 0.625 + 0.375 + 0.375}{9} = 0.28$$

Relationships among mean kinship, inbreeding, and genetic diversity

Average mean kinship within a population is a direct measure of the amount of genetic diversity that has been lost over time, and minimizing kinship equates to maximizing heterozygosity

The average mean kinship for a population ($\overline{mk}$) is the expected inbreeding coefficient in the next generation with random mating. Consequently, the average mean kinship is directly related to the proportion of genetic diversity lost since time 0 ($1 - [H_{t+1}/H_0]$), as follows (Ballou & Lacy 1995):

$$\overline{mk} = \overline{F}_{t+1} = 1 - \frac{H_{t+1}}{H_0} \qquad 13.3$$

Consequently, if kinship is minimized, inbreeding in the next generation is minimized, and proportional heterozygosity retained is maximized.

Kinship and relatedness are often confused and used interchangeably. However, relatedness (r_e) is directly proportional to kinship ($r_e = 2mk_i$) only when there is no inbreeding in the population (Wright 1922; Crow & Kimura 1970). Since we are concerned with small inbred population fragments, r_e is not an appropriate measure to use for quantifying relationships within and between the populations.

Thus, managing by minimizing mean kinship is superior to other methods for retaining genetic diversity.

Mean kinship between populations

Mean kinship between populations measures the genetic similarity of two different populations (1 − divergence)

The mean kinship between populations *A* and *B* (mk_{AB}) is the average of the pairwise kinships of all individuals in population *A* with all individuals in population *B*:

$$mk_{AB} = \left(\sum_{i=1}^{N_A} \sum_{j=1}^{N_B} k_{ij}\right) / (N_A N_B) \qquad 13.4$$

where k_{ij} is the kinship of individual *i* in population *A* (with size N_A) with individual *j* in population *B* (with size N_B). $N_A N_B$ represents the total number of pairwise kinships. In Example 13.1, the kinship between populations A and B, based on data in the nine unshaded cells in the upper right of the kinship matrix, is:

$$mk_{AB} = \frac{(0.25 + 0 + 0 + 0 + 0 + 0 + 0 + 0 + 0)}{9} = 0.028$$

(Note that $mk_{AB} = mk_{BA}$.) Later we present other methods for estimating kinships without known pedigree by using genetic marker data. However, the concepts of within and between population kinships are the same regardless of how the kinships are calculated.

Genetic structure of fragmented populations can be characterized as a matrix of within and between population mean kinships

We can characterize the genetic structure of a fragmented population as a matrix consisting of mean kinships within populations on the diagonal and mean kinships between populations off-diagonal, as shown in Fig. 13.2.

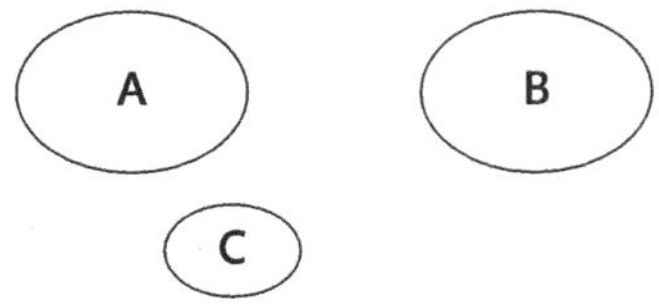

Population

Population		A	B	C
	A	MK_{AA}	MK_{AB}	MK_{AC}
	B	MK_{BA}	MK_{BB}	MK_{BC}
	C	MK_{CA}	MK_{CB}	MK_{CC}

Population

Population		A	B	C
	A	0.03	0.005	0.02
	B	0.005	0.02	0.01
	C	0.02	0.01	0.13

Fig. 13.2 The population mean kinship matrix for three hypothetical population fragments, illustrating within and between population kinships. Fragment C is smaller and more inbred than A and B. Within population kinships are shaded.

Restoring genetic diversity by gene flow based on minimizing mean kinship

Management of gene flow should be based on estimates of the mean kinships within and between individuals in different fragments, with gene flow directed into fragments with high mean kinships from fragments with low mean kinships with that population

In much the same way that kinships can recognize individuals of greater or lesser genetic value, mean kinships within and between population fragments can be used to identify those in need of additional genetic variation, and pinpoint the best sources of individuals for augmentation. Fragments in need of genetic rescue have high within population *mk*. Populations with individuals on average least related to that population are the most genetically valuable source of rescue, and will have the lowest between population *mk* with that fragment (Fig. 13.2), e.g. fragment C has high mean kinship ($mk_{CC} = 0.13$, mean kinship closer than half-sibs) and is likely in need of augmentation. Fragment C has a lower mean kinship with B ($mk_{BC} = 0.01$) than with A ($mk_{AC} = 0.02$) and thus B is a more suitable source of genetic rescuers for C.

How these mean kinships are estimated depends on the information available, as explained in what follows.

Population mean kinship

Population mean kinship can be estimated from historical population size and gene flow, from pedigrees, or from molecular markers.

Population mean kinships from knowledge of historical population sizes and migration

When detailed historical demographic information is available on fragments or individuals, computer simulations can be used to estimate kinships between and within fragments to guide management of gene flow more precisely

VORTEX software (Lacy & Pollak 2014) can be used to model kinship structure within and between populations and can model the effectiveness of using kinships to guide management of gene flow.

For example, genetic management of the Allegheny woodrats in Example 12.1 could be improved from random translocations of individuals among fragments, to more detailed management where the fragments with least diversity receive more immigrants from fragments with low mean kinships to the recipients and typically high genetic diversity.

Population mean kinships from molecular markers

Within and between population mean kinships can be estimated through molecular analyses of individuals sampled from all fragments

Molecular genetic data (microsatellites or especially SNPs) can provide more precise information on kinship structure (Fig. 13.1). A random selection of individuals needs to be sampled from each population or fragment and mean kinships calculated (Hoban et al. 2013; see Example 13.2). We consider sample size and power issues later.

Example 13.2 Calculating population mean kinships from allele frequency data

To illustrate the general approach for calculating kinships from allele frequencies, we will show the basic calculation for estimating the average probability that alleles that are shared between populations, ignoring, for the time-being, whether alleles are identical by descent or not. This measure of kinship is the "gene identity" (Jxy) that has long been used in population genetics as a measure of population similarity (Nei 1987):

$$J_{xy} = \sum (p_{xi}\, p_{yi})$$

with the sum being across the product of the frequencies of each allele in populations X and Y. In general, *mk*, the probability of shared alleles that are identical by descent, will be proportional to J_{XY}.

The following hypothetical example illustrates the method for calculating J_{XY} between four populations, based on allele frequencies (p_i):

	p_1	p_2	p_3	p_4
Population A	0.25	0.25	0.25	0.25
Population B	0.5	0.5	0	0
Population C	0	0.5	0.5	0
Population D	0	0	0.5	0.5

So, for example

$$J_{AA} = 4\ (0.25 \times 0.25) = 0.25$$

$$J_{AB} = (0.25 \times 0.5) + (0.25 \times 0.5) + (0.25 \times 0) + (0.25 \times 0) = 0.25$$

After calculating J_{XY} for each pair of populations, we can compile the matrix of pairwise gene identities:

Population	Population: A	B	C	D
A	0.25	0.25	0.25	0.25
B	0.25	0.5	0.25	0
C	0.25	0.25	0.5	0.25
D	0.25	0	0.25	0.5

Most kinship calculation methods from microsatellite or SNP data use a version of this simple approach to calculate both between and within population kinships (see Example 13.3).

Methods for estimating kinship from molecular markers are reviewed by Wang (2011), de Cara et al. (2013), and Toro et al. (2014), and a selection of software packages to calculate kinships from molecular data is given in the Software section at the end of the chapter.

Sampling strategies need to be sufficient to detect differences in kinships at a level of precision needed for genetic management. The probability of detecting kinship differences will depend on the number and kind of markers used, number of individuals sampled per population, and the degree to which populations differ in mean kinship. For our purposes, we need to be able to detect differences in mean kinship at a level of 0.10 or finer. Hoban et al. (2013) provides a tool for planning sampling strategies to detect population structure for different types of markers. In smaller populations with lower genetic diversity, more markers are needed for useful kinship estimation (e.g. for $N_e = 20$, 30–40 markers are required). Taylor et al. (2015) found that it is not possible to accurately estimate inbreeding in species with low genetic diversity from pedigrees based solely on a typical array of microsatellite markers, but SNPs provide better estimates. Specific cases should use computer models to help develop sampling strategies to ensure that estimates of genetic differences among fragments are sufficiently accurate to be used for management (e.g. Ryman & Palm 2006, Hoban et al. 2013). However, given the much higher precision for estimating kinships using SNPs (see below), we recommend that SNPs now be routinely used instead of microsatellites.

Example 13.3 illustrates the use of microsatellite data to estimate mean kinships within and among populations of the jellyfish tree. Similarly, kinships within and among giant panda populations have been estimated from data on 150,000 SNPs (Garbe et al. 2016).

Example 13.3 Within and between population kinships in the jellyfish tree

(Finger et al. 2011)

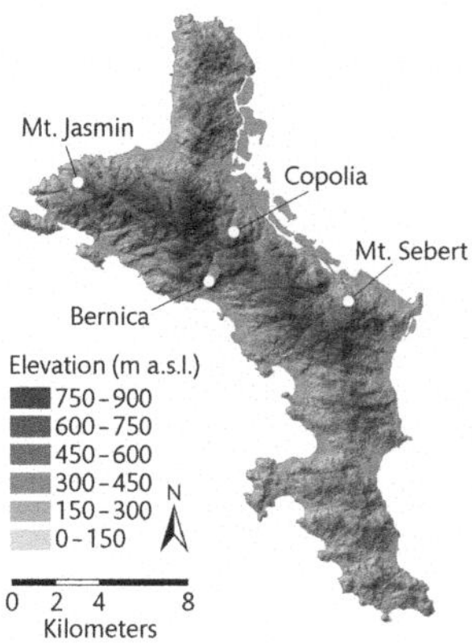

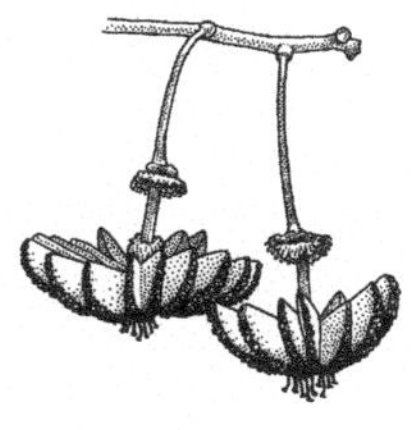

Jellyfish tree (Seychelles)

Finger et al. (2011) examined the population kinship structure of the critically endangered and endemic jellyfish tree based on genotyping all 90 existing trees for 10 microsatellite markers. The trees are located in four isolated fragments (three extremely small, one a little larger—see following table) on the island of Mahe in the Seychelles.

Finger et al. (2011) used the following method to calculate kinship (k_{ij}):

$$k_{ij} = \frac{\sum_{ij}(p_i - \bar{p})(p_j - \bar{p})}{k\bar{p}(1 - \bar{p})} + \frac{2}{(8k+1)^{0.5} - 1} \; for \; (i<j)$$

where the first term is the expected value of the correlation of frequencies of alleles in individuals i and j, k is the number of pairwise combinations ($n[n - 1]/2$), and the second term adjusts for bias due to finite sample size and standardizes k_{ij} to have an expected value of zero for populations in Hardy–Weinberg equilibrium. This was calculated for each locus, and they were then combined by weighting the result for each locus by its polymorphic index, $\sum p_i(1 - p_i)$ (see Loiselle et al. 1995). These kinships may be negative (as in the table below) if they have allelic correlations that are less than expected under Hardy–Weinberg equilibrium.

The between and within kinships in the four populations were:

Kinship matrix

Population (*N*)	**Mt. Jasmin**	**Copolia**	**Mt. Sebert**	**Bernica**
Mt. Jasmin (2)	0.57	0.07	0.02	–0.02
Copolia (3)	—	0.17	0.09	–0.02
Mt. Sebert (7)	—	—	0.41	–0.04
Bernica (78)	—	—	—	0.004

The largest population, Bernica, had much lower within population mean kinship than did the other populations and low between population kinships with the other populations. Thus, genetic rescue would best be accomplished by gene flow from Bernica to the other populations. Crosses between populations showed genetic rescue for fitness (Finger et al. 2011).

Kinships at the individual level

When kinship information is available on all individuals, more precise genetic management is possible, e.g. selecting specific individuals for moving between populations

Information on kinships between individuals within and among fragments allows managers to identify individuals best suited to transfer—those least related on average to the recipient population—in addition to identifying the best source population. It also allows estimation of how removing an individual from a population affects the genetic diversity of that population (see Fig. 13.3). Individual kinships can be calculated using pedigrees or genetic markers such as microsatellites or SNPs, but many markers are required to achieve adequate precision for management.

Individual kinships based on pedigrees

In the rare circumstance where pedigrees of wild populations are known, kinship calculations can be based on the pedigree

Occasionally, full or partial pedigrees of wild populations are known from observations of parentage each generation, as for example for several golden lion tamarin populations, and Chatham Island black robins (Mickelberg 2011; Kennedy et al. 2014).

Traditionally, the most reliable estimates of kinship were based on pedigree calculations. Most pedigrees however, have some proportion of their parentage as unknown or uncertain. RCL & JDB (pers. comm.) found that when more than about 20% of a managed population's pedigree is derived from unknown ancestry, mean kinship calculations become unreliable, as do those for inbreeding coefficients (Taylor 2015).

Inadequate pedigrees can have major impacts on inferences: in song sparrows, unobserved matings caused up to 550% underestimation of inbreeding depression (Reid et al. 2014). Further, pedigrees going back only a generation or two may not provide enough information to determine kinships originating from relationships several generations further back. Fortunately, many of these data deficiencies can be resolved using molecular methods.

Individual kinships based on molecular data

Many methods exist to calculate kinships based on molecular data

Increasingly, kinships and pedigrees are being estimated from molecular data (microsatellites and especially SNPs). For many threatened species, no amount of markers may be adequate to reconstruct pedigrees with useful accuracy (Taylor et al. 2015). Instead, direct estimation of kinships can be used. The precision of kinship estimates based on molecular markers depends on the type and number of markers used and on the numbers of markers per chromosome. For example, with microsatellites, Fernández et al. (2005) concluded that 10 loci each with 10 alleles were required per chromosome to approach management efficiency similar to that of using pedigrees. Thus, for the endangered salt marsh harvest mouse (*Reithrodontomys raviventris*) with 19 pairs of chromosomes (Hood et al. 1984), a minimum of 190 microsatellite markers each with 10 alleles would be required for all the potential parents if the inferred pedigree was to be used for genetic management. This level of coverage has rarely been feasible, due to the costs of developing and scoring microsatellites, and most studies have used fewer than 20 microsatellites. To partially compensate for the subsequent decrease in precision, low kinship values are ignored, instead focusing simply on first or second degree kinships (k = 0.25, 0.125) versus zero kinship. MOLKIN is one software program often used to estimate kinships based on microsatellite data (Gutiérrez et al. 2005). Taylor (2015) presents recommendations for standardized investigation and reporting of power. We recommend that individual level management in the absence of pedigrees be avoided unless many SNPs are used to estimate kinships.

Individual kinships from genomic information

If all individuals of a species, or a group of population fragments have their genomes sequenced, kinships can be estimated with high accuracy without distortions due to pedigree errors, selection, or segregation

It is now increasingly feasible to obtain full DNA sequence information on all individuals in captive and natural populations, yielding accurate genome-wide kinships estimates (see Manichaikul et al. 2010 for methods). Gómez-Romano et al. (2013) concluded that management based on kinships derived from pedigrees and genomic information gave similar results if SNP density (SNPs per Morgan—a unit of recombination) was three times the effective size of the population. The marker density in the simulations was the number of markers per chromosome. Thus, a species with N_e = 20 would require 60 SNP markers per chromosome, corresponding to at least 1,140 markers for the endangered salt marsh harvest mouse referred to above. Use of SNPs rather than microsatellite data to calculate kinships will improve accuracy and precision of estimates (Taylor et al. 2015).

SNPs can even provide more accurate estimates of kinship than pedigrees, because SNPs measure actual allele sharing while the pedigree methods are based on average expected allele sharing, assuming that there is no segregation distortion or allele frequency changes due to natural selection. Thus, full genomic sequence information should represent an improvement upon even an excellent pedigree (Curik et al. 2014; Bosse et al. 2015; Knief et al. 2015). However, if a pedigree is sufficiently deep and accurate, the very small further gain in kinship precision accessible with extensive SNP genotyping probably is rarely worth the cost and effort.

How does management differ when based on molecular versus pedigree kinships?

There are differences in goals and outcomes of genetic management based on markers versus pedigrees

Marker based kinships reflect any alleles that are shared, whether they are identical by descent (alleles inherited from a common ancestor), or identical by state (indistinguishable, but not from recent common ancestry) (Toro et al. 2014). Conversely, pedigree kinships only encompass alleles that are identical by descent.

Management with pedigree based kinships seeks to equalize contributions from each of the pedigree's founder individuals, while approaches using marker based kinships seek to equalize the frequency of all alleles. Thus, use of marker based kinships will promote breeding those individuals that have rare alleles, even if they come from inbred peripheral populations (with less common alleles), and especially those individuals that are actually hybrids or unrecognized members of another taxon. For example, marker based management of American bison (*Bison bison*) conducted without regard to the prior introgression from domesticated cattle would increase the percentage of cattle genes (Halbert & Derr 2007; Hedrick 2009).

Three approaches have been suggested to minimize the problems with marker based kinship. The first is to remove markers with low minor allele frequencies to avoid increasing the frequency of rare deleterious alleles close to such markers (de Cara 2013). The second uses kinship based on segments (haplotypes) of identity, a measure of the proportion of genome segments shared by two individuals from a common ancestor (de Cara 2013). These two should bridge much of the gap between pedigree and molecular estimates of kinship, and simulations indicate that they lie between pedigree and simple marker kinship in their characteristics and management consequences. Third, if introgression is known to exist, it can be selected against by minimizing the kinship between breeding candidates of the target population and the introgressing population, while still managing to maximize genetic diversity within the target population (Toro et al. 2014). If it is not known whether introgression is occurring, kinship outliers should be checked to determine whether they have morphological and other characteristics of hybrids. If so, they should be excluded.

Managing gene flow at the individual level

Individuals that are minimally inbred and minimally related to the recipient population are the best choice for genetic rescue

Optimal genetic rescue is identified by evaluating what each individual's mean kinship would be if they were transferred to each fragment, and then sending individuals to the recipient fragment where it has the lowest mean kinship, as illustrated for the California condor in Box 13.3.

California condor (USA)

Box 13.3 Managing gene flow among California condor populations based on individual mean kinships from pedigrees

Each year, chicks produced by the captive California condors are either retained in the captive population, or moved to one of the three reintroduced populations in California, Arizona, and Baja California (Mexico). To minimize inbreeding and mean kinships in the population, the chicks are placed into the population fragments to which they share the lowest kinship. In the following figure, each line represents the mean kinship of a potential chick from each of the 41 captive pairs of condors if it were placed in each of the three wild populations. While chicks are ideally placed in the population where they have the lowest mean kinship, considerations other than kinship also come into play, such as the chick's sex, the population sizes, maintaining genetically valuable chicks as future breeders in the more secure captive population, and logistics such as obtaining permits for export to Mexico.

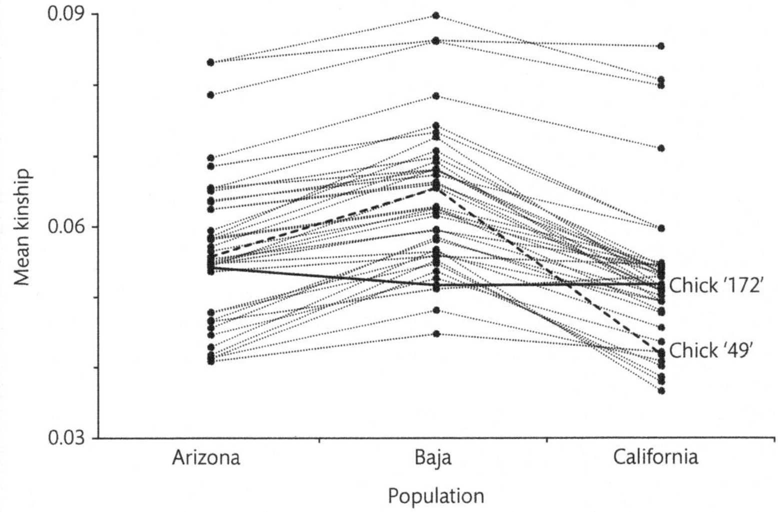

The bold dashed line identifies the chick of female 49 that has large differences in mean kinship in the three wild populations, while the bold solid line illustrates the chick of female 172 with little difference. The chick of female 49 would best go to the California population, while the chick of female 172 would be beneficial in Baja, as it has one of the lowest *mk* values to that population.

Where both donor and recipient populations are threatened, it is desirable to consider the effect of translocations on the genetic diversity of both the donor and recipient populations. Change in genetic diversity can be measured from the change in average mean kinship in the population when an individual is added or removed (using eqn 13.3). For example, the impact of removing individual golden lion tamarins from the captive population and adding them to the reintroduced wild population on expected heterozygosity for the two populations is shown in Fig. 13.3.

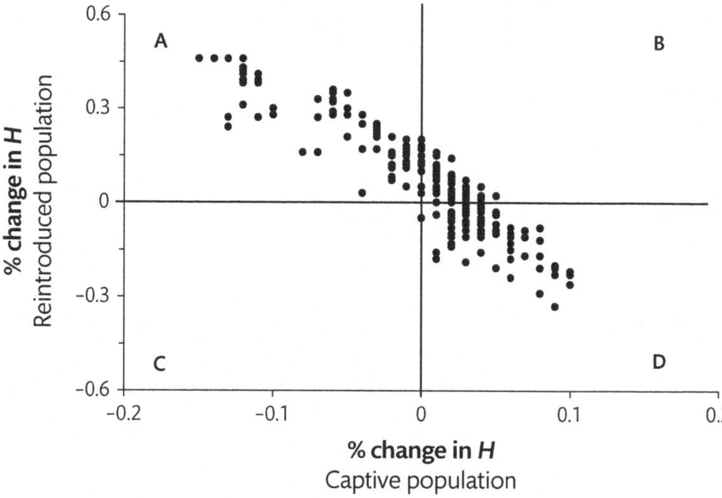

Fig. 13.3 Effect on retained heterozygosity in the captive and reintroduced populations of golden lion tamarins when individuals are transferred from captivity into the wild (Frankham et al. 2002, Fig. 18.6).

Individuals (points) in quadrant A are those whose addition would benefit the reintroduced population, but whose loss from the captive population would reduce its gene diversity. These are genetically valuable captive animals with few relatives in the reintroduced population. Moving quadrant B individuals to the wild will benefit both the reintroduced and the captive populations, as these are genetically over-represented captive animals with few reintroduced relatives. Reintroduction of individuals in quadrant C would be universally detrimental, as they are valuable in the captive population, but already have lots of reintroduced relatives. Reintroducing individuals in quadrant D would be beneficial to the captive population, but detrimental to the reintroduced population (individuals over-represented in both populations).

While this general strategy involves gene flow into populations with high mean kinship from populations that share low mean kinship, other strategies can be implemented by quantitative approaches that identify the optimal movements of individuals between fragments (Fernández et al. 2008). Such an approach was used to optimize the hypothetical movement of 10 individuals among four different fragments of golden lion tamarins (GLT) (Fig. 13.4). The individual mean kinship matrix data in the four different fragments were provided to the program METAPOP (Pérez-Figueroa et al. 2009). The movement regime decreased the mean kinship in each of the four population fragments.

(a)

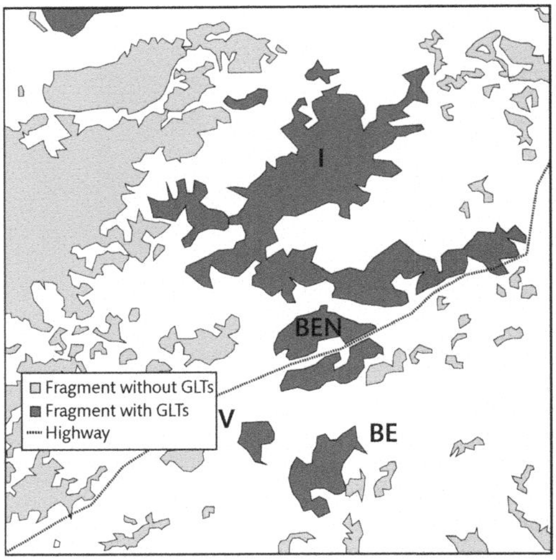

(b)

	I (63)	BEN (25)	V (12)	BE (15)
I	0.146	0.070	0.049	0.040
BEN	0.070	0.268	0.079	0.067
V	0.049	0.079	0.162	0.097
BE	0.040	0.067	0.097	0.205

(c)

From \ To	I	BEN	V	BE
I	—	2	2	5
BEN	0	—	0	0
V	0	1	—	0
BE	0	0	0	—

(d)

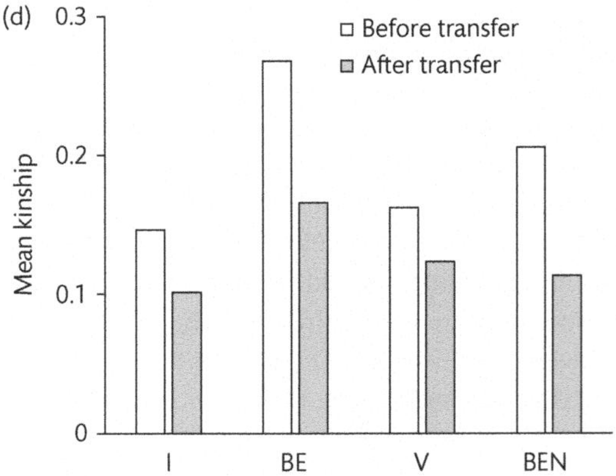

Fig. 13.4 The METAPOP program (Pérez-Figueroa et al. 2009) was used to determine the optimal exchange of 10 hypothetical individuals between four fragments of the wild golden lion tamarin population. (a) Map of the four populations. (b) The population mean kinship matrix of these fragments; individual kinships were also known from the pedigrees (Mickelberg 2011). From the individual kinship matrices, METAPOP was used to identify the best 10 individuals to move between populations to maximize within fragment genetic diversity. (c) The transfer matrix (e.g. move five individuals from "I" to "BE"). (d) The expected mean kinship of fragments after that transfer followed by breeding (by mean kinship).

Mean kinship has additional management use beyond managing gene flow. For example, some small wild populations require occasional culling to avoid exceeding the carrying capacity of their habitat. Culling individuals with the highest mean kinship values enables populations to retain higher levels of genetic diversity than when individuals are selected at random genetically. This approach has been suggested for both wild horses and bison (Eggert et al. 2010; Giglio et al. 2016).

When to cease?

Since we wish to restore normal levels of genetic diversity to population fragments and to reduce their mean inbreeding coefficients to < 0.10, these goals provide guidelines for when to cease augmentation of gene flow. Reasonable objectives would be to cease augmentations when genetic diversity is >95% of the maximum achievable for the target population (given the available donor sources), and its inbreeding coefficient is much less than 0.1 (in natural outbreeders). Additional augmentation may be required later if a fragment continues to have a small population size with consequent loss of genetic diversity (Chapter 12).

What genetic monitoring is needed?

The consequences of genetic management of gene flow should be monitored to ensure that it is proceeding as planned

It is important, if possible, to monitor the long-term genetic consequences of gene flow management between fragmented populations to quantify:

- levels of realized flow compared to targets (e.g. the difficulty in integrating captive individuals into the wild could lead to less gene flow than intended, or high reproductive success of outbred immigrants could lead to more (Harrisson et al. 2016)
- whether natural selection is distorting impacts (e.g. immigrant alleles are swamping the resident population; Adams et al. 2011)
- whether there is some natural gene flow from nearby populations or undocumented translocations (Adams et al. 2011; Hedrick et al. 2014)

Mauritius kestrel

Swedish adder

- whether loss of genetic diversity is faster than expected (Weeks et al. 2013)
- how rapidly inbreeding increases subsequent to a genetic rescue (Bouzat et al. 2009; Adams et al. 2011; Hedrick et al. 2014)
- the demographic responses of the populations to genetic augmentation (Johnson et al. 2010; Hedrick et al. 2014; Fitzpatrick et al. 2016).

There are many examples of monitoring the impacts of genetic management in animal populations, but such studies are rare in invertebrates and plants, especially those using nuclear DNA markers (Wandeler et al. 2007). Genetic determinations pre and post augmentation using microsatellites or SNPs have been done, for example in Mauritius kestrels, Florida panthers, greater prairie chickens, Isle Royale gray wolves, Scandinavian wolves, Swedish adders, and Trinidadian guppies (Madsen et al. 1996, 1999; Groombridge et al. 2000; Vilà et al. 2003; Bouzat et al. 2009; Johnson et al. 2010; Adams et al. 2011; Hedrick et al. 2014; Fitzpatrick et al. 2016). These examples encompass molecular genetic data on near neutral markers, adaptive variation, fitness, and demography.

Integrating genetic rescue with other management considerations

There are many other non-genetic or genetic issues not associated with fragmentation to consider in programs to augment gene flow. For example, there are often behavioral, ecological, and demographic considerations. When new individuals are introduced into a population, introduced animals can be killed or maimed by residents, or new males may kill juveniles, as in lions.

In the next chapter we address genetic management issues in fragmented populations in the context of global climate change.

Summary

1. With more detailed genetic information, more precise and effective management of gene flow into populations suffering genetic erosion can be instituted.
2. Minimizing mean kinship is the superior approach to genetic management of fragmented populations, while using F_{ST} is inappropriate.

3. Mean kinships within and between populations can be estimated at the population and individual level, either from modeling, pedigrees, or from multilocus genotypic data.
4. Minimizing pedigree based kinships will work towards equalizing contributions from each of the pedigree's founder individuals, while management with marker based kinships will work towards equalizing the frequency of all alleles.
5. Management by minimizing mean kinship can be instituted at the population level, or at the individual level. The latter generally allows higher precision of genetic management.
6. Individuals with low mean kinship to the recipient population have the best potential genetic impacts.
7. Populations should be monitored to ensure that movement of individuals has resulted in the desired levels of gene flow, that genetic diversity has been enhanced, and inbreeding reduced.

FURTHER READING

Ballou & Lacy (1995) Pivotal paper justifying the use of minimizing mean kinship to manage threatened populations in captivity based on analytical work and computer simulations.

de Cara et al. (2013) Simulation study comparing management by minimizing kinship based on pedigree versus SNP data.

Taylor (2015) Reviews the use and abuse of genetic marker-based estimates of relatedness and inbreeding.

Toro et al. (2014) Reviews the application of genomics to genetic management in conservation programs.

SOFTWARE

COANCESTRY: a program for simulating, estimating, and analyzing kinship and inbreeding coefficients (Wang 2011). https://www.zsl.org/science/software/coancestry

KING: calculates kinships from SNPs (Manichaikul et al. 2010). http://people.virginia.edu/~wc9c/KING/

METAPOP: software for the dynamic and flexible management of conserved subdivided populations (Pérez-Figueroa et al. 2009). http://anpefi.webs.uvigo.es/metapop/

MOLKIN: software for estimating kinship from microsatellite data (Gutiérrez et al. 2005). https://pendientedemigracion.ucm.es/info/prodanim/html/JP_Web.htm

PMx: software for demographic and genetic management of single and fragmented captive populations, based on pedigrees (Lacy et al. 2012). http://vortex10.org/PMx.aspx

POWSIM: simulation based software that estimates statistical power when testing the null hypothesis of $F_{ST} = 0.0$ (Ryman and Palm 2006). http://www.zoologi.su.se/~ryman.

VORTEX: software for species risk assessment, conservation planning, and population management (Lacy & Pollak 2014). www.vortex10.org/Vortex10.aspx

Global climate change increases the need for genetic management

CHAPTER 14

Adverse genetic impacts on fragmented populations are expected to accelerate under global climate change. Many populations and species may not be able to adapt *in situ*, or to move unassisted to suitable habitat. Management may reduce these threats by augmenting genetic diversity to improve the ability to adapt evolutionarily, by assisted translocation, or by ameliorating non-genetic threats. Global climate change amplifies the need for genetic management of fragmented populations.

TERMS

Assisted colonization, backcross, genetic stochasticity, *in situ*, invasive species, phenology, range contraction, range expansion

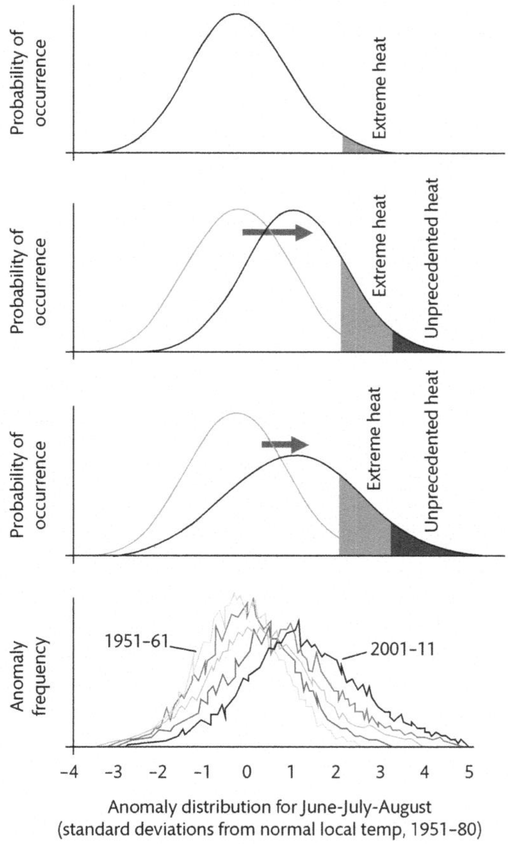

Increase in global temperatures and increase in heat anomalies over the last six decades. The top three panels illustrate the predicted impacts of global climate change from pre-industrial levels (1st panel), to impacts of an increase in mean temperature (2nd panel), to an increase in both mean and variability (3rd panel) on the mean temperature, and on the frequencies of extreme and unprecedented heat. The bottom panel shows the observed temperature distributions from 1951-1961 to 2001-2011 indicating increases in mean and frequency of extremes (NOAA 2016, http://www.ncdc.noaa.gov/cag/).

Genetic Management of Fragmented Animal and Plant Populations. R. Frankham, J. D. Ballou, K. Ralls, M. D. B. Eldridge, M. R. Dudash, C. B. Fenster, R. C. Lacy & P. Sunnucks. Oxford University Press (2017) DOI 10.1093/oso/9780198783398.001.0001

What is the problem?

Global climate change is adversely affecting many species and its impacts are projected to worsen

We have addressed the genetic management of fragmented populations as if they inhabit relatively stable environments. However, human activities are resulting in the warming of the planet, an increased frequency of climatic extremes, rising sea levels, and ocean acidification (IPCC 2014). For example, extreme high temperature events have increased from covering 0.1–0.2% of the planet in 1951–1980 to 10% in 2006–2010 (Hansen et al. 2012).

Innumerable terrestrial and marine species across the planet are adversely impacted by global climates change, or are projected to be (Welbergen et al. 2008; Pacifici et al. 2015; Urban 2015). For example, coral reef communities are at serious risk (Box 14.1). Further, population extinctions due to global climate change have already been reported: two populations of checkerspot butterfly (*Euphydras editha bayensis*) in California were extirpated due to increased variability in precipitation (an expected consequence of global climate change), while the Bramble Cay melomys (*Melomys rubicola*) population from Australia was extirpated by ocean inundation (McLaughlin et al. 2002; Gynther et al. 2016).

Brain coral (*Goniastrea favulus*) (Australia)

Box 14.1 Adverse impact of global climate change on coral reef communities

(Hoegh-Guldberg et al. 2007; De'ath et al. 2012; van Oppen et al. 2015)

Coral reef communities are already suffering from global climate change and many may be extirpated (Hoegh-Guldberg et al. 2007). Corals are suffering increasingly frequent episodes of mass bleaching (first recorded in 1980: Hoegh-Guldberg 1999) that kills them and reduces suitable habit for other reef organisms. For example, the 2016 bleaching on the Great Barrier Reef, Australia was the worst on record and similar impacts were recorded on many other reef systems across the planet (AIMS 2016; Markham 2016). Ocean acidification is also impacting corals and reef invertebrates, and is projected to worsen (Dove et al. 2013). Further, storms are damaging reefs weakened by bleaching and ocean acidification. Coral mass has halved on the Great Barrier Reef between 1985 and 2012 (De'ath et al. 2012), and 80% of its corals may be gone within 40 years without world action to reduce carbon pollution.

The ability of the species in the reef community to evolve to cope with global climate changes is generally unknown. However, the widely distributed reef-building Pacific stony coral (*Acropora millepora*) does exhibit genetic variation for thermal tolerance, primarily due to different symbiont zooxanthellae within its cells, rather than nuclear locus genetic diversity in the coral (Császár et al. 2010). This coral species has been observed to alter its symbiont to a more thermal tolerant type following bleaching (Berkelmans & van Oppen 2006). Further, some reef organisms are capable of evolutionary adaptation to acidification (see later).

The impacts of climate change on corals are so extreme that they are expected to outpace their evolutionary capacity to adapt to acidification and warming waters. Consequently, a series of innovative projects is underway in an attempt to save corals (van Oppen et al. 2015), namely:

- selective breeding to improve their ability to persist in the face of global warming and ocean acidification
- conditioning corals in the laboratory to make them more resilient to stress, and releasing them into the wild
- introducing zooxanthellae symbionts that confer enhanced coral survival and growth under projected environments
- hybridizing closely related species to determine whether progeny will be more resilient than the parents
- translocating better performing corals from regions where they are doing well to regions where corals are doing poorly.

Many species will require assistance to persist under climate change

Many populations and species (especially those with fragmented distributions) will require human intervention to cope with global climate change, because they will be unable to adapt genetically to the changed conditions, or move to a more suitable location

In many cases, changes in environments are projected to occur at rates greater than movements of species recorded in paleoecological records (Huntley 1991; Jump & Peñuelas 2005; Corlett & Westcott 2013; IPCC 2014).

To cope, populations and species can adapt, move to more suitable locations, or do both. Adaptation may be due to phenotypic plasticity, evolutionary adaptation, or a combination of these. However, only evolutionary adaptation will suffice in the face of large environmental change (Hoffmann et al. 2015). Some species are evolving too slowly to keep pace with environmental changes, for example great tits and European beech, with extinction the likely outcome (Bradshaw & Holzapfel 2006; Reusch & Wood 2007).

Fragmented populations often have inhospitable habitat (matrix) separating them from more suitable sites, so individuals often cannot move without human assistance (translocation) to localities where conditions may better match their requirements. This is especially the case for species living on mountain tops, or islands, and ones with limited dispersal capabilities.

In this chapter, we consider how the challenges from climate change alter our genetic management recommendations for fragmented populations. First, we discuss the implications of global climate change for biodiversity conservation. Second, we consider genetic management options to alleviate these effects. Third, we discuss approaches for improving the ability of species to adapt to environmental change in their current location. Fourth, we discuss genetic issues involved in translocations to cope with climate change in fragmented populations, including movement of species outside their historical range (assisted colonization).

We conclude that the genetic management principles discussed in previous chapters still apply, but the need for active genetic management of fragmented populations is becoming ever more urgent under global climate change.

Why does global climate change increase the need for genetic management?

Global change is accelerating extinction risk for fragmented populations

With global climate change, many species are faced with altered environments to which they are not adapted, with consequent reductions in reproduction and survival. Demographic and environmental stochasticity and catastrophes are worsening, thus accelerating adverse genetic impacts

The path to species extinction typically involves both deterministic and stochastic factors (Gilpin & Soulé 1986; Shaffer et al. 2000; Brook et al. 2008). While the direct effects of climate change on population fitness (Brook et al. 2008) and to a lesser degree on evolutionary adaptation to climate change have been discussed (Hoffmann & Sgrò 2011), increased impacts of stochastic factors and interactions among deterministic and stochastic factors are also expected. For example, the Arctic fox (*Alopex lagopus*) is facing multiple threats to its persistence associated with climate change (Box 14.2), as are many other species. Further, laboratory experiments in rotifer populations showed that habitat fragmentation, over-harvesting, and environmental warming stressors resulted in populations declining up to 50 times more rapidly when combined than when acting alone (Mora et al. 2007).

Box 14.2 The Arctic fox is experiencing multiple threats associated with global climate change

(Sillero-Zubir & Aungerbjorn 2009)

The Arctic fox inhabits the Arctic tundra and sea ice. Populations in Sweden, Finland, and Norway were decimated by hunting for fur in the early 20th century and have failed to recover despite legal protection since 1940. European populations now number only ~ 150 and are threatened with extinction. Although the Arctic fox in the rest of its range is relatively abundant, it has been disappearing from the southern edge of the tundra, raising concerns over the species' long-term future.

The Arctic fox faces multiple threats from climate change: its tundra habitats are shrinking, the sea ice is disappearing, its lemming prey are becoming less abundant in some areas, and it is being displaced by the larger red fox (*Vulpes vulpes*), which is moving northward as temperatures warm.

Arctic fox

The extinction vortex is accelerating

The extinction vortex is accelerating due to the worsening deterministic and stochastic factors caused by global climate change

Global climate change is expected to accelerate the extinction vortex by (Brook et al. 2008):

- reducing species' fitness and population sizes as they become less well adapted to their environment
- increasing environmental stochasticity
- increasing the frequency of catastrophes
- increasing demographic stochasticity
- increasing genetic stochasticity
- stronger interactions among these effects.

Climate change causes habitat loss, even if fragment size is unchanged, because some parts of the fragment are no longer suitable habitat for any individuals in the population. Thus, population size is smaller. As the habitat is less suitable, reproduction and survival decline. Increased climate variability results in greater fluctuations in population size, so N_e becomes a smaller proportion of N. These result in reduced genetic diversity and inbreeding, further reducing the reproductive rate. As each of the components of the extinction vortex worsens, the decline towards extinction accelerates. In addition, other aspects of species' environments may be affected by global climate change, such as movement of diseases, pests, parasites, and invasive species or worsening of their impacts. For example, the last known population of timber rattlesnake (*Crotalus horridus*) in New Hampshire, USA is declining due to the interactions among climate change, disease, and low genetic diversity (Clark et al. 2011).

These factors combine to reduce population size and increase its variability, thus worsening genetic stochasticity.

Effects of genetic stochasticity are worsening

The impacts of genetic stochasticity are becoming greater with climate change from four effects (Pertoldi et al. 2007):

- decline in fitness in species that cannot evolve fast enough to keep up with climate change, reducing N_e for many species (Selwood et al. 2015)
- reduction in N_e/N ratios due to increasing climatic extremes, as fewer individuals breed and population sizes fluctuate more (Beaumont et al. 2011; Hansen et al. 2012; IPCC 2014)
- worsening inbreeding depression due to more rapid inbreeding and more stressful environments (Armbruster & Reed 2005; Fox & Reed 2011; Enders & Nunney 2012)
- reduced ability of many populations to adapt genetically due to diminished N_e, loss of genetic diversity, reduced fitness, and increased environmental variance (Lynch & Lande 1993; Lande & Shannon 1996; Chevin et al. 2010).

These impacts are even more serious in fragmented habitats (Leimu et al. 2010).

How have species responded to global climate change?

Many species have moved in response to climate change, others have adapted via phenotypic plasticity, and some have undergone evolutionary adaptation

Examples of species' responses to climate change are provided in what follows (Dawson et al. 2011).

Moved

Many species have already moved in response to climate change, with range shifts averaging 16.9 km per decade towards the poles (Parmesan & Yohe 2003; Root et al. 2003; Menzel et al. 2006; Chen et al. 2011; Poloczanska et al. 2013). Examples range from mammals to mollusks, and grasses to trees. In 80% of cases, movement was in the predicted elevational or compass direction.

Adapted via phenotypic plasticity

Species often respond to changed environments via phenotypic plasticity (Agrawal 2001; Gienapp et al. 2008; Nicotra et al. 2010). In plants, this may involve shifts in timing of annual events, or changes in leaf size, shape, thickness, or pigmentation, height at maturity, water use efficiency, seed size and number, chemical defenses, etc. Similarly, there are many plastic responses to environmental changes in animals, including behavioral modifications, altered timing of migration, changed breeding time, body weight, morphs in *Daphnia* in response to predator cues, etc. (Agrawal 2001; Réale et al. 2003; Charmantier et al. 2008; Gardner et al. 2011).

Many species have shown changed timing of annual events (phenological shifts), such as earlier flowering and fruiting in plants, emergence of butterflies, breeding in frogs, and migration in some birds (Kearney et al. 2010; Primack 2014; Kullberg et al. 2015). For example, 43 plant species are flowering on average seven days earlier than in the 1850s at Concord, Massachusetts, USA in response to warming (Fig. 14.1).

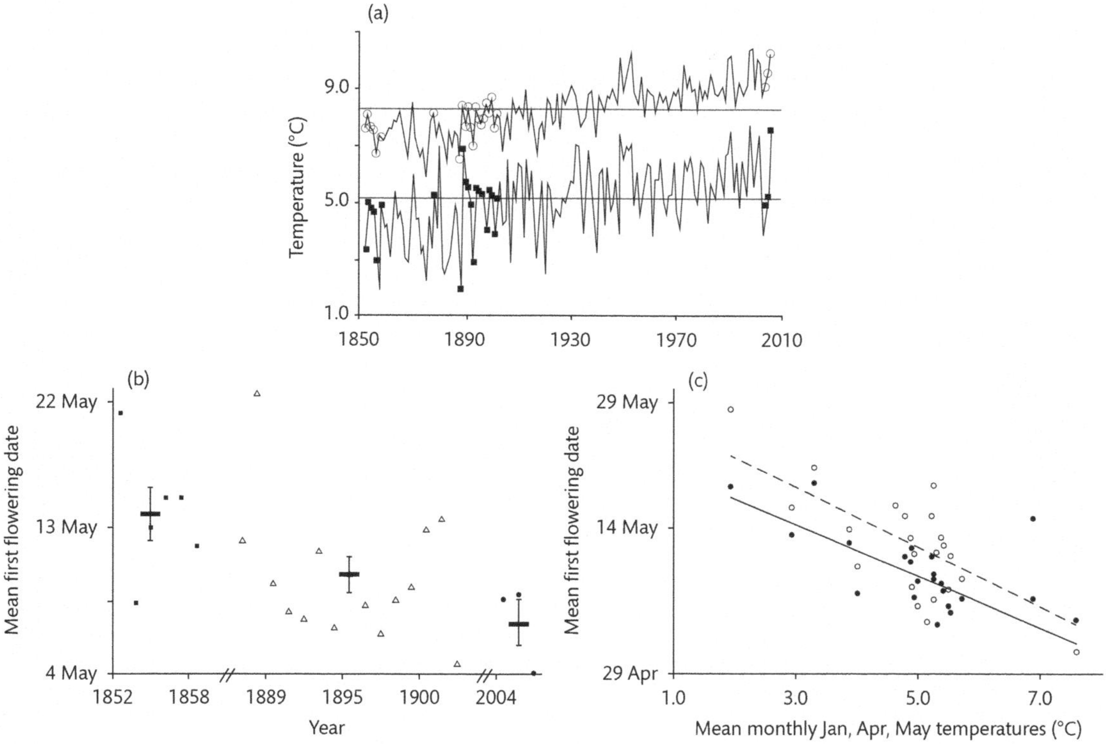

Fig. 14.1 Earlier first flowering dates for 43 plant species associated with rising temperatures near Concord, MA, USA from 1852 to 2006 (Miller-Rushing & Primack 2008). (a) Temperatures: the upper line is the mean annual temperature, while the lower one is the mean monthly temperatures for January, April, and May; this is most closely associated with flowering time. Circles and squares show years with flowering data. (b) First flowering dates: the solid horizontal bars with standard error bars represent the means for the different observers over time (note the split X-axis). The filled squares, triangles, and filled circles are the yearly means for them. (c) First flowering dates are much earlier with higher temperatures. The solid line and solid circles are for 33 native species, and dotted line and open circles for 10 non-native species.

Most documented cases of adaptation have probably involved phenotypic plasticity, though tests for evolutionary changes have only rarely been done. Changes in egg laying date with climate change in great tit birds in Wytham, UK are attributable to phenotypic plasticity, rather than evolution (Charmantier et al. 2008).

Evolved

Some species have evolved to cope with climate change

Species known to have genetically adapted to their altered climatic conditions include Canadian red squirrels (*Tamiasciurus hudsonicus*), blackcap warblers (*Sylvia atricapilla*), several species of *Drosophila*, pitcher plant mosquitoes (*Wyomia smithii*), Mediterranean wild thyme (*Thymus vulgaris*), canola (*Brassica rapa*), and thale cress (Bradshaw & Holzapfel 2006; Reusch & Wood 2007; Franks & Hoffmann 2012; Thompson et al. 2013; Wilczek et al. 2014). For example, Anderson et al. (2012) found that Drummond's rockcress plants (*Boechera stricta*) in the Rocky Mountains (USA) flowered earlier with increasing temperature, due to evolutionary adaptation and phenotypic plasticity. Further, Nevo et al. (2012) found that average flowering times of wild populations of barley (*Hordeum spontaneum*) and emmer wheat (*Triticum dicoccoides*) in Israel were 8.5 and 10.9 days earlier, respectively in plants derived from seed collected in 2008 than in those from 1980 seed (Fig. 14.2).

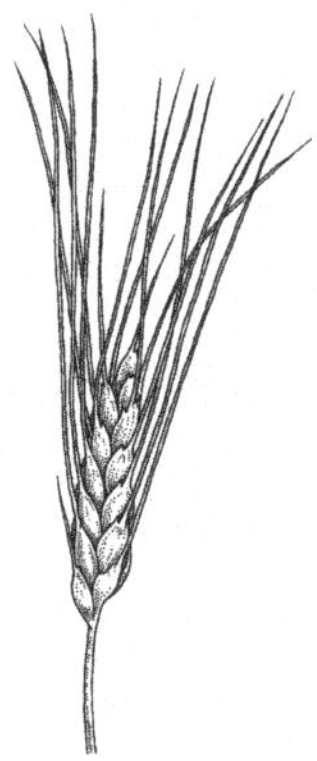

Emmer wheat (Israel)

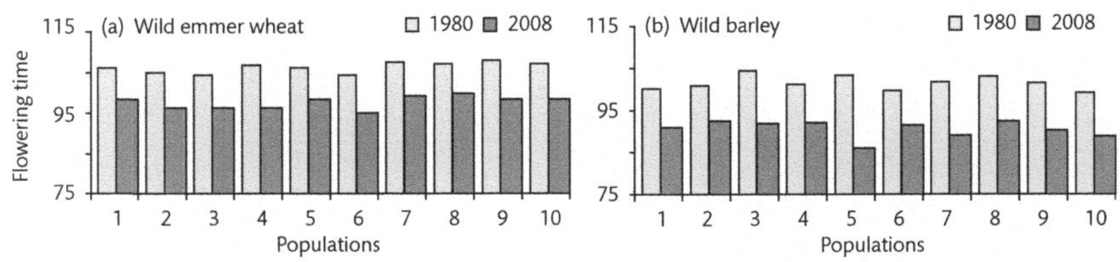

Fig. 14.2 Evolution of earlier flowering time in wild populations of (a) emmer wheat and (b) wild barley in Israel in response to global climate change. Differences in flowering time for 10 populations of each species derived from seed collected from the wild in 1980 and 2008, and grown contemporaneously in a greenhouse under the same conditions (Nevo et al. 2012).

We next consider what management options we have if a species cannot currently adapt to climate change in its current location or move itself to another suitable location.

What genetic management options do we have for populations that cannot move or adapt sufficiently?

Given the effects of climate change on species, active management will often be needed to stabilize them and prevent population or species extinctions by (Dawson et al. 2011):

- improving the ability of species to adapt evolutionarily (Chapters 4, 6, and 11)
- translocation to more suitable habitats
- both.

A decision tree for assessing management options to improve the ability of a population to cope with climate change is given in Fig. 14.3.

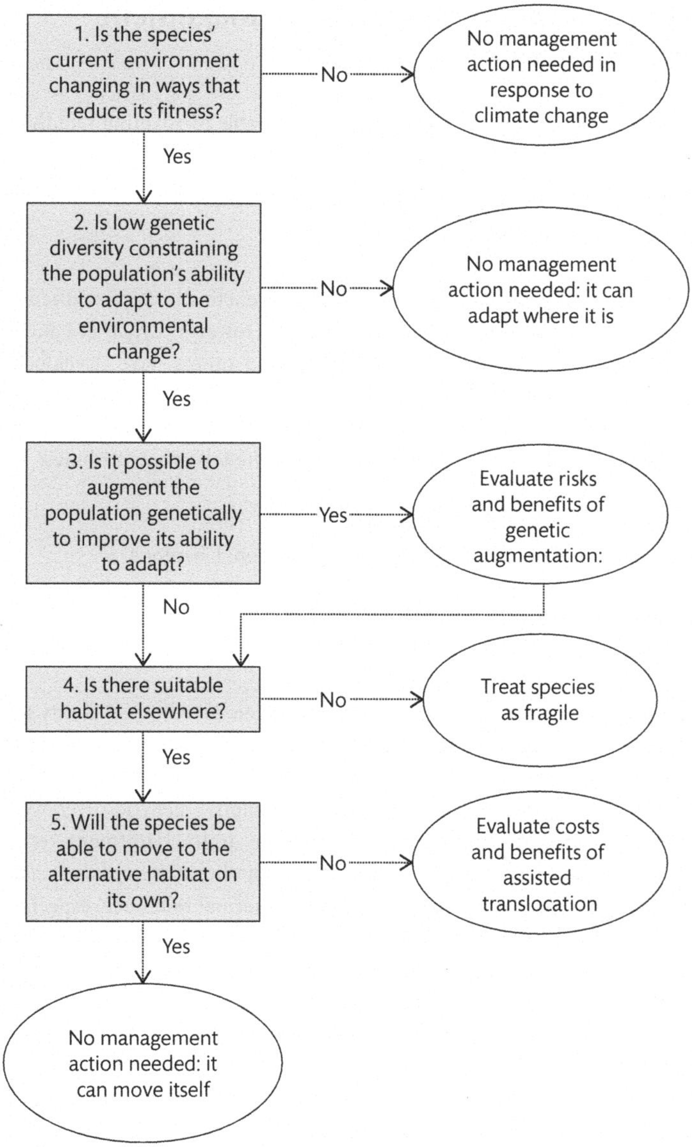

Fig. 14.3 Decision tree for assessing genetic management options to improve the ability of populations and species to cope with global climate change.

We begin with the ability to adapt evolutionarily, first by considering how we can identify populations that are incapable of adapting rapidly enough to keep up with climate change, second by identifying where new genetic diversity can be obtained, and third by implementing programs to improve the ability to adapt.

How do we implement genetic management to assist adaptation?

The first step is to identify populations that are incapable of adapting rapidly enough.

How can we identify populations that are incapable of adapting rapidly enough?

We do not know what proportion of species lack the genetic diversity required to evolve to cope with current and projected climate change. However, Lynch & Lande (1993), Lande & Shannon (1996), Chevin et al. (2010), and others have developed theory to predict whether species have the ability to adapt sufficiently rapidly in response to environmental change to avoid extinction. Populations that are unlikely to be able to adapt evolutionarily will have some combination of the following attributes:

- be experiencing rapid climate change
- low genetic diversity (especially adaptive variation; Chapter 4)
- low N_e
- low reproductive rates
- long generation intervals.

For example, the European beech and other trees often have limited ability to rapidly adapt, because of long generation times (Petit & Hampe 2006; but see Alberto et al. 2013). Further, tropical *Drosophila* species with short generation lengths cannot evolve desiccation resistance due to the lack of adaptive genetic diversity, despite high levels of genetic diversity for morphological traits (Hoffmann et al. 2003; Kellermann et al. 2006). Even the widespread cosmopolitan *Drosophila melanogaster* does not always exhibit evolutionary potential for thermal tolerance, especially when tests are conducted in ecologically relevant ways, despite the ability to evolve for other traits and high microsatellite genetic diversity (Schou et al. 2014). Further, wild mustard (*Brassica juncea*) was unable to adapt to projected global climate change conditions through either phenotypic plasticity or evolutionary change (Potvin & Tousignant 1996).

Conversely, the ability to evolve in response to increased ocean acidification has been reported in several calcifying organisms (presumed to have high population sizes, ample genetic diversity, and short generation intervals), including sea urchins, bryozoans, and coccolithophores (unicellular algae) (Pistevos et al. 2011; Sunday et al. 2011; Foo et al. 2012; Lohbeck et al. 2012; Pespeni et al. 2013).

Practical means for measuring the evolutionary potential of populations were considered in Chapter 4 (see also Visser 2008; Sunday et al. 2014). If these are not feasible, PVA models, run with and without evolutionary adaptation, provide an important means for identifying populations that are incapable of keeping pace with climate change (Dunlop et al. 2007; Reed et al. 2011; Kuparinen & Hutchings 2012; Piou & Prévost 2012). For example, with a simulated increase in summer temperatures of 2°C by 2100 for sockeye salmon (*Oncorhynchus nerka*) in the Fraser River, Canada, adult migration from the ocean to the river advanced by ~10 days due to evolutionary change when $h^2 = 0.5$. This substantially reduced the risk of population extinction compared to what it was without evolutionary change (Reed et al. 2011). Few such studies have been conducted, but we recommend them, if feasible. An even more comprehensive approach has been advocated by Dawson et al. (2011) that includes paleoecological data, long term monitoring, experimental manipulations, etc., which we describe in more detail later. However, some of the additional items will not be feasible, given the urgency to act.

The ability of species to adapt evolutionarily may be improved by augmenting their adaptive genetic diversity

Species and populations with low genetic variation may need genetic augmentation if they are to survive. Empirical studies with *Tribolium* flour beetles have verified the effectiveness of gene flow from other populations in enhancing fitness and evolutionary adaptation under changing environments (Hufbauer et al. 2015a).

From where can adaptive genetic diversity be sourced?

The potential sources of genetic diversity to improve evolutionary potential in a population are:

- other populations within the species
- other species and sub-species (introgression or gene transfer).

Up to this point we have focused our discussion on within species management (Chapters 4 and 6); however, we now consider sourcing new genetic diversity from other taxa.

Introgression as a source of adaptive genetic variation

Given the rapid changes expected with global climate change, genetic diversity from other sub-species or species may sometimes be the only potential source of new adaptive genetic variation

Introgression from other species may seem a radical suggestion, as presently even genetic augmentation from other sub-species is generally avoided by authorities, such as the US Fish and Wildlife Service. However, some sub-species designations have little scientific basis (Chapter 9), so such categorical restrictions are not justified. In the Florida panther, augmentation from Texas cougars, considered at that time to be a different sub-species, has been beneficial (Culver et al. 2000).

If there are no other conspecifics to cross with the recipient population, adding new adaptive material by introgression from a related species demands serious consideration (Hamilton & Miller 2016). First, gene exchange in some species is equivalent to population crosses in others, depending upon which species concept has been employed, and whether classifications have been done by splitters or lumpers (Chapter 9). Second, many species show low natural levels of introgression (Hausdorf 2011).

Darwin's medium ground finches (Ecuador)

Third, natural utilization of adaptive variation derived from a related species has been reported in butterflies, birds, primates, and plants (Arnold et al. 2008; Grant & Grant 2008). For example, the ability to adapt to climatic cycles by changes in beak size in Darwin's medium ground and cactus finches (*Geospiza fortis* and *G. scandens*) on Daphne Major Island in the Galápagos archipelago, Ecuador largely utilizes alleles originating by introgression from each other (Grant & Grant 2008). Further, Lewontin & Birch (1966) inferred that the Queensland fruit fly (*Bactrocera* [*Dacus*] *tyroni*) in Australia extended its range using genetic variation for thermal tolerance derived by hybridization with a closely related species. Fourth, some species have evolved by hybridization between species, but remained diploid and adapted to new environmental niches, as for example in several sunflowers (Rieseberg et al. 2003; Schwarz et al. 2005; Mavárez et al. 2006).

Fifth, multiple species composites have often been used by animal and plant breeders to develop populations for human use in food and horticulture. For example, several edible bananas (*Musa*) have genetic contributions from different species (Sardos et al. 2016). Further, several cattle breeds designed for harsh tropical environments (e.g. Santa Gertrudis, Brangus, and Australian milking zebu) have been synthesized with contributions from *Bos indicus* and *B. taurus* (Hayman 1974; Frankham 2009b).

Gene transfer

Genetic engineering offers the potential to introduce new beneficial alleles into target species from any other species, even those from which it is reproductively isolated

Genetic engineering can source new beneficial alleles (especially ones of large effect) from any species, as has been done in plant and animal breeding (Forabosco et al. 2013; Prado et al. 2014). Whilst there are reservations about using gene transfer in conservation contexts, there are well-established regulations and protocols governing the creation and release of genetically modified organisms (e.g. Hamilton 2001; Lee 2009). In the conservation context, a wheat gene has been inserted into the American chestnut (*Castanea dentata*) to confer resistance to an introduced fungus that almost eliminated this iconic tree (Newhouse et al. 2014). The precision of such manipulations has recently been greatly improved by application of the CRISPR-Cas9 gene editing technology (Doudna & Charpentier 2014).

How does one find DNA sequences with large beneficial effects that will aid adaptation to global climate change? Many candidate loci affecting a variety of traits have already been identified as a result of genome sequencing, especially in humans and domestic animals and plants, and many more will be found with the rapid growth of sequencing and annotation of genomes of wild species (Grossman et al. 2013; Vitti et al. 2013; Hoban et al. 2016). In corals, DNA sequences conferring increased tolerance to temperature or acidification are candidates for gene transfer.

Given the severity of the problems that global climate change poses, gene transfer offers worthwhile prospects and is a much more precise method than introgressive hybridization, because a single allele, rather than a whole haploid genome is added to a species. If a particular gene transfer does not improve fitness in the wild, natural selection will usually eliminate it. Nevertheless, care needs to be exercised with gene transfers to evaluate the consequence of gene insertion or replacement on other aspects of fitness. Trade-offs may occur, as with the harmful effects on health and reproduction of adding bovine growth hormone genes into pigs (*Sus scrofa domesticus*) (Pursel et al. 1989).

Human assisted evolution

Artificial selection, manipulation of symbionts and other associated microbes, and hybridizations have the potential to improve adaptation in species affected by global climate change

Rates and magnitudes of improvement are typically greater with artificial selection than with natural selection (Futuyma 1998). Consequently, corals are being artificially selected for increased tolerance to warming and acidity (van Oppen et al. 2015), and related options exist in other species. However, this approach has two shortcomings—genetic adaptations to captivity that are harmful on return to the wild, and trade-offs between fitness components (Frankham 2008, 2009b; van der Werf et al. 2009). The use of *in situ* coral nurseries avoids genetic adaptation to captivity and has been used to establish better adapted forms of asexual species in new locations (van Oppen et al. 2015).

In corals, attempts are also being made to replace symbiotic zooxanthellae that confer low heat tolerance with ones that confer increased heat tolerance (van Oppen et al. 2015). Further, interspecific hybrids between species that naturally hybridize are being produced and tested for adaptation to global climate change effects (van Oppen et al. 2015).

The other management options to cope with global climate change are to translocate to a new environment, or if this is not possible, to reduce other stresses on the population.

How do we decide what populations and species need translocation to avoid extinction?

If a species or population is incapable of genetically adapting sufficiently rapidly, and/or unable to move itself rapidly enough, it is a potential candidate for translocation to cope with climate change

The Florida torreya tree is an example of a species likely to benefit from translocation, as fewer than 1,000 individuals of this long-lived tree exist in a restricted area, it is not reproducing in its native habitat, is unlikely to move naturally, and has low genetic diversity (Schwartz 1993; McLachlan et al. 2007).

Methods for assessing the ability of species to cope with climate change have been reviewed by Pacifici et al. (2015), and the IUCN Climate Change Specialist Group recently published guidelines for assessing the vulnerability of species to climate change (Foden & Young 2016). As mentioned previously, the most comprehensive means for assessing whether a species needs assisted translocation is to conduct a PVA using a spatially explicit model that incorporates projected global climate change and genetic factors. For example, Landguth et al. (2014) identified populations of bull trout

(*Salvelinus confluentus*) that were vulnerable to climate change. However, detailed individual species modeling typically requires extensive species-specific biological and ecological data that are often unavailable.

Consequently, simpler climate envelope models have been used to identify species at risk from climate change (Pearson & Dawson 2003; Schwartz et al. 2006; Phillips & Dudík 2008; Garnett & Franklin 2014). Using this method, Thomas et al. (2004) estimated that 35% of species on the planet would be committed to extinction under the global climate change scenario that the planet is currently tracking or exceeding.

In plants, Keith et al. (2008) linked dynamic habitat suitability and spatially explicit stochastic population models to determine how variations in life history, disturbance regime, and distribution patterns influence the viability of populations under stable and changing climate, and applied it to South African fynbos species. Plant species' responses to climate change exhibited complex dependencies on all of the factors examined. Ideally the preceding models should include evolutionary adaptation to determine whether adaptation is a realistic mitigation strategy, but few do (Pierson et al. 2015; see earlier).

The most comprehensive approach to identifying species at risk from climate change combines evidence from (Dawson et al. 2011):

- direct observations
- paleoecological records
- ecophysiological models
- climate-envelope models
- population models
- experimental manipulations.

We endorse the approach of using all available evidence, but emphasize the need for prompt action to address threats to species persistence (Soulé 1985). Gaps in knowledge can then be progressively filled and incorporated into an adaptive management process.

Below we address genetic management of translocations to cope with climate change.

How should we genetically manage translocations to cope with climate change?

The objective of translocation is to successfully establish or augment populations that have a high probability of persisting over the long term. The questions we need to answer are:

- To where should the translocatees be moved?
- What genotypes/populations should be moved?

Populations that are not well adapted to their current location should be moved to locations where they are likely to be better adapted now and in the future (Weeks et al. 2011). This may involve movement to locations outside their historical range, but such translocations are controversial (Ricciardi & Simberloff 2009; Seddon et al. 2009; Minteer & Collins 2010; Kreyling et al. 2011; Schwartz et al. 2012). Given that assisted colonization may be the only option to avert extinction under environmental change, we endorse its use. Its potential use should be carefully evaluated in cost-benefit analyses for each particular case.

Founding translocated populations outside the historical range

With large projected environmental changes due to global climate change, translocations to sites outside species' historical ranges should be contemplated when there are no suitable sites within the historical range

The movement of species beyond their historical range has been suggested as an option for species at immediate risk of extinction (Hoegh-Guldberg et al. 2008; Weeks et al. 2011), and has been undertaken as a last resort in some instances (Jamieson et al. 2006; Grueber & Jamieson 2008; Miskelly et al. 2009). For example, Willis et al. (2009) used "species climate" models to choose suitable sites for introductions ~65 and ~35 km, respectively, beyond the then range margins of marbled white (*Melanargia galathea*) and small skipper (*Thymelicus sylvestris*) butterflies in northern England. Assisted translocations were done to these sites in 1999–2000. Both introduced populations grew and expanded their ranges over six years (2001–2006) and were still thriving in 2008.

Choice of populations for translocations

The choice of one or more populations to use in translocations depends on the genetic diversity in the available populations, the extent of genetic differentiation among them, their isolation, and their adaptation to the environment of the translocation site

When a decision is made to translocate a population to rescue it from climate change, the genetic considerations discussed throughout this book become especially important. Unfortunately, species and populations at immediate risk of extinction are often already suffering from a genetic viewpoint (inbreeding depression, loss of genetic diversity, maladaptation, etc.: Frankham et al. 2010; Frankham 2012). The likely success of a translocation will be greatest when the source population(s) have high reproductive fitness, high genetic variation, and are already adapted to the environment of the translocation site, characteristics that are rarely found in likely candidates for translocation, unless population crosses are used (Wolf et al. 1996).

Care must be taken with translocations to ensure that the translocated population is representative of the genetic diversity in the source population(s) (Weeks et al. 2011). An example of poor choice of source population is provided by the koala (*Phascolarctos cinereus*) in southeastern Australia (Box 14.3). In general, care should be taken when island populations are being considered as source populations for translocation, because they typically have low genetic diversity and are inbred compared to mainland populations (Frankham 1997, 1998; Eldridge et al. 1999, 2004). However, F_1 progeny of between island or island–mainland population crosses should fare much better.

Koala (Australia)

Box 14.3 Translocation of koalas in southeastern Australia: a poorly designed program with adverse genetic impacts

(Houlden et al. 1996; Sherwin et al. 2000; Seymour et al. 2001; after Frankham et al. 2010)

(Frankham et al. 2010, Box 17.3 map)

The koala's range and numbers in Eastern Australia (the historical distribution of the koala is shaded on the map above) were drastically reduced by hunting, habitat loss, and disease by the 1930s, having disappeared from South Australia and declined markedly in Victoria. The fur trade ceased in the 1930s, following legislation that gave koalas legal protection. They were still common in Queensland and subsequently recovered without major assistance, as they did in New South Wales.

Beginning in 1923, a population founded from as few as 2–3 individuals on French Island (FI), Victoria was used to directly or indirectly supply 10,000 animals for translocations to 70 locations in Victoria and South Australia (Houlden et al. 1996). Consequently, the populations in Victoria and South Australia now possess about half the genetic diversity found in less-perturbed populations further north, and they also have an elevated frequency of males with testicular aplasia (missing testicles), this being worst in the most bottlenecked population on the South Australian mainland (Seymour et al. 2001). While this example does not involve climate change, it illustrates the problems that can arise when genetic factors are not considered in translocations.

From where should individuals for translocations be sourced?

The ideal population(s) to use for assisted colonizations depend upon the genetic diversity in the population fragments, extent of fragmentation, adaptive differentiation among fragments, and their adaptation to the new site. Figure 14.4 illustrates a range of situations reflecting these issues in the face of global climate change. Where there is evidence of adaptive genetic differentiation among extant populations, the translocated individuals should come from populations most likely to be adapted to the reintroduction habitat. Scenarios 3 and 4 are most relevant to our concerns about fragmented populations. If the source populations have low genetic diversity and are fragmented, then genetic material for translocation should be sourced from multiple populations (Broadhurst et al. 2008; Breed et al. 2013; Prober et al. 2015) and matched ecologically with the intended translocation site (Fig. 14.4 scenario 4).

	Genetic diversity of source population	Distribution of source population	Solution	Schematic
1	High	Continuous	Take from entire distribution, or tip if this matches target area ecologically	
2	High	Fragmented along cline	Take from fragment(s) that match target area ecologically; this might be closest population (particularly in altitudinal series)	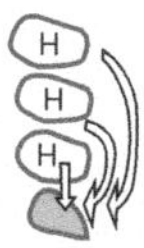
3	High or low	2+ populations- large disjunction- no population adapted to site	Cross and pool (especially if low genetic diversity)	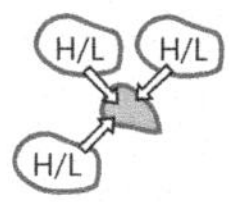
4	Low	Fragmented along cline	Take from multiple populations to augment diversity and adaptation; match ecologically if possible	

Fig. 14.4 Genetic considerations in establishing populations outside their current or historical species range. Relevant scenarios depend particularly on whether levels of genetic variability in the populations are high (H) or low (L), whether populations are fragmented, and whether populations are distributed along environmental gradients (after Weeks et al. 2011). The arrows indicate the recommended sources of individuals to found the population in the new location (gray).

If the new site is significantly different from the current environment in several attributes (e.g. it is heavily altered by human activities), then establishing the new population with maximum genetic diversity and minimum inbreeding by crossing between several available populations should maximize the probability of success (Fig. 14.4 scenario 3). In many cases the translocation habitats will only partially match the environmental requirements of the species, so augmenting genetic diversity to enhance the ability of the population to adapt to its new environment will often be the desired option.

Strategies for choosing sources for translocations that are adjusted to account for evolutionary genetic considerations are beginning to be implemented, particularly in revegetation (e.g. Prober et al. 2015; Garner et al. 2016). Further, Moll et al. (1965) found that F_1 and F_2 crosses of maize varieties adapted to different environments generally exceeded the grain yield of the mid-parent, and often exceeded that of the better performing parent, except for the most divergent crosses.

Risk assessment framework for genetic translocations

Risk assessment should be undertaken for genetic translocations, because they may fail, or have adverse effects on the target, or on other species, and they need to be compared to the risk from doing nothing

Weeks et al. (2011) devised a simple risk assessment framework for genetic translocations (Fig. 14.5), based on whether the translocation is planned within the historical range of the species, or outside, genetic structure and isolation of populations, and the likelihood of hybridization to other species or sub-species. An application of this decision tree is given in Box 14.4.

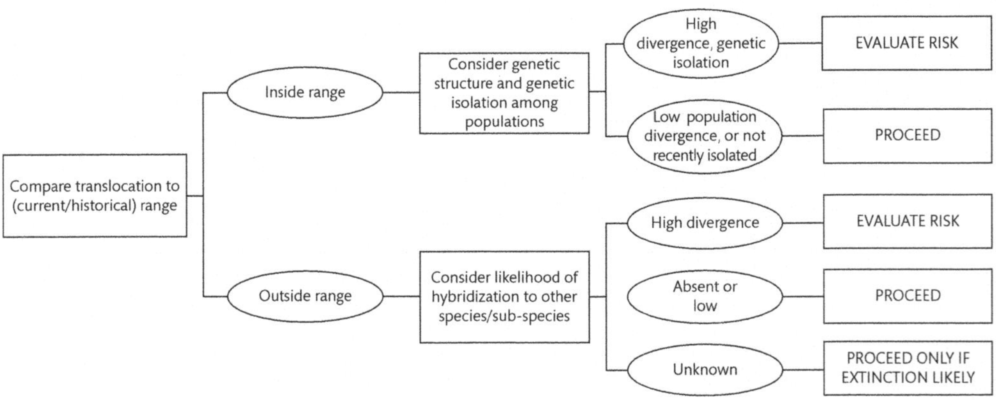

Fig. 14.5 Simplified decision tree for determining whether to proceed or assess risk in translocations (Weeks et al. 2011).

Box 14.4 Risk assessment for restoration of genetic diversity in the mountain pygmy possum (*Burramys parvus*)

(after Weeks et al. 2011, 2015)

Mountain pygmy possum (Australia)

The endangered mountain pygmy possum is restricted to three genetically distinct populations in the Snowy Mountains of southeastern Australia (map) (Mitrovski et al. 2007). One of the threats to this mountain-restricted species comes from climate change. The southern population (Mt Buller) crashed in the 10 years subsequent to 1996, losing ~ 70% of its microsatellite genetic diversity and becoming inbred (Weeks et al. 2015). Consequently, Mitrovski et al. (2007) proposed that individuals from the central region be translocated to the Mt Buller population to alleviate the effects of inbreeding, restore genetic diversity, increase adaptive potential, and enhance the long-term persistence of the population.

Using the decision tree (Fig. 14.5), the proposed augmentation of the Mt Buller population is inside the species' current range, genetic structure is high among populations, and therefore we should evaluate risk. The four risks to evaluate are outbreeding depression, loss of local adaptation, replacement of the Mt Buller genome with the Central genome, and disease transmission. The risk of outbreeding depression is moderate given that these populations have been evolving independently in similar environments for ~ 20,000 years (Mitrovski et al. 2007). A mitigation strategy would be to evaluate preliminary crosses, either in the field or in captivity. The risk of losing local adaptation is moderate (in the absence of direct knowledge) and a mitigation strategy would be to backcross to the recipient population. The likelihood of replacing the Mt Buller genome with the Central genome is low and can be prevented if only males are translocated, and the frequency of gene flow into the population is monitored genetically. Introduction of a disease could cause extinction of the Mt Buller population, but the likelihood of this is minimal given that there are no known diseases present in pygmy possum populations. A mitigation strategy would be veterinary examination of the translocated animals and quarantine if appropriate.

If the outcome of the decision tree had been to abandon the translocation, the risks for the Mt Buller population would have been continued loss of genetic diversity, increasing inbreeding depression, and a high extinction threat, given the current population size was ~ 30 individuals. The only mitigation strategy would have been to undertake captive breeding to improve the chances of short-term persistence, but genetic adaptation to captivity becomes a problem over generations (Frankham 2008).

In 2010 the Mt Buller population was augmented with males from the central population. Fertile F_1 hybrid individuals have been produced, they have produced offspring, the genetic diversity has more than doubled, and the population size has increased (Weeks et al. 2015). There has been no evidence of outbreeding depression to date.

Species with diverse mating systems and modes of inheritance (not outbreeding diploids) often have altered risks and benefits associated with genetic translocations, and justify different decisions and modified management. Their characteristics are described in Chapter 8 and additional detail is given in Weeks et al. (2011).

Take home messages for the readers and Final messages for managers of wild animal and plant populations are presented in Chapter 15.

Summary

1. Climate change is accelerating genetic erosion and elevating extinction risks for many species, increasing the urgency of genetic management.
2. Many populations and species will be unable to adapt to new environmental conditions, move to more suitable locations (including outside of their historical range), or both.
3. If populations or species have low adaptive genetic diversity, it may be possible to improve their evolvability by gene flow from other populations, sub-species, or species.
4. Species needing translocations are those that have (a) an inability to move without assistance due to poor dispersal, inhospitable matrix, or long distances to new suitable habitat, (b) inability to adapt genetically due to low genetic diversity, low fitness, and small N_e, or (c) combinations of these.
5. Individuals translocated to cope with climate change should result in a population with ample genetic diversity and low inbreeding that is well adapted to the new environment.
6. If the species consists of small isolated fragments with high inbreeding and low genetic diversity, and especially when the new site is rather different in several attributes to the populations' current environment, the individuals for translocation should be taken from several populations and ideally crossed such that F_1 individuals are translocated.

FURTHER READING

Hoffmann & Sgrò (2011) Review of climate change and evolutionary adaptation.

Hoffmann et al. (2015) Provides a framework for incorporating evolutionary genomics into conservation management under climate change.

IPCC (2014) *Report of the Intergovernmental Panel on Climate Change*: Addresses likely scenarios for climate change and their consequences.

Menzel et al. (2006) Meta-analysis of phenological changes with global climate change for 542 species of plants and 19 species of animals in Europe.

van Oppen et al. (2015) Review of innovative genetic approaches being used in an attempt to save corals under global climate change.

Weeks et al. (2011) Review of genetic translocations of animals and plants to cope with climate change.

SOFTWARE

MAXENT: Software to define environmental envelopes for species, sub-species, and populations, as part of predicting the impact of global climate change on persistence (Phillips & Dudick 2008). http://www.cs.princeton.edu/~schapire/maxent/

VORTEX: Population viability analysis software that tracks the predicted heterozygosity and inbreeding coefficients of individuals (Lacy & Pollak 2014). www.vortex10.org/

CHAPTER 15

Take home messages

Having addressed the major issues concerning genetic management of fragmented animal and plant populations, we now summarize the material into take home messages. They are:

1. Genetic management of fragmented populations is one of the most important issues in conservation biology, but is very rarely addressed in a satisfactory fashion.
2. Due to human activities, most species have fragmented distributions, many with small isolated population fragments that will experience loss of genetic diversity, become increasingly inbred, be unable to adapt to future environments, and have elevated extinction risks.
3. If populations of naturally outbreeding species are inbred, they should be presumed to be experiencing inbreeding depression and managed appropriately without waiting for specific evidence of inbreeding depression.
4. Inbreeding depression, loss of genetic diversity, and loss of evolutionary potential can be reversed by augmenting gene flow from a genetically different population.
5. Crossing between populations is occasionally harmful (outbreeding depression), but this is largely predictable and a less serious problem than inbreeding depression.
6. The expression of outbreeding depression is often temporary because natural selection removed such harmful effects in all investigated cases.
7. Some species definitions are unsuitable for conservation purposes (especially the Phylogenetic and General Lineage Species Concepts) and many species delineations lack robust scientific support. For conservation purposes we recommend that species delineations be based on reproductive isolation in the broad sense.
8. We strongly urge that standardized species delineation protocols be devised for conservation purposes, including appropriate species concepts, geographic sampling regimes, sample sizes, characters, habitat characteristics, and statistical analyses.
9. When population differentiation is detected within a species, it is important to distinguish whether this is due to drift (where augmentation of gene flow should be evaluated) or differential adaptation (where separate management is usually indicated).
10. We recommend augmentation of gene flow for isolated population fragments of outbreeding species that are suffering inbreeding and low genetic diversity, provided the proposed population cross has a low risk of outbreeding depression and the predicted benefits justify the cost.

Genetic Management of Fragmented Animal and Plant Populations. R. Frankham, J. D. Ballou, K. Ralls, M. D. B. Eldridge, M. R. Dudash, C. B. Fenster, R. C. Lacy & P. Sunnucks. Oxford University Press (2017) © R. Frankham, J. D. Ballou, K. Ralls, M. D. B. Eldridge, M. R. Dudash, C. B. Fenster, R. C. Lacy & P. Sunnucks 2017. DOI 10.1093/oso/9780198783398.001.0001

11. Fitness benefits from crossing for selfing species do not persist over generations (but evolutionary rescue should), so genetic rescue attempts for them are less likely to be justified. Fitness benefits from crossing in mixed mating species do not persist as well as they do in outcrossing species.
12. Choosing among genetic management actions (and inactions) needs to assess the overall risks and benefits of different scenarios. Doing nothing is a choice that is often harmful to the persistence of populations and species.
13. We recommend managing gene flow among isolated population fragments by minimizing mean kinship. If kinship analyses are not feasible, we advocate management of gene flow to maximize genetic diversity, based on conservation genetics principles.
14. Species will need to be even more adaptable to cope with projected global environmental change, increasing the need for genetic management.
15. Threatened species need integrated management across populations, disciplines (including genetics), institutions, and political boundaries, as exemplified by the One Plan approach.

Final messages for managers of wild animal and plant populations

- The persistence of species with fragmented distributions is heavily dependent upon active genetic management.
- It is indefensible to passively manage small isolated populations to extinction (without augmenting gene flow) if there are other populations of the same species adapted to similar environments and without fixed chromosomal differences, from which gene flow can be augmented.
- Almost any regime of augmented gene flow is likely to be beneficial in such circumstances, thereby reducing unnecessary population extinctions.
- Augmenting gene flow is likely to be a highly cost-effective management option, improving prospects for species to persist in the face of other threats.
- Management practices will have to facilitate ongoing adaptation to rapidly changing environments.
- Should you require assistance to implement genetic management, there are conservation and evolutionary geneticists who can assist you.

We trust that this book will contribute to more effective genetic management of fragmented populations in particular, and more informed consideration of the importance of conserving genetic diversity in general.

Glossary

Revised and updated from Frankham et al. (2010, 2014).

Adaptive evolution: Genetic change due to natural selection that improves the fitness of a population to its environment.

Additive: Locus where the heterozygote has a mean phenotype exactly intermediate between the two homozygotes. Also referred to as additivity.

Additive variance: The proportion of the quantitative genetic variation in a population due to variance in the average effects of alleles.

AFLP: See amplified fragment length polymorphism.

Allelic diversity: Average number of alleles per locus, a measure of genetic diversity within a population.

Allopatric: Populations or species whose geographic distributions do not overlap.

Allopolyploid: A species whose chromosomal complement derives from the full chromosome complements of two or more species, e.g. allotetraploid. Many plant species have evolved via this mechanism of instantaneous speciation.

Allotetraploid: A species whose chromosomal complement derives from the combination of two separate diploid species (compare autotetraploid).

Allozymes: Alternative forms of a protein detected by electrophoresis and protein staining that are due to alternative alleles at a single locus.

Amphidiploid: An allopolyploid species that shows chromosomal segregation as if it were a diploid, e.g. bread wheat has 42 chromosomes, with 7 pairs derived from each of three progenitor species, but normally produces gametes with 21 chromosomes.

Amplified fragment length polymorphism (AFLP): Method for detecting genetic variation across the genome by cutting genomic DNA with an enzyme that cleaves particular DNA sequences, adding short synthetic adaptor DNA fragments of known sequence to the cut ends, carrying out PCR using primers that complement the adapter sequence, running amplified DNA on a gel, and visualizing the fragments. A multilocus DNA fingerprint is produced.

Asexual: Reproduction without fusion of gametes resulting in offspring identical to parents (clonal), apart from new mutations.

Assignment test: Assigning an individual to one of several populations, based upon highest relative probability of individual's genotype deriving from that population.

Assisted colonization: Human assisted movement of wild plant or animal individuals to a location outside their historical distribution.

Autopolyploid: A species derived by meiotic error that combines full sets of chromosomes from the same diploid species (compare allopolyploid), e.g. autotetraploid.

Autotetraploid: A species derived by the combining of two full sets of chromosomes from the same species (compare allotetraploid). One form of autopolyploidy.

Backcross: Cross of F_1 progeny to one of the parental genotypes.

Balancing selection: Selection that maintains genetic variation in a population, encompassing heterozygote advantage (overdominance), rare-advantage selection, and particular forms of selection that vary over space or time.

Base population: Population from which studied population(s) were founded.

Bayesian: Methods of statistical analysis that incorporate other (prior) information, originally devised by Thomas Bayes.

Binomial distribution: The distribution describing the number of occurrences of two (or more) independent events in a sample of size *n*, e.g. the number of heads and tails in 50 tosses of a coin.

Biological Species Concept (BSC): Concept that defines reproductively isolated units as species. Gene flow is possible within species, but weak or absent between them.

Biparental inbreeding: Inbreeding due to mating of relatives more remote than self (full-sibs, half-sibs, cousins, etc.).

Bottleneck: A sudden restriction in population size, resulting in loss of genetic diversity, often followed by recovery of population size, but without recovery of pre-bottleneck genetic diversity.

Breeding system: Characteristics of a flower that influence gamete transfer among conspecifics, e.g. the spatial and temporal arrangement of male and female reproductive organs within a flower.

Catastrophe: An extreme environmental fluctuation that has a devastating impact on a population, e.g. hurricane, drought, extreme winter, disease epidemic, etc.

Census population size: Number of individuals in a population, sometimes potentially breeding adults (compare effective population size).

Centric fusion: Chromosome rearrangement where two rod shaped chromosomes (with centromeres at or near one end) fuse, resulting in V or J shaped chromosomes.

Centromere: Region of a chromosome with a constriction that attaches to spindle fibers during mitosis and meiosis.

Chloroplast DNA (cpDNA): Circular DNA molecules found in the chloroplasts of plants. They are usually maternally inherited in angiosperms (sometimes biparentally), but paternally inherited in gymnosperms.

Chromosomal translocation: A chromosomal variant where segments of non-homologous chromosomes swap locations.

Cline: Continuous change in genetic composition of a population over a region (often an environmental gradient), such as a latitudinal or an altitudinal cline.

Clones: Individuals with identical genotypes (bar mutations), e.g. cuttings deriving from a single plant, individuals produced by asexual reproduction from a single parent, or multiple individual animals derived from a single animal by nuclear transplantation.

Coadapted: Combination of alleles that have higher fitness in combination than expected from the sum of their average effects.

Coalesce: At the point in a gene tree where DNA sequence lineages converge on a common ancestor, they are said to coalesce.

Coalescence: The study of the convergence of DNA sequence lineages on common ancestors, to make biological inferences based on the numerical properties of gene trees.

Coancestry: The coancestry of two individuals is the probability that two alleles, one from each individual, are identical by descent. Synonymous with kinship.

Common ancestor: An individual that is an ancestor of both parents of an individual.

Common garden experiment: Comparison of different genotypes contemporaneously in the same environment to distinguish genetic differences from environmental ones, usually for quantitative characters.

Conspecific: Belonging to the same species.

Corridor: Ribbon of habitat between population fragments.

cpDNA: See chloroplast DNA.

Critically Endangered: A species with a very high probability of extinction within a short time, e.g. 50% probability of extinction within 10 years, or 3 generations, whichever is longer.

Decision tree: A diagram with a series of connected questions with dichotomous answers (yes or no) to aid in making decisions on some issue.

Demographic stochasticity: Random fluctuation in numbers of births and deaths, sex ratio, immigration and emigration that may drive a small population to extinction.

Demography: The study of how vital rates, such as fecundity, survival, and migration influence population growth and persistence.

Differential Fitness Species Concept (DFSC): Defines species as groups of individuals that are reciprocally characterized by features that would have negative fitness effects in other groups, and that cannot be regularly exchanged between groups upon contact.

Dioecious: Individuals within a population having separate sexes, especially used to describe plant reproductive systems (compare hermaphrodite).

Diploid: A genome with two doses of each chromosome (apart from sex-chromosomes, if present).

Directional selection: Selection in which the most extreme high (or low) ranked individuals on some trait(s) from a population are most successful as parents of the next generation.

Disomic: A pattern of inheritance whereby homologous chromosomes pair and segregate into haploid copies in gametes (the typical pattern of pairing and segregation observed in diploids and amphidiploids).

Disruptive selection: Selection of varying direction within the range of a species, e.g. favored melanic peppered moths in polluted areas, but non-melanic forms in non-polluted areas.

DNA fingerprint: The "bar code" produced by visualizing minisatellite regions within the genome of an individual, obtained by cutting DNA with a restriction enzyme, size separating by electrophoresis on a gel, probing with a minisatellite DNA sequence, and visualizing areas of complementary base pairing. Also called Variable Number Tandem Repeats (VNTR).

Dominance (*d*): Deviation of the heterozygote phenotype from the mean of the homozygotes phenotypes at a locus (typically a quantitative trait one).

Ecotypes: Populations within a species that are genetically adapted to different ecological conditions, often of soil and/or climate.

Effectively neutral: The situation where the selective forces on an allele are so weak that it behaves as if it were not subject to natural selection. Occurs when the selection coefficient is less than ~ $1/(2N_e)$, where N_e is the effective population size.

Effective number of alleles (n_e): The number of alleles that if equally frequent would result in the observed heterozygosity.

Effective population size: (N_e) The number of individuals that would result in the same loss of genetic diversity, inbreeding, genetic drift, or coalescence if they behaved in the manner of an idealized population.

Electrophoresis: A method for separating proteins or DNA fragments in a gel according to one or more of their net electrical charge, shape, and size.

Endangered: A species or population with a high probability of extinction within a short time, e.g. a 20% probability of extinction within 20 years or 10 generations, whichever is longer.

Endemic: A population or species found in only one geographic area or country.

Environmental stochasticity: Effects of natural fluctuations in environmental conditions, such as rainfall, food supply, competitors, winter temperatures, etc., on demographic parameters of a species and that may drive small populations to extinction.

Epigenetic: Heritable changes in gene activity that are not the result of changes in DNA sequence, but of DNA methylation, histone modification, etc. Often last only a single generation, but some may persist for a few generations, especially in plants.

Epistasis: Non-additive interactions among gene loci in their effects on a phenotype.

Equilibrium: State where a population or other entity has no tendency to change from its present condition across time.

ESUs: See evolutionarily significant units.

Evolution: Change in the genetic composition of a population.

Evolutionary potential: The ability of a population to evolve to cope with environmental changes. Often equated with genetic diversity, especially for quantitative characters such as fitness.

Evolutionary rescue: Recovery by adaptive evolution of a population or species that is threatened by environmental change, by utilizing pre-existing genetic diversity and/or new mutations.

Evolutionary Species Concept (ESC): A species definition based on delineating species on the basis of distinct lineages of ancestral descent that maintain their identity from other such lineages.

Evolutionarily significant units (ESUs): Partially genetically differentiated populations that justify management as separate units. Definition may be for neutral markers and/or adaptive differences according to diverse definitions.

Exchangeability: A measure of the ability of genotypes to successfully replace each other in their home environments, especially in having similar reproductive fitness.

Expected heterozygosity (H_e): The heterozygosity expected for a random mating population with given allele frequencies according to the Hardy–Weinberg equilibrium.

Extinction: Permanent disappearance of a population or species.

Extinction vortex: Describes the likely adverse interactions between human impacts, inbreeding, and demographic fluctuation that result in a reinforcing feedback loop and spiral downwards in population size towards extinction.

***F* statistics:** Measures of total inbreeding in a population (F_{IT}), partitioned into that due to inbreeding within fragments (F_{IS}) and that due to differentiation among them (F_{ST}).

Fitness: Reproductive fitness, the number of fertile offspring that survive to reproductive age, contributed by an individual (lifetime reproductive success).

Fixation: All individuals in a population being identically homozygous at a locus, e.g. all A_1A_1.

Fixed gene differences: Populations homozygous for different alleles or haplotypes at one or more loci, or alternatively sharing no alleles or haplotypes at one or more loci.

Forensics: Application of science to the law, including detection by scientific means of illegal activities (including wildlife poaching).

Full-sib mating: A mating between a brother and a sister having both parents in common.

Gene diversity: The extent of genetic variation at a locus in a population; Hardy–Weinberg expected heterozygosity.

Gene flow: Movement of alleles between populations via migrants or gametes.

Gene genealogies: Trees showing the relationships between different copies of a single locus or DNA segment (gene trees).

General Lineage Species Concept: Concept that defines species as components of separately evolving metapopulation lineages.

Genetic distance: A measure of the genetic difference between allele frequencies in two populations or species, e.g. Nei's genetic distance.

Genetic diversity: The extent of genetic variation in a population, or species, or across a group of species, e.g. heterozygosity, or allelic diversity, or heritability.

Genetic drift: Changes in the genetic composition of a population due to random sampling in finite populations. Also referred to as random genetic drift.

Genetic erosion: The process whereby small populations lose genetic diversity and become inbred, leading to inbreeding depression, reduced ability to evolve, and elevated extinction risk.

Genetic load: The amount of deleterious alleles in a population, some due to mutation-selection balance (mutation load), and others to heterozygote advantage and other forms of balancing selection (balanced load).

Genetic rescue: Improvement in reproductive fitness and increase in genetic diversity due to crossing a population previously suffering from inbreeding and low genetic diversity to another distinct population.

Genetic stochasticity: Genetic consequences of chance effects in small populations, including inbreeding, loss of genetic diversity due to genetic drift, and chance fixations of deleterious mutations that reduce fitness, and may drive a population or species to extinction.

Genetic swamping: Substantial gene flow from another population, especially a related introduced species, resulting in a dilution of native genetic composition and potentially reduced local adaptation.

Gene trees: Trees showing the relationships between different copies of a single locus (gene genealogies), typically derived on the basis of DNA sequences.

Genome: The entire DNA or all of the chromosomes in an individual, species, or organelle.

Genotype × environment interaction: Differential performance of diverse genotypes in dissimilar environments, e.g. the relative performance of two genotypes switched between alternative environments.

Gynodioecious: Species containing some individual plants with female flowers and others with hermaphroditic flowers.

Haplodiploid: Chromosome numbers in a species whereby females are diploid and males haploid, e.g. Hymenoptera. Females result from fertilized eggs and males typically from unfertilized ones.

Haploid: Gamete or individual with one dose of each chromosome or locus.

Haplotype: Haploid allelic composition for several different loci or DNA bases in a chromosomal region, e.g. $A_1B_3C_2$.

Haplotype diversity: A measure of genetic diversity among haplotypes that is a DNA sequence analogue of heterozygosity.

Hardy-Weinberg equilibrium: The equilibrium genotype frequencies achieved in a random mating population with no perturbing forces from mutation, migration, selection, or chance. If two alleles A_1 and A_2 have frequencies of p and q, the Hardy–Weinberg equilibrium frequencies for the A_1A_1, A_1A_2, and A_2A_2 genotypes are p^2, $2pq$, and q^2, respectively, and are attained after one generation of random mating.

Heritability (h^2): Proportion of the variation for a quantitative character due to (additive) genetic causes, estimated from resemblances between relatives.

Hermaphrodite: An animal or plant with both sexes present in single individuals. In hermaphroditic plants, male and female functions occur within each flower.

Heterochromatin: Blocks of highly repeated DNA sequences that may be differentially stained. Often located around centromeres.

Heterosis: Hybrid vigor. Superior performance of hybrid genotypes for a quantitative character, usually indicating superiority to both parents.

Heterozygosity: Proportion of heterozygous individuals for a locus (or set of loci) in a population.

Heterozygote advantage: A form of selection where the heterozygote has a higher reproductive fitness than the homozygotes. Also referred to as overdominance.

Heterozygote disadvantage: An uncommon form of selection where the heterozygote has a lower reproductive fitness than the homozygotes, as may be found at some loci when distinct species are crossed.

Idealized population: A conceptual random mating population with equal numbers of hermaphrodite individuals breeding in each generation and Poisson variation in family sizes (mean = variance). Used as a standard against which other populations are equated when defining effective population sizes.

Identity by descent: Alleles that are identical copies of an allele present in a common ancestor.

Inbreeding: The production of offspring from mating of individuals related by descent, e.g. self-fertilization, brother × sister, cousin × cousin matings, etc.

Inbreeding coefficient (*F*): The probability that two alleles at a locus in an individual are identical by descent. Used to measure the magnitude of inbreeding.

Inbreeding depression: Reduction in mean for a quantitative trait due to inbreeding, especially manifest in reproductive fitness traits.

***In situ* conservation:** Conservation of a species or population in its normal wild habitat.

Integron: A mobile DNA element found in bacteria that can capture and carry gene loci.

Introgression: Introduction of genetic material from another species or subspecies into a population. A threat to the genetic integrity of some canid, fish, plant, etc. species (high levels of introgression result in genetic swamping).

Invasive species: A species introduced into a new range, that establishes and spreads, typically invading large areas and adversely affecting many native species.

Inversion: A chromosome aberration involving two breaks within a chromosome, turning the middle section through 180 degrees, and re-joining the ends such that gene order is changed, say from ABCDE to ACBDE.

Isolation by distance: Describes a distribution of population genetic composition where individuals show increasing genetic differentiation with increasing geographic distances.

Isolation by environment: Describes a distribution of population genetic composition where individuals show increasing genetic differentiation with increasing differences in their environments.

IUCN: International Union for Conservation of Nature.

Kinship coefficient (k_{ij}): The probability that two alleles, one from each of two individuals, are identical by descent (also termed coancestry). The inbreeding coefficient of any progeny of the two individuals.

Kinship matrix: Table of pairwise kinship coefficients between different individuals or populations.

Landscape genetics: Study of patterns of genetic diversity, gene flow, and adaptation across landscapes.

Lethal: Inconsistent with survival, as in a recessive lethal allele that results in death when homozygous.

Lethal equivalents (*B*): A measure for comparing the extent of inbreeding depression in different populations. A group of detrimental alleles that would cause death if homozygous, e.g. one lethal allele, two alleles each with a 50% probability of causing death, etc. Typically estimated from the slope of the regression of the natural logarithm of survival on the inbreeding coefficient *F*.

Lineage sorting: Random loss of genetic variants by genetic drift in different lineages deriving from a polymorphic common ancestral species (or population), resulting over generations in differentiated lineages.

Linkage disequilibrium: Non-random association of alleles at different loci (usually loci co-located on a chromosome).

Locus: Originally used for functional DNA unit coding for a protein, but now often used for a segment of DNA on a chromosome, as in a microsatellite locus.

Major histocompatibility complex (MHC): A large family of loci that play an important role in the vertebrate immune system and in fighting disease.

Maternal effect: Effect of maternal environment on phenotype of an offspring, e.g. effect on offspring weight and survival of nutrient supply from mother. Can also occur across more than one generation.

Mating system: In plants, refers to the proportion of mating events resulting from selfing and/or mating between related individuals compared to mating events between unrelated individuals (outcrossing). In animals, refers to which males and females mate, degree of promiscuity, etc.

Matrix (of habitat): Habitat separating fragments of a population, as in cleared pasture land between forest fragments.

Maximum likelihood: Statistical method used to estimate a parameter that maximizes the probability of the observed result.

Mean kinship (*mk*): The average kinship of an individual with all individuals in a population, including itself. It may also be used to measure the similarity between different populations. Minimizing mean kinship is the recommended method for genetically managing endangered species in captivity and we recommend its use in management of fragmented populations in the wild.

Meta-analysis: A statistical analysis based on the combined information from many different studies.

Metapopulation: A spatially distributed group of partially isolated population fragments of the same species that undergo local extinctions and re-colonizations.

MHC: See major histocompatibility complex.

Microsatellite: A locus with a short tandemly repeated DNA sequence (typically 1–5 bases in length), such as the sequence AC repeated 10 times. Typically shows variable number of repeats and high heterozygosities in populations.

Migration: Movement of individuals or gametes between populations, as in gene flow between populations.

Mitochondrial DNA (mtDNA): The circular DNA molecules contained within mitochondria that code for several proteins involved in energy metabolism and their expression machinery. Usually maternally inherited in animals and plants.

Mixed mating: Populations or species that exhibit both self-fertilization and outcrossing, with selfing within the range of 20–80%.

Monomorphic: The presence of only one allele at a locus, generally taken to mean the most common allele is at a frequency of greater than 99% or 95%. Contrast with polymorphic.

Monophyletic: A group of species (or DNA sequences) that derive from the same common ancestral species (or DNA sequence). Converse is polyphyletic.

mtDNA: See mitochondrial DNA.

Mutation: A sudden genetic change, i.e. parents lack the allele, but it is present in one or more of their offspring or a cell lacks the allele, but it is found in a daughter cell.

Mutational meltdown: Decline in reproductive rate and consequent downward spiral in population size towards extinction, due to chance fixation of new mildly deleterious mutations in small populations.

Mutation load: The load of harmful alleles carried in a population's genome.

Mutational variance: Increase in quantitative genetic variation per generation due to new mutations in the genome.

Mutation-selection balance: The equilibrium between the spontaneous occurrence of harmful mutations and natural selection removing them, resulting in low frequencies (typically $< 1\%$) of harmful alleles at individual loci in populations. However, this occurs at many loci across the genome.

Natural selection: Differential mortality and/or reproduction among individuals in a population due to natural environmental processes that alters the genetic composition of a population if the differences are heritable.

Nei's genetic distance (D_N): The most widely used measure of genetic difference between allele frequencies in two populations or species (devised by Masatoshi Nei).

Neutral mutation: A mutation that is equivalent in effects on reproductive fitness to the existing allele.

Nucleotide diversity (π): A measure of genetic diversity at the nucleotide level. Heterozygosity at the nucleotide level in a random mating population.

Outbred: An individual whose parents are unrelated.

Outbreeding: Not inbreeding. Approximately random mating within a population.

Outbreeding depression: A reduction in reproductive fitness compared to either parent due to crossing between two genetically divergent populations (or subspecies or species) caused, for instance by some combination of fixed chromosomal differences, and/or disruption of local adaptation and long isolation.

Outcrossing: Proportion of matings in hermaphroditic species that are not-self. Primarily outcrossing species have outcrossing rates of > 0.8 (< 20% selfing). In animals, may be used for crossing of an individual or population to another that is not closely related.

Outcrossing rate: Proportion of successful fertilizations in a plant population between non-related individuals within the same population or between other populations.

Overdominance: Heterozygote advantage; a form of selection where heterozygotes have a higher fitness than homozygotes.

Parapatric: A geographic distribution of populations where they abut each other.

Parthenogenesis: Reproduction without mating or fertilization, as in virgin birth. May result in offspring identical to their parent (bar mutations), or may involve meiotic division with segregation and recombination.

Partial dominance: Where the heterozygote has a phenotype or gene expression closer to one homozygote than the other, e.g. heterozygotes for most deleterious alleles are nearly, but not completely normal (compare dominant, recessive, and additive).

Pedigree: A chart specifying lines of descent and relationship among individuals.

Peripheral trait: A character with limited relationship to reproductive fitness, e.g. bristle number in fruit flies or tail length in rodents.

Phenological: Concerned with the timing of periodic events in the life cycle of an animal or plant population, in relation to climatic conditions, e.g. change with temperature in first flowering date in plant populations, or arrival date of the first migrants in migratory birds.

Phenotypic plasticity: Variation in the phenotype of individuals with the same genotype, as a response to different environments, e.g. hemoglobin levels will change if one lives at high altitude for several months, or plants of the same genotype grown with different nutrient levels will produce different biomass.

Phylogenetic Species Concepts (PSC): Defines species as diagnosably different population segments (variously based on fixed gene differences, lack of shared alleles, or reciprocal monophyly), irrespective of whether they are inherently reproductively isolated (compare Biological Species Concept).

Phylogenetic tree: A tree representing the closeness of relationship between species or populations.

Phylogeography: Study of the geographic distribution of genealogical lineages, especially within species.

Polyandry: A mating system whereby a female mates with more than one male at approximately the same time.

Polymorphic: The presence of more than one allele at a locus, generally taken to mean the most common allele is at a frequency of less than 99% or 95% (compare with monomorphic).

Polyphyletic: A group of DNA sequences, individuals, or species that derives from more than one ancestral DNA sequence, individual, or species.

Polyploid: Having more than two doses of each chromosome, e.g. tetraploid (4n).

Population: A group of individuals of the same species that could potentially interbreed with each other.

Population fragmentation: Destruction of habitat, resulting in a population's continuous distribution being converted into separate spatial fragments, as in clearing paths for roads through a forest or in partially clearing a forest for agriculture.

Population genomics: The study of variation and evolution of genomes.

Population viability analysis (PVA): A systems modeling approach for predicting the fate of a population (including risk of extinction) due to the combined effects of all systematic and stochastic threats faced by a population. Typically, population size, means and standard deviation of birth and death rates, density feedbacks, plus risks and severity of catastrophes, inbreeding depression, etc. are input into a software package and many replicates projected over multiple generations using stochastic computer simulation. Used as a management and research tool in conservation biology.

Private allele: Allele found in only a single population.

Purging: Reducing the frequency of deleterious alleles (frequently partial recessives) by natural selection, especially in populations that have suffered size bottlenecks and/or inbreeding.

PVA: See population viability analysis.

QTL: See quantitative trait loci.

Quantitative character: Typically a trait with a continuous distribution influenced by genetic variation (typically at many loci) and environmental variation, e.g. fecundity, survival, height, behavior, and weight.

Quantitative genetic variation: Genetic variation affecting a quantitative character. Presumed to be due to the cumulative effects of variation at many quantitative trait loci (QTL).

Quantitative trait loci (QTL): Loci affecting quantitative characters (also referred to as polygenes).

Random genetic drift: Changes in the genetic composition of a population due to random sampling in finite populations.

Random mating: A pattern of mating where the chances of two genotypes or phenotypes breeding is determined by their frequencies in the population (synonym panmictic).

Range contraction: Reduction in the geographic extent of distribution of a species or population.

Range expansion: Increase in the geographic extent of distribution of a species or population.

Reciprocal monophyly: Two groups of populations (or DNA sequences) with all members of the first more closely related to each other than to the second group, and vice versa.

Reintroduction: Returning a species or population to part of its former range where it had become extinct, using individuals from captive populations.

Relatedness: Proportion of alleles in two individuals that are identical by descent, equal to twice the inbreeding coefficient in their offspring (if they could have them) when the individuals are not inbred.

Relative fitness: The fitness of a genotype compared to that of another genotype, usually at the same locus, e.g. if fitnesses of genotypes at a locus conferring warfarin resistance are 30%, 80%, and 54% for RR, RS, and SS genotypes, then their relative fitnesses are 30/80 = 0.375, 80/80 = 1, and 54/80 = 0.68, respectively.

Reproductive fitness: The number of fertile offspring surviving to reproductive age, contributed by an individual. Encompasses mating ability, fertilization capacity, fecundity, and survival (lifetime reproductive success). Often referred to as fitness.

Restriction fragment length polymorphism (RFLP): Genetic diversity detected by cutting DNA with an enzyme that identifies and cuts at particular sequences, followed by electrophoretic separation and visualization of the resulting fragments.

RFLP: See restriction fragment length polymorphism.

Selection coefficient (*s*): Difference in relative fitness between a genotype at a locus and the genotype with the highest fitness, e.g. if three genotypes A_1A_1, A_1A_2, and A_2A_2 at a locus have relative fitnesses of 1, 1, and 0.9, the selection coefficient for the A_2A_2 genotype is $s = 1 - 0.9 = 0.1$.

Selection differential (*Sd*): A measure of the magnitude of selection on a quantitative character; difference in mean between the (selected) parents and the mean of the total population from which they derived.

Selective sweep: Action of natural selection driving a single allele to fixation and at the same time reducing genetic diversity for surrounding DNA (often not under selection itself). Common in genome regions with low genetic recombination.

Self-incompatible: Genetically based inability of an individual (usually plant) to produce offspring following attempted self-fertilization. Many plant species have loci that control self-incompatibility.

Selfing: Self-fertilizing.

Self-sterility: See self-incompatible.

Single large or several small (SLOSS): The concept that compares the consequences of a single large population versus several small populations of equivalent total size, especially in terms of their loss of genetic variation and extinction risk.

Single nucleotide polymorphism (SNP): A position in DNA of a species or population at which two or more alternative bases occur at appreciable frequency (> 1%).

SLOSS: See single large or several small.

SNP: See single nucleotide polymorphism.

Source-sink: A population structure where one population, the source, is permanent and supplies individuals to restart one or more transient (sink) populations.

Speciation: The processes by which populations diverge and become reproductively isolated, turning into different species.

Species: Mayr defined species as "groups of actually or potentially interbreeding natural populations which are reproductively isolated from other such groups" according to the Biological Species Concept. There are many other definitions of species.

Stabilizing selection: Selection favoring phenotypic intermediates for a quantitative character at the expense of phenotypic extremes.

Statistical power: The ability to reject an erroneous null hypothesis.

Stochastic: Having a chance element. Having variable outcomes described by a probability distribution, e.g. environmental and demographic stochasticity, genetic drift.

Sub-species: Taxonomic units within species that are partially genetically differentiated populations, often located in separate geographic regions.

Sympatric: Populations that share the same or largely overlapping geographic distributions.

Tandem repeats: Multiple copies of the same DNA sequence lying one after another in a series, as in microsatellite repeats, or ribosomal RNA loci.

Taxa: Several populations belonging to a taxonomic unit, e.g. several species or several sub-species, etc. (Singular taxon.)

Taxonomic Species Concept (TSC): A method of delineating species based on the determination of a taxonomist without specifying a defined species concept.

Telomere: Short tandem DNA repeats at the ends of eukaryotic chromosomes that stabilize them.

Tetraploid: Species with four copies of each chromosome.

Tetrasomic: A pattern of inheritance where there are four copies of a chromosome (or allele) that pair at random and segregate to give diploid gametes (compare disomic), e.g. selfing of a plant with genotype $A_1A_1A_1A_2$ results in equal numbers of A_1A_1 and A_1A_2 gametes, and a 1:2:1 ratio of $A_1A_1A_1A_1$: $A_1A_1A_1A_2$: $A_1A_1A_2A_2$ genotypes in the offspring. Exhibited by recently evolved autotetraploid species.

Threatened: A population or species that has a finite risk of extinction within a relatively short time frame, say a greater than 10% risk of extinction within 100 years. Under the IUCN system this is the sum of the Critically Endangered, Endangered, and Vulnerable categories.

Translocation (physical movement): The movement of an individual animal or plant from one wild location to another as a result of human actions (distinguish from a chromosomal translocation).

Transposons: Mobile genetic elements found in species from bacteria to plants and animals.

Variance: The most commonly used measure of dispersion among quantitative measurements. The average of the squared deviations from the mean. The square of the standard deviation.

VNTR: Variable number tandem repeat; see DNA fingerprint.

Vulnerable: A species or population with a tangible risk of extinction within a moderate time, e.g. a 10% probability within 100 years.

Wahlund effect: Reduction in heterozygosity, compared to Hardy–Weinberg expectations, in a population split into partially isolated population fragments (named after its discoverer).

References

Abplanalp, H., 1990. Inbreeding, In Poultry Breeding and Genetics. ed. R.D. Crawford, pp. 955–984. Elsevier, Amsterdam.

Adams, J.R., Vucetich, L.M., Hedrick, P.W., et al., 2011. Genomic sweep and potential genetic rescue during limiting environmental conditions in an isolated wolf population. Proceedings of the Royal Society B: Biological Sciences 278, 3336–3344.

Agapow, P.M., Bininda-Emonds, O.R., Crandall, K.A., et al., 2004. The impact of species concept on biodiversity studies. Quarterly Review of Biology 79, 161–179.

Agrawal, A.A., 2001. Phenotypic plasticity in the interactions and evolution of species. Science 294, 321–326.

Aguilar, R., Quesada, M., Ashworth, L., et al., 2008. Genetic consequences of habitat fragmentation in plant populations: susceptible signals in plant traits and methodological approaches. Molecular Ecology 17, 5177–5188.

AIMS, 2016. Coral bleaching events. Australian Institute of Marine Science. http://www.aims.gov.au/docs/research/climate-change/coral-bleaching/bleaching-events.html.

Aitken, S.N., Whitlock, M.C., 2013. Assisted gene flow to facilitate local adaptation to climate change. Annual Review of Ecology, Evolution, and Systematics 44, 367–388.

Alberto, F.J., Aitken, S.N., Alía, R., et al., 2013. Potential for evolutionary responses to climate change—evidence from tree populations. Global Change Biology 19, 1645–1661.

Allendorf, F.W., Luikart, G., 2007. Conservation and the Genetics of Populations. Blackwell, Malden, MA.

Allendorf, F.W., Waples, R.S., 1996. Conservation and genetics of salmonid fishes, In Conservation Genetics: Case Histories from Nature. eds J.C. Avise, J.L. Hamrick, pp. 238–280. Chapman & Hall, New York.

Allendorf, F.W., Luikart, G., Aitken, S.N., 2013. Conservation and the Genetics of Populations, 2nd edition. Wiley-Blackwell, Oxford, UK.

Amberger, J.S., Bocchini, C.A., Schiettecatte, F., et al., 2015. OMIM.org: Online Mendelian Inheritance in Man (OMIM®), an online catalog of human genes and genetic disorders. Nucleic Acids Research 43, D789–D798.

American Rivers, 2016. https://www.americanrivers.org/.

Amos, J.N., Harrisson, K.A., Radford, J.Q., et al., 2014. Species- and sex-specific connectivity effects of habitat fragmentation in a suite of woodland birds. Ecology 95, 1556–1568.

Anderson, J.T., Inouye, D.W., McKinney, A.M., et al., 2012. Phenotypic plasticity and adaptive evolution contribute to advancing flowering phenology in response to climate change. Proceedings of the Royal Society of London B: Biological Sciences 279, 3843–3852.

Anderson, J.T., Willis, J.H., Mitchell-Olds, T., 2011. Evolutionary genetics of plant adaptation. Trends in Genetics 27, 258–266.

Angeloni, F., Ouborg, N.J., Leimu, R., 2011. Meta-analysis on the association of population size and life history with inbreeding depression in plants. Biological Conservation 144, 35–43.

Anonymous, 2003. Wolves under threat as rabies outbreak bites. Nature 425, 892.

Anonymous, 2014. No genetic rescue for Island wolves. Science 344, 240–241.

Armbruster, P., Reed, D.H., 2005. Inbreeding depression in benign and stressful environments. Heredity 95, 235–242.

Arnason, U., Gullberg, A., Janke, A., et al., 1996. Pattern and timing of evolutionary divergences among hominoids based on analyses of complete mtDNAs. Journal of Molecular Evolution 43, 650–661.

Arnold, M.L., Hodges, S.A., 1995. Are natural hybrids fit or unfit relative to their parents? Trends in Ecology and Evolution 10, 67–71.

Arnold, M.L., Sapir, Y., Martin, N.H., 2008. Genetic exchange and the origin of adaptations: prokaryotes to primates. Philosophical Transactions of the Royal Society of London B: Biological Sciences 363, 2813–2820.

Arntzen, J.W., Jehle, R., Bardakci, F., et al., 2009. Asymmetric viability of reciprocal-cross hybrids between crested and marbled newts (*Triturus cristatus* and *T. marmoratus*). Evolution 63, 1191–1202.

Aslan, C.E., Zavaleta, E.S., Tershy, B., et al., 2013. Mutualism disruption threatens global plant biodiversity: a systematic review. PLOS ONE 8, e66993.

Avise, J.C., Ball, R.M.J., 1990. Principles of genealogical concordance in species concepts and biological taxonomy. Oxford Surveys in Evolutionary Biology 7, 45–67.

Baden, A.L., Holmes, S.M., Johnson, S.E., et al., 2014. Species-level view of population structure and gene flow for a critically endangered primate (*Varecia variegata*). Ecology and Evolution 4, 2675–2692.

Baker, R.J., Bickham, J.W., 1986. Speciation by monobrachial centric fusions. Proceedings of the National Academy of Sciences 83, 8245–8248.

Balkenhol, N., Cushman, S., Storfer, A., et al., 2015. Landscape Genetics: Concepts, Methods, Applications. John Wiley & Sons, Chichester, UK.

Ball, S.J., Adams, M., Possingham, H.P., et al., 2000. The genetic contribution of single male immigrants to small, inbred populations: a laboratory study using *Drosophila melanogaster*. Heredity 84, 677–684.

Ballard, J.W.O., Rand, D.M., 2005. The population biology of mitochondrial DNA and its phylogenetic implications. Annual Review of Ecology, Evolution and Systematics 36, 621–642.

Ballard, J.W.O., Whitlock, M.C., 2004. The incomplete natural history of mitochondria. Molecular Ecology 13, 729–744.

Ballou, J.D., 1983. Calculating inbreeding coefficients from pedigrees, In Genetics and Conservation: A Reference for Managing Wild Animal and Plant Populations. eds C.M. Schonewald-Cox, S.M. Chambers, B. MacBryde, L. Thomas, pp. 509–520. Benjamin/Cummings, Menlo Park, CA.

Ballou, J.D., 1995. Genetic Management, Inbreeding Depression and Outbreeding Depression in Captive Populations. Ph.D. thesis, University of Maryland, College Park, MD.

Ballou, J.D., 1997. Effects of ancestral inbreeding on genetic load in mammalian populations. Journal of Heredity 88, 169–178.

Ballou, J.D., Lacy, R.C., 1995. Identifying genetically important individuals for management of genetic diversity in pedigreed populations, In Population Management for Survival and Recovery: Analytical Methods and Strategies in Small Population Conservation. eds J. Ballou, M. Gilpin, T. Foose, pp. 76–111. Columbia University Press, New York.

Ballou, J.D., Ralls, K., 1982. Inbreeding and juvenile mortality in small populations of ungulates: a detailed analysis. Biological Conservation 24, 239–272.

Ballou, J.D., Lacy, R.C., Pollak, J.P., 2011. PMx: software for demographic and genetic analysis and management of pedigreed populations (version 1.0). Chicago Zoological Society, Brookfield, IL.

Ballou, J.D., Lees, C., Faust, L.J., et al., 2010. Demographic and genetic management of captive populations, In Wild Mammals in Captivity, 2nd edition. eds D.G. Kleiman, K.V. Thompson, C. Kirk-Bauer, pp. 219–252. University of Chicago Press, Chicago, IL.

Balloux, F., Goudet, J., 2002. Statistical properties of population differentiation estimators under stepwise mutation in a finite island model. Molecular Ecology 11, 771–783.

Banes, G.L., Galdikas, B.M.F., Vigilant, L., 2016. Reintroduction of confiscated and displaced mammals risks outbreeding and introgression in natural populations, as evidenced by orang-utans of divergent subspecies. Scientific Reports 6, 22026, doi: 10.1038/srep22026.

Bangert, R.K., Turek, R.J., Martinsen, G.D., et al., 2005. Benefits of conservation of plant genetic diversity to arthropod diversity. Conservation Biology 19, 379–390.

Barber, H.N., 1955. Adaptive gene substitution in Tasmanian eucalypts. I. Evolution 9, 1–14.

Barber, H.N., Jackson, W.D., 1957. Natural selection in action in *Eucalyptus*. Nature 179, 1267–1269.

Barner, A.K., Pfister, C.A., Wootton, J.T., 2011. The mixed mating system of the sea palm kelp *Postelsia palmaeformis*: few costs to selfing. Proceedings of the Royal Society of London B: Biological Sciences 278, 1347–1355.

Barr, C.M., Neiman, M., Taylor, D.R., 2005. Inheritance and recombination of mitochondrial genomes in plants, fungi and animals. New Phytologist 168, 39–50.

Barrett, S., Jackson, T., 2007. Translocation proposal: feather-leaved banksia *Banksia brownii* R. Br (Proteaceae). Department of Environment and Conservation, Kensington, Western Australia.

Barrett, S.C., Charlesworth, D., 1991. Effect of a change in the level of inbreeding on the genetic load. Nature 352, 522–524.

Barrett, S.C.H., Kohn, J.R., 1991. Genetic and evolutionary consequences of small population size in plants: implications for conservation, In Genetics and Conservation of Rare Plants. eds D.A. Falk, K.E. Holsinger, pp. 3–61. Oxford University Press, New York.

Barringer, B.C., Geber, M.A., 2008. Mating system and ploidy influence levels of inbreeding depression in *Clarkia* (Onagraceae). Evolution 62, 1040–1051.

Barton, N., Whitlock, M.C., 1997. The evolution of metapopulations, In Metapopulation Biology: Ecology, Genetics, and Evolution. eds I.A. Hanski, M.E. Gilpin, pp. 183–210. Academic Press, San Diego.

Beaumont, L.J., Pitman, A., Perkins, S., et al., 2011. Impacts of climate change on the world's most exceptional ecoregions. Proceedings of the National Academy of Sciences 108, 2306–2311.

Becker, C., Hagmann, J., Müller, J., et al., 2011. Spontaneous epigenetic variation in the *Arabidopsis thaliana* methylome. Nature 480, 245–249.

Becker, W.A., 1984. Manual of Quantitative Genetics, 4th edition. Academic Enterprises, Pullman, Wash.

Beckmann, J.P., Clevenger, A.P., Huijser, M., et al., 2010. Safe Passages: Highways, Wildlife, and Habitat Connectivity. Island Press, Washington, DC.

Beerli, P., Felsenstein, J., 2001. Maximum likelihood estimation of a migration matrix and effective population sizes in *n* subpopulations by using a coalescent approach. Proceedings of the National Academy of Sciences 98, 4563–4568.

Beheregaray, L.B., Sunnucks, P., Alpers, D.L., et al., 2000. A set of microsatellite loci for the hairy-nosed wombats (*Lasiorhinus krefftii* and *L. latifrons*). Conservation Genetics 1, 89–92.

Beier, P., Spencer, W., Baldwin, R.F., et al., 2011. Toward best practices for developing regional connectivity maps. Conservation Biology 25, 879–892.

Beilharz, R.G., 1982. The effect of inbreeding on reproduction in mice. Animal Production 34, 49–54.

Beissinger, S.R., McCullough, D.R., 2002. Population Viability Analysis. University of Chicago Press, Chicago, IL.

Benson, J.F., Hostetler, J.A., Onorato, D.P., et al., 2011. Intentional genetic introgression influences survival of adults and subadults in a small, inbred felid population. Journal of Animal Ecology 80, 958–967.

Bergl, R.A., Vigilant, L., 2007. Genetic analysis reveals population structure and recent migration within the highly fragmented range of the Cross River gorilla (*Gorilla gorilla diehli*). Molecular Ecology 16, 501–516.

Berkelmans, R., van Oppen, M.J.H., 2006. The role of zooxanthellae in the thermal tolerance of corals: a "nugget of hope" for coral reefs in an era of climate change. Proceedings of the Royal Society B: Biological Sciences 273, 2305–2312.

Berthold, P., Helbig, A.J., Mohr, G., et al., 1992. Rapid microevolution of migratory behaviour in a wild bird species. Nature 360, 668–670.

Bever, J.D., Felber, F., 1994. The theoretical population genetics of autopolyploids. Oxford Surveys in Evolutionary Biology 8, 185–217.

Bickford, D., Lohman, D.J., Sodhi, N.S., et al., 2007. Cryptic species as a window on diversity and conservation. Trends in Ecology and Evolution 22, 148–155.

Bidau, C.J., 1991. Multivalents resulting from monobrachial homologies within a hybrid zone in *Dichroplus pratensis* (Acrididae): meiotic orientation and segregation. Heredity 66, 219–232.

Bieri, J., Kawecki, T.J., 2003. Genetic architecture of differences between populations of cowpea weevil (*Callosobruchus maculatus*) evolved in the same environment. Evolution 57, 274–287.

Bijlsma, R., Bundgaard, J., Boerema, A.C., 2000. Does inbreeding affect the extinction risk of small populations?: Predictions from *Drosophila*. Journal of Evolutionary Biology 13, 502–514.

Bijlsma, R., Bundgaard, J., Van Putten, W.F., 1999. Environmental dependence of inbreeding depression and purging in *Drosophila melanogaster*. Journal of Evolutionary Biology 12, 1125–1137.

Bijlsma, R., Westerhof, M.D.D., Roekx, L.P., et al., 2010. Dynamics of genetic rescue in inbred *Drosophila melanogaster* populations. Conservation Genetics 11, 449–462.

Bilde, T., Maklakov, A.A., Schilling, N., 2007. Inbreeding avoidance in spiders: evidence for rescue effect in fecundity of female spiders with outbreeding opportunity. Journal of Evolutionary Biology 20, 1237–1242.

Bingaman Lackey, L., 1999. International Giraffe Studbook. ISIS, Apple Valley, MN.

Binks, R.M., Prince, J., Evans, J.P., et al., 2012. More than bindin divergence: reproductive isolation between sympatric subspecies of a sea urchin by asynchronous spawning. Evolution 66, 3545–3557.

Birchler, J.A., Auger, D.L., Riddle, N.C., 2003. In search of the molecular basis of heterosis. The Plant Cell 15, 2236–2239.

Birky, C.W., Fuerst, P., Maruyama, T., 1989. Organelle gene diversity under migration, mutation, and drift: equilibrium expectations, approach to equilibrium, effects of heteroplasmic cells, and comparison to nuclear genes. Genetics 121, 613–627.

Biscotti, M.A., Olmo, E., Heslop-Harrison, J.S., 2015. Repetitive DNA in eukaryotic genomes. Chromosome Research 23, 415–420.

Bishop, J.A., Cook, L.M., 1975. Moths, melanism and clean air. Scientific American 232, 90–99.

Bittles, A.H., Neel, J.V., 1994. The costs of human inbreeding and their implications for variations at the DNA level. Nature Genetics 8, 117–121.

Bittner, T.D., King, R.B., 2003. Gene flow and melanism in garter snakes revisited: a comparison of molecular markers and island vs. coalescent models. Biological Journal of the Linnean Society 79, 389–399.

Blair, C., Weigel, D.E., Balazik, M., et al., 2012. A simulation-based evaluation of methods for inferring linear barriers to gene flow. Molecular Ecology Resources 12, 822–833.

Blomqvist, D., Pauliny, A., Larsson, M., et al., 2010. Trapped in the extinction vortex? Strong genetic effects in a declining vertebrate population. BMC Evolutionary Biology 10, 33.

Boakes, E.H., Wang, J., Amos, W., 2007. An investigation of inbreeding depression and purging in captive pedigreed populations. Heredity 98, 172–182.

Bogart, J., Bi, K., 2013. Genetic and genomic interactions of animals with different ploidy levels. Cytogenetic and Genome Research 140, 117–136.

Bohonak, A.J., 1999. Dispersal, gene flow and population structure. Quarterly Review of Biology 74, 21–45.

Bolnick, D.I., Near, T.J., 2005. Tempo of hybrid inviability in centrarchid fishes (Teleostei: Centrarchidae). Evolution 59, 1754–1767.

Borlase, S.C., Loebel, D.A., Frankham, R., et al., 1993. Modeling problems in conservation genetics using captive *Drosophila* populations: consequences of equalization of family sizes. Conservation Biology 7, 122–131.

Bosse, M., Megens, H-J., Madsen, O., et al., 2015. Using genome-wide measures of coancestry to maintain diversity and fitness in endangered and domestic pig populations. Genome Research 25, 970–981.

Bossuyt, B., 2007. Genetic rescue in an isolated metapopulation of a naturally fragmented plant species, *Parnassia palustris*. Conservation Biology 21, 832–841.

Bousalem, M., Viader, V., Mariac, C., et al., 2010. Evidence of diploidy in the wild Amerindian yam, a putative progenitor of the endangered species *Dioscorea trifida* (Dioscoreaceae). Genome 53, 371–383.

Bouzat, J.L., Cheng, H.H., Lewin, H.A., et al., 1998a. Genetic evaluation of a demographic bottleneck in the greater prairie chicken. Conservation Biology 12, 836–843.

Bouzat, J.L., Johnson, J.A., Toepfer, J.E., et al., 2009. Beyond the beneficial effects of translocations as an effective tool for the genetic restoration of isolated populations. Conservation Genetics 10, 191–201.

Bouzat, J.L., Lewin, H.A., Paige, K.N., 1998b. The ghost of genetic diversity past: historical DNA analysis of the greater prairie chicken. American Naturalist 152, 1–6.

Bradshaw, A.D., 1984. Adaptation of plants to soils containing toxic metals—a test for conceit, In Origins and Development of Adaptation. eds D. Evered, G.N. Collins, pp. 4–19. Pitman, London.

Bradshaw, A.D., 1991. Genostasis and the limits to evolution. Philosophical Transactions of the Royal Society of London B: Biological Sciences 333, 289–305.

Bradshaw, C.J.A., Isagi, Y., Kaneko, S., et al., 2007. Low genetic diversity in the bottlenecked population of endangered non-native banteng in northern Australia. Molecular Ecology 16, 2998–3008.

Bradshaw, W.E., Holzapfel, C.M., 2006. Evolutionary response to rapid climate change. Science 312, 1477–1478.

Brazeau, D.A., Gleason, D.F., Morgan, M.E., 1998. Self-fertilization in brooding hermaphrodite Caribbean corals: evidence from molecular markers. Journal of Experimental Marine Biology and Ecology 231, 225–238.

Breed, M.F., Stead, M.G., Ottewell, K.M., et al., 2013. Which provenance and where? Seed sourcing strategies for revegetation in a changing environment. Conservation Genetics 14, 1–10.

Bremner-Harrison, S., Cypher, B.L., 2007. Feasibility and strategies for reintroducing San Joaquin kit foxes to vacant or restored habitats. California State University—Stanislaus, Endangered Species Recovery Program, Fresno, CA.

Briggs, D., Walters, S.M., 2016. Plant Variation and Evolution, 4th edition. Cambridge University Press, Cambridge, UK.

Briscoe, D.A., Tait, N.N., 1995. Allozyme evidence for extensive and ancient radiations in Australian Onychophora. Journal of Heredity 114, 91–102.

Briscoe, D.A., Malpica, J.M., Robertson, A., et al., 1992. Rapid loss of genetic variation in large captive populations of *Drosophila* flies: implications for the genetic management of captive populations. Conservation Biology 6, 416–425.

Broadhurst, L.M., Lowe, A., Coates, D.J., et al., 2008. Seed supply for broadscale restoration: maximizing evolutionary potential. Evolutionary Applications 1, 587–597.

Brommer, J.E., Kirkpatrick, M., Qvarnström, A., et al., 2007. The intersexual genetic correlation for lifetime fitness in the wild and its implications for sexual selection. PLOS ONE 2, e744.

Brook, B.W., Sodhi, N.S., Bradshaw, C.J.A., 2008. Synergies among extinction drivers under global change. Trends in Ecology and Evolution 23, 453–460.

Brook, B.W., Tonkyn, D.W., O'Grady, J.J., et al., 2002. Contribution of inbreeding to extinction risk in threatened species. Conservation Ecology 6(1), 16 [online] http://www.consecol.org/vol16/iss11/art16.

Brookfield, J.F.Y., Sharp, P.M., 1994. Neutralism and selectionism face up to DNA data. Trends in Genetics 10, 109–111.

Brooks, T.M., Helgen, K.M., 2010. Biodiversity: a standard for species. Nature 467, 540–541.

Brown, A.H.D., 1989. Genetic characterisation of plant mating systems, In Plant Population Genetics, Breeding and Genetic Resources. eds A.H.D. Brown, M.T. Clegg, A.L. Kahler, B.S. Weir, pp. 145–162. Sinauer, Sunderland, MA.

Brown, A.H.D., Young, A.G., 2000. Genetic diversity in tetraploid populations of the endangered daisy *Rutidosis leptorrhynchoides* and implications for its conservation. Heredity 85, 122–129.

Bryant, E.H., Backus, V.L., Clark, M.E., et al., 1999. Experimental tests of captive breeding for endangered species. Conservation Biology 13, 1487–1496.

Brys, R., Jacquemyn, H., 2016. Severe outbreeding and inbreeding depression maintain mating system differentiation in *Epipactis* (Orchidaceae). Journal of Evolutionary Biology 29, 352–359.

Bull, J.K., Sands, C.J., Garrick, R.C., et al., 2013. Environmental complexity and biodiversity: the multi-layered evolutionary history of a log-dwelling velvet worm in montane temperate Australia. PLOS ONE 8, e84559.

Bult, A., Lynch, C., 2000. Breaking through artificial selection limits of an adaptive behavior in mice and the consequences for correlated responses. Behavior Genetics 30, 193–206.

Bunce, M., Szulkin, M., Lerner, H.R.L., et al., 2005. Ancient DNA provides new insights into the evolutionary history of New Zealand's extinct giant eagle. PLOS Biology 3, e9.

Buri, P., 1956. Gene frequency in small populations of mutant *Drosophila*. Evolution 10, 367–402.

Burt, A., Trivers, R., 2006. Genes in Conflict: The Biology of Selfish Genetic Elements. Belknap Press, Cambridge, MA.

Busch, J., 2005. Inbreeding depression in self-incompatible and self-compatible populations of *Leavenworthia alabamica*. Heredity 94, 159–165.

Busch, J.W., 2006. Heterosis in an isolated, effectively small, and self-fertilizing population of the flowering plant *Leavenworthia alabamica*. Evolution 60, 184–191.

Byers, D.L., 1998. Effect of cross proximity on progeny fitness in a rare and a common species of Eupatorium (Asteraceae). American Journal of Botany 85, 644–653.

Byers, D.L., Meagher, T.R., 2005. Consequences of prairie fragmentation on the progeny sex ratio of a gynodioecious species, *Lobelia spicata* (Campanulaceae). Heredity 95, 69–75.

Byers, D.L., Waller, D.M., 1999. Do plant populations purge their genetic load? Effects of population size and mating history on inbreeding depression. Annual Review of Ecology and Systematics 30, 479–513.

Byers, O., Lees, C., Wilcken, J., et al., 2013. The "One Plan" approach: the philosophy and implementation of CBSG's approach to integrated species conservation planning. WAZA Magazine 14, 2–5.

Byrne, M., Steane, D.A., Joseph, L., et al., 2011. Decline of a biome: evolution, contraction, fragmentation, extinction and invasion of the Australian mesic zone biota. Journal of Biogeography 38, 1635–1656.

Cahill, A.E., Aiello-Lammens, M.E., Fisher-Reid, M.C., et al., 2013. How does climate change cause extinction? Proceedings of the Royal Society B: Biological Sciences 280, 20121890.

Cale, G.H., Gowen, J.G., 1956. Heterosis in the honey bee (*Apis mellifera*). Genetics 41, 292–303.

Campeau-Péloquin, A., Kirsch, J.A.W., Eldridge, M.D.B., et al., 2001. Phylogeny of the rock-wallabies (Marsupialia: Macropodidae) based on DNA/DNA hybridisation. Australian Journal of Zoology 49, 463–486.

Carbone, L., Nergadze, S.G., Magnani, E., et al., 2006. Evolutionary movement of centromeres in horse, donkey, and zebra. Genomics 87, 777–782.

Cardoso, M.J., Eldridge, M.D.B., Oakwood, M., et al., 2009. Effects of founder events on the genetic variation of translocated island populations: implications for conservation management of the northern quoll. Conservation Genetics 10, 1719–1733.

Carney, S.E., Gardner, K.A., Rieseberg, L.H., 2000. Evolutionary changes over the fifty-year history of a hybrid population of sunflowers (*Helianthus*). Evolution 54, 462–474.

Carr, R.L., Carr, G.D., 1983. Chromosome races and structural heterozygosity in *Calycadenia ciliosa* Greene (Asteraceae). American Journal of Botany 70, 744–755.

Carrington, M., Nelson, G.W., Martin, M.P., et al., 1999. HLA and HIV-1: heterozygote advantage and *B*35-Cw*04* disadvantage. Science 283, 1748–1752.

Carroll, S.P., Loye, J.E., Dingle, H., et al., 2005. And the beak shall inherit—evolution in response to invasion. Ecology Letters 8, 944–951.

Carstens, B.C., Dewey, T.A., 2010. Species delimitation using a combined coalescent and information-theoretic approach: an example from North American *Myotis* bats. Systematic Biology 59, 400–414.

Carstens, B.C., Knowles, L.L., 2007. Estimating species phylogeny from gene-tree probabilities despite incomplete lineage sorting: an example from *Melanoplus* grasshoppers. Systematic Biology 56, 400–411.

Castillo, D.M., Burger, M.K., Lively, C.M., et al., 2015. Experimental evolution: assortative mating and sexual selection, independent of local adaptation, lead to reproductive isolation in the nematode *Caenorhabditis remanei*. Evolution 69, 3141–3155.

Castric, V., Vekemans, X., 2004. Plant self-incompatibility in natural populations: a critical assessment of recent theoretical and empirical advances. Molecular Ecology 13, 2873–2889.

Catalán, P., Segarra-Moragues, J.G., Palop-Esteban, M., et al., 2006. A Bayesian approach for discriminating among alternative inheritance hypotheses in plant polyploids: the allotetraploid origin of genus *Borderea* (Dioscoreaceae). Genetics 172, 1939–1953.

Ceballos, G., Ehrlich, P.R., 2002. Mammal population losses and the extinction crisis. Science 296, 904–907.

Chang, C.-C., Rodriguez, J., Ross, J., 2016. Mitochondrial-nuclear epistasis impacts fitness and mitochondrial physiology of interpopulation *Caenorhabditis briggsae* hybrids. G3: Genes| Genomes| Genetics 6, 209–219.

Charlesworth, B., 2009. Effective population size and patterns of molecular evolution and variation. Nature Reviews Genetics 10, 195–205.

Charlesworth, B., Charlesworth, D., 1999. The genetic basis of inbreeding depression. Genetical Research 74, 329–340.

Charlesworth, B., Charlesworth, D., 2010. Elements of Evolutionary Genetics. Roberts and Company, Greenwood Village, CO.

Charlesworth, D., 2006. Balancing selection and its effects on sequences in nearby genome regions. PLOS Genetics 2, e64.

Charlesworth, D., Charlesworth, B., 1987. Inbreeding depression and its evolutionary consequences. Annual Review of Ecology and Systematics 18, 237–268.

Charlesworth, D., Charlesworth, B., 1995. Quantitative genetics in plants: the effect of the breeding system on genetic variability. Evolution 49, 911–920.

Charlesworth, D., Willis, J.H., 2009. The genetics of inbreeding depression. Nature Reviews Genetics 10, 783–796.

Charlesworth, D., Morgan, M.T., Charlesworth, B., 1993. Mutation accumulation in finite outbreeding and inbreeding populations. Genetical Research 61, 39–56.

Charmantier, A., Garant, D., 2005. Environmental quality and evolutionary potential: lessons from wild populations. Proceedings of the Royal Society of London B: Biological Sciences 272, 1415–1425.

Charmantier, A., McCleery, R.H., Cole, L.R., et al., 2008. Adaptive phenotypic plasticity in response to climate change in a wild bird population. Science 320, 800–803.

Charron, G., Leducq, J.B., Landry, C.R., 2014. Chromosomal variation segregates within incipient species and correlates with reproductive isolation. Molecular Ecology 23, 4362–4372.

Chen, C., Durand, E., Forbes, F., et al., 2007. Bayesian clustering algorithms ascertaining spatial population structure: a new computer program and a comparison study. Molecular Ecology Notes 7, 747–756.

Chen, I.-C., Hill, J.K., Ohlemüller, R., et al., 2011. Rapid range shifts of species associated with high levels of climate warming. Science 333, 1024–1026.

Chenuil, A., Galtier, N., Berrebi, P., 1999. A test of the hypothesis of an autopolyploid vs. allopolyploid origin for a tetraploid lineage: application to the genus *Barbus* (Cyprinidae). Heredity 82, 373–380.

Cheptou, P.O., Donohue, K., 2011. Environment-dependent inbreeding depression: its ecological and evolutionary significance. New Phytologist 189, 395–407.

Chevin, L.-M., Lande, R., Mace, G.M., 2010. Adaptation, plasticity, and extinction in a changing environment: towards a predictive theory. PLOS Biology 8, e1000357.

Christie, M.R., Knowles, L.L., 2015. Habitat corridors facilitate genetic resilience irrespective of species dispersal abilities or population sizes. Evolutionary Applications 8, 454–463.

Churikov, D., Gharrett, A.J., 2002. Comparative phylogeography of the two pink salmon broodlines: an analysis based on mitochondrial DNA genealogy. Molecular Ecology 11, 1077–1101.

Clark, R.W., Marchand, M.N., Clifford, B.J., et al., 2011. Decline of an isolated timber rattlesnake (*Crotalus horridus*) population: interactions between climate change, disease, and loss of genetic diversity. Biological Conservation 144, 886–891.

Clarke, A., Robertson, M.A., Pieck, A., 2013. Recovery Plan for *Acanthocladium dockeri* (Spiny Daisy). Department of Environment, Water and Natural Resources, Clare, South Australia.

Coates, D., Monks, L., Hickman, E., 1998. Translocation proposal: round leaf honey-suckle *Lambertia orbifolia* CA Gardner (Proteaceae). Department of Conservation and Land Management, Kensington, Western Australia.

Coates, D.J., Hamley, V.L., 1999. Genetic divergence and the mating system in the endangered and geographically restricted species, *Lambertia orbifolia* Gardner (Proteaceae). Heredity 83, 418–427.

Coates, F., Walsh, N.G., James, E.A., 2002. Threats to the survival of the Grampians pincushion lily (*Borya mirabillis*, Lilliaceae)—a short-range endemic from western Victoria. Australian Systematic Botany 15, 477–485.

Cody, M.L., Overton, J.M., 1996. Short-term evolution of reduced dispersal in island plant populations. Journal of Ecology 84, 53–61.

Coghlan, A., Eichler, E.E., Oliver, S.G., et al., 2005. Chromosome evolution in eukaryotes: a multi-kingdom perspective. Trends in Genetics 21, 673–682.

Cohan, F.M., Perry, E.B., 2007. A systematics for discovering the fundamental units of bacterial diversity. Current Biology 17, R373–R386.

Coleman, R.A., Weeks, A.R., Hoffmann, A.A., 2013. Balancing genetic uniqueness and genetic variation in determining conservation and translocation strategies: a comprehensive case study of threatened dwarf galaxias, *Galaxiella pusilla* (Mack) (Pisces: Galaxiidae). Molecular Ecology 22, 1820–1835.

Collet, M., Vayssade, C., Auguste, A., et al., 2016. Diploid male production correlates with genetic diversity in the parasitoid wasp *Venturia canescens*: a genetic approach with new microsatellite markers. Ecology and Evolution 6, 6721–6734.

Collevatti, R.G., Grattapaglia, D., Hay, J.D., 2001. High resolution microsatellite based analysis of the mating system allows the detection of significant biparental inbreeding in *Caryocar brasiliense*, an endangered tropical tree species. Heredity 86, 60–67.

Colosimo, P.F., Hosemann, K.E., Balabhadra, S., et al., 2005. Widespread parallel evolution in sticklebacks by repeated fixation of ectodysplasmin alleles. Science 307, 1928–1933.

Coltman, D.W., Pilkington, J.G., Smith, J.A., et al., 1999. Parasite-mediated selection against inbred Soay sheep in a free-living island population. Evolution 53, 1259–1267.

Cook, J.M., Crozier, R.H., 1995. Sex determination and population biology in the Hymenoptera. Trends in Ecology and Evolution 10, 281–286.

Cook, L.M., Sutton, S.L., Crawford, T.J., 2005. Melanic moth frequencies in Yorkshire, an old English industrial hot spot. Journal of Heredity 96, 522–528.

Corander, J., Marttinen, P., Sirén, J., et al., 2008. Enhanced Bayesian modelling in BAPS software for learning genetic structures of populations. BMC Bioinformatics 9, 1.

Corbett-Detig, R.B., Zhou, J., Clark, A.G., et al., 2013. Genetic incompatibilities are widespread within species. Nature 504, 135–137.

Corlett, R.T., Westcott, D.A., 2013. Will plant movements keep up with climate change? Trends in Ecology and Evolution 28, 482–488.

Coulon, A., Fitzpatrick, J.W., Bowman, R., et al., 2008. Congruent population structure inferred from dispersal behaviour and intensive genetic surveys of the threatened Florida scrub-jay (*Aphelocoma cœrulescens*). Molecular Ecology 17, 1685–1701.

Coyne, J.A., 1984. Correlation between heterozygosity and rate of chromosomal evolution in animals. American Naturalist 123, 725–729.

Coyne, J.A., Orr, H.A., 2004. Speciation. Sinauer, Sunderland, MA.

Coyne, J.A., Meyers, W., Crittenden, A.P., et al., 1993. The fertility effects of pericentric inversions in *Drosophila melanogaster*. Genetics 134, 487–496.

Cracraft, J., 1983. Species concepts and speciation analysis, In Current Ornithology. ed. R.F. Johnston, pp. 159–187. Plenum Press, New York.

Cracraft, J., 1992. The species of the birds-of-paradise (Paradisaeidae): applying the phylogenetic species concept to a complex pattern of diversification. Cladistics 8, 1–43.

Cracraft, J., 1997. Species concepts in systematics and conservation biology—an ornithological viewpoint, In Species: The Units of Biodiversity. eds M.F. Claridge, H.A. Dawah, M.R. Wilson, pp. 325–339. Chapman & Hall, London.

Crandall, K.A., Bininda-Emonds, O.R.P., Mace, G.M., et al., 2000. Considering evolutionary processes in conservation biology: an alternative to "evolutionary significant units." Trends in Ecology and Evolution 15, 290–295.

Cremer, K.W., 1966. Dissemination of seed from *Eucalyptus regnans*. Australian Forestry 30, 33–37.

Crispo, E., Hendry, A., 2005. Does time since colonization influence isolation by distance? A meta-analysis. Conservation Genetics 6, 665–682.

Crnokrak, P., Barrett, S.C.H., 2002. Purging the genetic load: a review of evidence. Evolution 56, 2347–2358.

Crnokrak, P., Roff, D.A., 1999. Inbreeding depression in the wild. Heredity 83, 260–270.

Cronin, M.A., Cánovas, A., Bannasch, D.L., et al., 2015. Single nucleotide polymorphism (SNP) variation of wolves (*Canis lupus*) in Southeast Alaska and comparison with wolves, dogs, and coyotes in North America. Journal of Heredity 106, 26–36.

Crow, J.F., Kimura, M., 1970. An Introduction to Population Genetics Theory. Harper and Rowe, New York.

Császár, N.B.M., Ralph, P.J., Frankham, R., et al., 2010. Estimating the potential for adaptation of corals to climate warming. PLOS ONE 5, e9751.

Culver, M., Hedrick, P.W., Murphy, K., et al., 2008. Estimation of the bottleneck size in Florida panthers. Animal Conservation 11, 104–110.

Culver, M., Johnson, W.E., Pecon-Slattery, J., et al., 2000. Genomic ancestry of the American puma (*Puma concolor*). Journal of Heredity 91, 186–197.

Curik, I., Ferenčaković, M., Sölkner, J., 2014. Inbreeding and runs of homozygosity: a possible solution to an old problem. Livestock Science 166, 26–34.

Cutter, A.D., 2012. The polymorphic prelude to Bateson–Dobzhansky–Muller incompatibilities. Trends in Ecology and Evolution 27, 209–218.

Dalton, R.M., Koski, M.H., Ashman, T-L., 2013. Maternal sex effects and inbreeding depression under varied environmental conditions in gynodioecious *Fragaria vesca* subsp. *bracteata*. Annals of Botany 112, 613–621.

Daniels, S.J., Priddy, J.A., Walters, J.R., 2000. Inbreeding in small populations of red-cockaded woodpeckers: insights from a spatially-explicit individual-based model, In Genetics, Demography and Viability in Fragmented Populations. eds A.G. Young, G.M. Clarke, pp. 128–147. Cambridge University Press, Cambridge, UK.

Darwin, C., 1859. On the Origin of Species by Means of Natural Selection, or the Preservation of Favoured Races in the Struggle for Life. John Murray, London.

Darwin, C., 1868. The Variation of Animals and Plants under Domestication. John Murray, London.

Darwin, C., 1876. The Effects of Cross and Self Fertilisation in the Vegetable Kingdom. John Murray, London.

Dasmahapatra, K.K., Lacy, R.C., Amos, W., 2008. Estimating levels of inbreeding using AFLP markers. Heredity 100, 286–295.

Davies-Mostert, H., Mills, G., McDonald, D.W., 2009. A critical assessment of South Africa's managed metapopulation recovery strategy for African wild dogs, In Reintroduction of Top-Order Predators—Using Science to Improve Conservation Management. eds M.W. Hayward, M. Somers, pp. 10–42. Wiley-Blackwell, Oxford, UK.

Davis, J., 1992. The Wildlands Project, Wild Earth, Special Issue edition. Cenozoic Society, Inc, Canton, NY.

Dawson, T.P., Jackson, S.T., House, J.I., et al., 2011. Beyond predictions: biodiversity conservation in a changing climate. Science 332, 53–58.

Dayrat, B., 2005. Towards integrative taxonomy. Biological Journal of the Linnean Society 85, 407–415.

De'ath, G., Fabricius, K.E., Sweatman, H., et al., 2012. The 27-year decline of coral cover on the Great Barrier Reef and its causes. Proceedings of the National Academy of Sciences 109, 17995–17999.

de Boer, L.E.M., 1982. Karyological problems in breeding owl monkeys *Aotus trivirgatus*. International Zoo Yearbook 22, 119–124.

de Cara, M.Á.R., Villanueva, B., Toro, M.Á., et al., 2013. Using genomic tools to maintain diversity and fitness in conservation programmes. Molecular Ecology 22, 6091–6099.

Degnan, S.M., 1993. The perils of single gene trees—mitochondrial versus single-copy nuclear DNA variation in white-eyes (Aves: Zosteropidae). Molecular Ecology 2, 219–225.

Del Hoyo, J., Collar, N.J., Christie, D.A., et al., 2014. Illustrated Checklist of the Birds of the World. Lynx Edicions, Barcelona.

Demauro, M.M., 1993. Relationship of breeding system to rarity in the Lakeside daisy (*Hymenoxys acaulis* var. *glabra*). Conservation Biology 7, 542–550.

de Oliveira, A.K., Cordeiro, A.R., 1980. Adaptation of *Drosophila willistoni* experimental populations to extreme pH medium. II. Development of incipient reproductive isolation. Heredity 44, 123–130.

de Queiroz, K., 1998. The general lineage concept of species, species criteria, and the process of speciation. In Endless Forms: Species and Speciation. eds D.J. Howard, S.H. Berlocher, pp. 55–57. Oxford University Press, New York.

de Queiroz, K., 2007. Species concepts and species delimitation. Systematic Biology 56, 879–886.

de Roode, J., 2007. The moths of war. New Scientist 196(2633), 46–49.

DeRose, M.A., Roff, D.A., 1999. A comparison of inbreeding depression in life-history and morphological traits in animals. Evolution 53, 1288–1292.

Dettman, J.R., Sirjusingh, C., Kohn, L.M., et al., 2007. Incipient speciation by divergent adaptation and antagonistic epistasis in yeast. Nature 447, 585–588.

de Villemereuil, P., Gimenez, O., Doligez, B., 2013. Comparing parent–offspring regression with frequentist and Bayesian animal models to estimate heritability in wild populations: a simulation study for Gaussian and binary traits. Methods in Ecology and Evolution 4, 260–275.

Dickerson, G.E., Lush, J.L., Culbertson, C.C., 1946. Hybrid vigor in single crosses between inbred lines of Poland China swine. Journal of Animal Science 5, 16–24.

Dietz, J.M., Baker, A.J., Ballou, J.D., 2000. Demographic evidence of inbreeding depression in wild golden lion tamarins, In Genetics, Demography and the Viability of Fragmented Populations. eds A.G. Young, G.M. Clarke, pp. 203–211. Cambridge University Press, Cambridge, UK.

Dietz, J.M., Dietz, L.A., Nagagata, E.Y., 1994. The effective use of flagship species for conservation of biodiversity: the example of lion tamarins in Brazil, In Creative Conservation. eds P.J.S. Olney, G.M. Mace, A.T.C. Feistner, pp. 32–49. Chapman & Hall, London.

Do, C., Waples, R.S., Peel, D., et al., 2014. NEESTIMATOR v2: re-implementation of software for the estimation of contemporary effective population size (N_e) from genetic data. Molecular Ecology Resources 14, 209–214.

Dobrynin, P., Liu, S., Tamazian, G., et al., 2015. Genomic legacy of the African cheetah, *Acinonyx jubatus*. Genome Biology 16, 1–20.

Dobzhansky, T., 1937. Genetics and the Origin of Species. Columbia University Press, New York.

Dobzhansky, T., Ayala, F.J., Stebbins, G.L., et al., 1977. Evolution. W.H. Freeman, San Francisco.

Dodd, D.M.B., 1989. Reproductive isolation as a consequence of adaptive divergence in *Drosophila pseudoobscura*. Evolution 43, 1308–1311.

Dolan, R.W., Yahr, R., Menges, E.S., et al., 1999. Conservation implications of genetic variation in three rare species endemic to Florida rosemary scrub. American Journal of Botany 86, 1556–1562.

Dolgin, E.S., Charlesworth, B., Baird, S.E., et al., 2007. Inbreeding and outbreeding depression in *Caenorhabditis* nematodes. Evolution 61, 1339–1352.

Donohue, K., 2014. The epigenetics of adaptation: focusing on epigenetic stability as an evolving trait. Evolution 68, 617–619.

Doudna, J.A., Charpentier, E., 2014. The new frontier of genome engineering with CRISPR-Cas9. Science 346, 1258096.

Dove, S.G., Kline, D.I., Pantos, O., et al., 2013. Future reef decalcification under a business-as-usual CO_2 emission scenario. Proceedings of the National Academy of Sciences 110, 15342–15347.

Driscoll, C.A., Menotti-Raymond, M., Nelson, G., et al., 2002. Genomic microsatellites as evolutionary chronometers: a test in wild cats. Genome Research 12, 414–423.

Dudash, M.R., 1990. Relative fitness of selfed and outcrossed progeny in a self-compatible, protandrous species, *Sabatia angularis* L. (Gentianaceae): a comparison of three environments. Evolution 44, 1129–1139.

Dudash, M.R., Carr, D.E., 1998. Genetics underlying inbreeding depression in *Mimulus* with contrasting mating systems. Nature 393, 682–684.

Dudash, M.R., Fenster, C.B., 2000. inbreeding and outbreeding in fragmented populations, In Genetics, Demography and Viability of Fragmented Populations. eds A.G. Young, G.M. Clarke, pp. 35–53. Cambridge University Press, Cambridge, UK.

Dudash, M.R., Fenster, C.B., 2001. The role of breeding system and inbreeding depression in the maintenance of an outcrossing mating strategy in *Silene virginica* (Caryophyllaceae). American Journal of Botany 88, 1953–1959.

Dudash, M.R., Murren, C.J., 2008. The influence of breeding systems and mating systems on conservation genetics and conservation decisions, In Conservation Biology: Evolution in Action. eds S.P. Carroll, C.W. Fox, pp. 68–80. Oxford University Press, New York.

Dudash, M.R., Carr, D.E., Fenster, C.B., 1997. Five generations of enforced selfing and outcrossing in *Mimulus guttatus*: inbreeding depression variation at the population and family level. Evolution 51, 54–65.

Dudley, J.W., 1977. 76 generations of selection for oil and protein percentage in maize, In Proceedings of the International Conference on Quantitative Genetics. eds E. Pollak, O. Kempthorne, T.B. Bailey, pp. 459–473. Iowa State University Press, Ames, IA.

Dunlop, E.S., Shuter, B.J., Dieckmann, U., 2007. Demographic and evolutionary consequences of selective mortality: predictions from an eco-genetic model for smallmouth bass. Transactions of the American Fisheries Society 136, 749–765.

Ebert, D., Haag, C., Kirkpatrick, M., et al., 2002. A selective advantage to immigrant genes in a *Daphnia* metapopulation. Science 295, 485–488.

Eckert, C.G., 2002. The loss of sex in clonal plants, In Ecology and Evolutionary Biology of Clonal Plants. eds J.F. Stuefer, B. Erschbamer, H. Huber, J.I. Suzuki, pp. 279–298. Springer, Dordrecht.

Edmands, S., 1999. Heterosis and outbreeding depression in interpopulation crosses spanning a wide range of divergence. Evolution 53, 1757–1768.

Edmands, S., 2002. Does parental divergence predict reproductive compatibility? Trends in Ecology and Evolution 17, 520–527.

Edmands, S., 2007. Between a rock and a hard place: evaluating the relative risks of inbreeding and outbreeding depression for conservation and management. Molecular Ecology 16, 463–475.

Edmands, S., Feaman, H.V., Harrison, J.S., et al., 2005. Genetic consequences of many generations of hybridization between divergent copepod populations. Journal of Heredity 96, 114–123.

Egan, S.P., Funk, D.J., 2009. Ecologically dependent postmating isolation between sympatric host forms of *Neochlamisus bebbianae* leaf beetles. Proceedings of the National Academy of Sciences 106, 19426–19431.

Eggert, L.S., Powell, D.M., Ballou, J.D., et al., 2010. Pedigrees and the study of the wild horse population of Assateague Island National Seashore. Journal of Wildlife Management 74, 963–973.

Ehrman, L., 1969. Genetic divergence in M. Vetukhiv's experimental populations of *Drosophila pseudoobscura*. 5. A further study of rudiments of sexual isolation. American Midland Naturalist 82, 272–276.

Eisen, E.J., 1975. Population size and selection intensity effects on long-term selection response in mice. Genetics 79, 305–323.

Eldredge, N., Cracraft, J., 1980. Phylogenetic Patterns and the Evolutionary Process: Method and Theory in Comparative Biology. Columbia University Press, New York.

Eldridge, M.D.B., Close, R.L., 1992. Taxonomy of rock-wallabies, *Petrogale* (Marsupialia: Macropodidae). I. A revision of the eastern *Petrogale* with a description of three new species. Australian Journal of Zoology 40, 605–625.

Eldridge, M.D.B., Close, R.L., 1993. Radiation of chromosome shuffles. Current Opinions in Genetics and Development 3, 15–22.

Eldridge, M.D.B., Close, R.L., 1997. Taxonomy of rock wallabies, *Petrogale* (Marsupialia: Macropodidae). II. An historical review. Australian Mammalogy 19, 113–122.

Eldridge, M.D.B., King, J.M., Loupis, A.K., et al., 1999. Unprecedented low levels of genetic variation and inbreeding depression in an island population of the black-footed rock-wallaby. Conservation Biology 13, 531–541.

Eldridge, M.D.B., Kinnear, J.E., Onus, M.L., 2001. Source population of dispersing rock-wallabies (*Petrogale lateralis*) identified by assignment tests on multilocus genotypic data. Molecular Ecology 10, 2867–2876.

Eldridge, M.D.B., Kinnear, J.E., Zenger, K.R., et al., 2004. Genetic diversity in remnant mainland and "pristine" island populations of three endemic Australian Macropodids (Marsupialia): *Macropus eugenii*, *Lagorchestes hirsutus* and *Petrogale lateralis*. Conservation Genetics 5, 325–338.

Eldridge, M.D.B., Wilson, A.C.C., Metcalfe, C.J., et al., 2001. Taxonomy of rock-wallabies, *Petrogale* (Marsupialia: Macropodidae). III. Molecular data confirms the species status of the purple-necked rock-wallaby (*Petrogale purpureicollis* Le Souef). Australian Journal of Zoology 49, 323–343.

Elias, B.A., Shipley, L.A., McCusker, S., et al., 2013. Effects of genetic management on reproduction, growth, and survival in captive endangered pygmy rabbits (*Brachylagus idahoensis*). Journal of Mammalogy 94, 1282–1292.

El-Kassaby, Y.A., Yanchuk, A.D., 1994. Genetic diversity, differentiation, and inbreeding in Pacific yew from British Columbia. Journal of Heredity 85, 112–117.

Ellison, C.K., Burton, R.S., 2010. Cytonuclear conflict in interpopulation hybrids: the role of RNA polymerase in mtDNA transcription and replication. Journal of Evolutionary Biology 23, 528–538.

Ellstrand, N.C., Roose, M.L., 1987. Patterns of genotypic diversity in clonal plant species. American Journal of Botany 74, 123–131.

Ellstrand, N.C., Schierenbeck, K.A., 2000. Hybridization as a stimulus for the evolution of invasiveness in plants? Proceedings of the National Academy of Sciences 97, 7043–7050.

Ellstrand, N.C., Biggs, D., Kaus, A., et al., 2010. Got hybridization? A multidisciplinary approach for informing science policy. Bioscience 60, 384–388.

Elsworth, P., Kovaliski, J., Cooke, B., 2012. Rabbit haemorrhagic disease: are Australian rabbits (*Oryctolagus cuniculus*) evolving resistance to infection with Czech CAPM 351 RHDV? Epidemiology and Infection 140, 1972–1981.

Enders, L.S., Nunney, L., 2012. Seasonal stress drives predictable changes in inbreeding depression in field-tested captive populations of *Drosophila melanogaster*. Proceedings of the Royal Society of London B: Biological Sciences 279, 3756–3764.

Endler, J.A., 1977. Genetic Variation, Speciation and Clines. Princeton University Press, Princeton, NJ.

Endler, J.A., 1986. Natural Selection in the Wild. Princeton University Press, Princeton, NJ.

England, P.R., Osler, G.H.R., Woodworth, L.M., et al., 2003. Effect of intense versus diffuse population bottlenecks on microsatellite genetic diversity and evolutionary potential. Conservation Genetics 4, 595–604.

Erickson, D.L., Fenster, C.B., 2006. Intraspecific hybridization and the recovery of fitness in the native legume *Chamaecrista fasciculata*. Evolution 60, 225–233.

Etterson, J.R., Shaw, R.G., 2001. Constraint to adaptive evolution in response to global warming. Science 294, 151–154.

Evans, S.R., Sheldon, B.C., 2008. Interspecific patterns of genetic diversity in birds: correlations with extinction risk. Conservation Biology 22, 1016–1025.

Evenson, R.E., Gollin, D., 2003. Assessing the impact of the Green Revolution, 1960 to 2000. Science 300, 758–762.

Ewen, J.G., Armstrong, D.P., Parker, K.A., et al., 2012. Reintroduction Biology: Integrating Science and Management. Wiley-Blackwell, Oxford, UK.

Excoffier, L., Heckel, G., 2006. Computer programs for population genetics data analysis: a survival guide. Nature Reviews Genetics 7, 745–758.

Excoffier, L., Smouse, P.E., Quattro, J.M., 1992. Analysis of molecular variance inferred from metric distances among DNA haplotypes: application to human mitochondrial DNA restriction data. Genetics 131, 479–491.

Fagan, W.F., Holmes, E.E., 2006. Quantifying the extinction vortex. Ecology Letters 9, 51–60.

Fairfax, K., 2016. Northern Australia mangrove dieback "unprecedented." Green Left Weekly 1102 (19 July 2016), 12.

Falconer, D.S., 1977. Some results of the Edinburgh selection experiments with mice, In Proceeding of the 1st International Conference on Quantitative Genetics. eds E. Pollak, O. Kempthorne, T.B. Bailey, pp. 101–115. Iowa State University Press, Ames, IA.

Falconer, D.S., King, J.W.B., 1953. A study of selection limits in the mouse. Journal of Genetics 51, 561–579.

Falconer, D.S., Mackay, T.F.C., 1996. Introduction to Quantitative Genetics, 4th edition. Longman, Harlow, UK.

Falush, D., Stephens, M., Pritchard, J.K., 2003. Inference of population structure using multilocus genotype data: linked loci and correlated allele frequencies. Genetics 164, 1567–1587.

Falush, D., Stephens, M., Pritchard, J.K., 2007. Inference of population structure using multilocus genotype data: dominant markers and null alleles. Molecular Ecology Notes 7, 574–578.

Faubet, P., Waples, R.S., Gaggiotti, O.E., 2007. Evaluating the performance of a multi-locus Bayesian method for the estimation of migration rates. Molecular Ecology 16, 1149–1166.

Fenner, F., Ratcliffe, F.N., 1965. Myxomatosis. Cambridge University Press, Cambridge, UK.

Fenster, C.B., 1991. Gene flow in *Chamaecrista fasciculata* (Leguminosae) II. Gene establishment. Evolution 45, 410–422.

Fenster, C.B., Dudash, M.R., 1994. Genetic considerations for plant population restoration and conservation, In Restoration of Endangered Species: Conceptual Issues, Planning and Implementation. eds M.L. Bowles, C.J. Whelan, pp. 34–62. Cambridge University Press, Cambridge, UK.

Fenster, C.B., Galloway, L.F., 2000. Inbreeding and outbreeding depression in natural populations of *Chamaecrista fasciculata* (Fabaceae). Conservation Biology 14, 1406–1412.

Fenster, C.B., Vekemans, X., Hardy, O.J., 2003. Quantifying gene flow from spatial genetic structure data in a metapopulation of *Chamaecrista fasciculata* (Leguminosae). Evolution 57, 995–1007.

Ferchaud, A.L., Hansen, M.M., 2016. The impact of selection, gene flow and demographic history on heterogeneous genomic divergence: three-spine sticklebacks in divergent environments. Molecular Ecology 25, 238–259.

Fernández, J., Toro, M.A., 2006. A new method to estimate relatedness from molecular markers. Molecular Ecology 15, 1657–1667.

Fernández, J., Toro, M.A., Caballero, A., 2004. Managing individuals' contributions to maximize the allelic diversity maintained in small, conserved populations. Conservation Biology 18, 1358–1367.

Fernández, J., Toro, M.A., Caballero, A., 2008. Management of subdivided populations in conservation programs: development of a novel dynamic system. Genetics 179, 683–692.

Fernández, J., Villanueva, B., Pong-Wong, R., et al., 2005. Efficiency of the use of pedigree and molecular marker information in conservation programs. Genetics 170, 1313–1321.

Fiedler, P.L., Laven, R.D., 1996. Selecting reintroduction sites, In Restoring Diversity. eds D.A. Falk, C.I. Millar, M. Olwell, pp. 157–169. Island Press, Washington, DC.

Finger, A., Kettle, C.J., Kaiser-Bunbury, C.N., et al., 2011. Back from the brink: potential for genetic rescue in a critically endangered tree. Molecular Ecology 20, 3773–3784.

Finlayson, G., Finlayson, S., Dickman, C., 2010. Returning the rat-kangaroos: translocation attempts in the family Potoroidae (superfamily Macropodoidea) and recommendations for conservation, In Macropods: The Biology of Kangaroos, Wallabies and Rat-Kangaroos. eds G. Coulson, M. Eldridge, pp. 245–262. CSIRO Publishing, Melbourne, Vic., Australia.

Finstermeier, K., Zinner, D., Brameier, M., et al., 2013. A mitogenomic phylogeny of living primates. PLOS ONE 8, e69504.

Fischer, M., Matthies, D., 1998. Effects of population size on performance in the rare plant *Gentianella germanica*. Journal of Ecology 86, 195–204.

Fisher, M.C., Henk, D.A., 2012. Sex, drugs and recombination: the wild life of *Aspergillus*. Molecular Ecology 21, 1305–1306.

Fitch, A.J., Goodman, A.E., Donnellan, S.C., 2005. Isolation and characterisation of microsatellite markers for the Australian monitor lizard, *Varanus acanthurus* (Squamata: Varanidae) and their utility in other selected varanid species. Molecular Ecology Notes 5, 521–523.

Fitzpatrick, S.W., Gerberich, J.C., Angeloni, L.M., et al., 2016. Gene flow from an adaptively divergent source causes rescue through genetic and demographic factors in two wild populations of Trinidadian guppies. Evolutionary Applications 9, 879–891.

Flight, P.A., 2010. Phylogenetic comparative methods strengthen evidence for reduced genetic diversity among endangered tetrapods. Conservation Biology 24, 1307–1315.

Flinn, K.M., 2006. Reproductive biology of three fern species may contribute to differential colonization success in post-agricultural forests. American Journal of Botany 93, 1289–1294.

Florida Ziziphus Recovery Team, 2012. Florida Ziziphus Recovery Team Report. http://www.archbold-station.org/html/research/plant/flziz.html.

Foden, W., Young, B. eds., 2016. IUCN SSC Guidelines for Assessing Species' Vulnerability to Climate Change. IUCN, Gland, Switzerland.

Foerster, K., Coulson, T., Sheldon, B.C., et al., 2007. Sexually antagonistic genetic variation for fitness in red deer. Nature 447, 1107–1110.

Foo, S.A., Dworjanyn, S.A., Poore, A.G.B., et al., 2012. Adaptive capacity of the habitat modifying sea urchin *Centrostephanus rodgersii* to ocean warming and ocean acidification: performance of early embryos. PLOS ONE 7, e42497.

Foose, T.J., 1986. Riders of the last ark: the role of captive breeding in conservation strategies, In The Last Extinction. eds L. Kaufman, K. Mallory, pp. 141–165. MIT Press, Cambridge, MA.

Forabosco, F., Löhmus, M., Rydhmer, L., et al., 2013. Genetically modified farm animals and fish in agriculture: a review. Livestock Science 153, 1–9.

Forbes, S.H., Hogg, J.T., 1999. Assessing population structure at high levels of differentiation: microsatellite comparisons of bighorn sheep and large carnivores. Animal Conservation 2, 223–233.

Fox, C.W., Reed, D.H., 2011. Inbreeding depression increases with environmental stress: an experimental study and meta-analysis. Evolution 65, 246–258.

François, O., Durand, E., 2010. Spatially explicit Bayesian clustering models in population genetics. Molecular Ecology Resources 10, 773–784.

Frankel, O.H., Soulé, M.E., 1981. Conservation and Evolution. Cambridge University Press, Cambridge, UK.

Frankham, R., 1983. Origin of genetic variation in selection lines, In Proceedings of the Thirty-Second Annual Breeders' Roundtable, St Louis, MO. pp. 1–18.

Frankham, R., 1995a. Effective population size/adult population size ratios in wildlife: a review. Genetical Research 66, 95–107.

Frankham, R., 1995b. Conservation genetics. Annual Review of Genetics 29, 305–327.

Frankham, R., 1995c. Inbreeding and extinction: a threshold effect. Conservation Biology 9, 792–799.

Frankham, R., 1996. Relationship of genetic variation to population size in wildlife. Conservation Biology 10, 1500–1508.

Frankham, R., 1997. Do island populations have lower genetic variation than mainland populations? Heredity 78, 311–327.

Frankham, R., 1998. Inbreeding and extinction: island populations. Conservation Biology 12, 665–675.

Frankham, R., 2000. Genetics and conservation: commentary on Elgar & Clode. Australian Biologist 13 (3), 45–54.

Frankham, R., 2005. Genetics and extinction. Biological Conservation 126, 131–140.

Frankham, R., 2008. Genetic adaptation to captivity in species conservation programs. Molecular Ecology 17, 325–333.

Frankham, R., 2009a. Genetic considerations in reintroduction programmes for top-order, terrestrial predators, In Reintroduction of Top-Order Predators. eds M.W. Hayward, M.J. Somers, pp. 371–387. Wiley-Blackwell, Oxford, UK.

Frankham, R., 2009b. Genetic architecture of reproductive fitness and its consequences, In Adaptation and Fitness in Animal Populations: Evolutionary and Breeding Perspectives on Genetic Resource Management. eds J. van der Werf, H.-U. Graser, R. Frankham, C. Gondro, pp. 15–39. Springer, Dordrecht.

Frankham, R., 2010a. Where are we in conservation genetics and where do we need to go? Conservation Genetics 11, 661–663.

Frankham, R., 2010b. Challenges and opportunities of genetic applications in biological conservation. Biological Conservation 143, 1919–1927.

Frankham, R., 2012. How closely does genetic diversity in finite populations conform to predictions of neutral theory? Large deficits in regions of low recombination. Heredity 108, 167–178.

Frankham, R., 2015. Genetic rescue of small inbred populations: meta-analysis reveals large and consistent benefits of gene flow. Molecular Ecology 24, 2610–2618.

Frankham, R., 2016. Genetic rescue benefits persist to at least the F3 generation, based on a meta-analysis. Biological Conservation 195, 33–36.

Frankham, R., Loebel, D.A., 1992. Modeling problems in conservation genetics using captive *Drosophila* populations: rapid genetic adaptation to captivity. Zoo Biology 11, 333–342.

Frankham, R., Ballou, J.D., Briscoe, D.A., 2002. Introduction to Conservation Genetics. Cambridge University Press, Cambridge, UK.

Frankham, R., Ballou, J.D., Briscoe, D.A., 2004. A Primer of Conservation Genetics. Cambridge University Press, Cambridge, UK.

Frankham, R., Ballou, J.D., Briscoe, D.A., 2010. Introduction to Conservation Genetics, 2nd edition. Cambridge University Press, Cambridge, UK.

Frankham, R., Ballou, J.D., Dudash, M.R., et al., 2012. Implications of different species concepts for conserving biodiversity. Biological Conservation 153, 25–31.

Frankham, R., Ballou, J.D., Eldridge, M.D.B., et al., 2011. Predicting the probability of outbreeding depression. Conservation Biology 25, 465–475.

Frankham, R., Bradshaw, C.J.A., Brook, B.W., 2014a. Genetics in conservation management: revised recommendations for the 50/500 rules, Red List criteria and population viability analyses. Biological Conservation 170, 56–63.

Frankham, R., Jones, J.P., Barker, J.S.F., 1968. The effect of population size and selection intensity in selection for a quantitative character in *Drosophila*. I. Short-term response to selection. Genetical Research 12, 237–248.

Frankham, R., Lacy, R.C., Ballou, J.D., et al., 2014b. Species concepts for conservation—reply to Russello and Amato. Biological Conservation, 170, 334–335.

Frankham, R., Lees, K., Montgomery, M.E., et al., 1999. Do population size bottlenecks reduce evolutionary potential? Animal Conservation 2, 255–260.

Franklin, J.F., Lindenmayer, D.B., 2009. Importance of matrix habitats in maintaining biological diversity. Proceedings of the National Academy of Sciences 106, 349–350.

Franks, S.J., Hoffmann, A.A., 2012. Genetics of climate change adaptation. Annual Review of Genetics 46, 185–208.

Franks, S.J., Sim, S., Weis, A.E., 2007. Rapid evolution of flowering time by an annual plant in response to a climate fluctuation. Proceedings of the National Academy of Sciences 104, 1278–1282.

Fraser, D.J., Bernatchez, L., 2001. Adaptive evolutionary conservation: towards a unified concept for defining conservation units. Molecular Ecology 10, 2741–2752.

Fredrickson, R.J., Siminski, P., Woolf, M., et al., 2007. Genetic rescue and inbreeding depression in Mexican wolves. Proceedings of the Royal Society B: Biological Sciences 274, 2365–2371.

Freeman, S., Herron, J.C., 2013. Evolutionary Analysis, 4th edition. Pearson Higher Education, Harlow, UK.

Fujita, M.K., Leaché, A.D., Burbrink, F.T., et al., 2012. Coalescent-based species delimitation in an integrative taxonomy. Trends in Ecology and Evolution 27, 480–488.

Funk, D.J., Nosil, P., Etges, W.J., 2006. Ecological divergence exhibits consistently positive associations with reproductive isolation across disparate taxa. Proceedings of the National Academy of Sciences 103, 3209–3213.

Funk, W.C., Lovich, R.E., Hohenlohe, P.A., et al., 2016. Adaptive divergence despite strong genetic drift: genomic analysis of the evolutionary mechanisms causing genetic differentiation in the island fox (*Urocyon littoralis*). Molecular Ecology 25, 2176–2194.

Funk, W.C., McKay, J.K., Hohenlohe, P.A., et al., 2012. Harnessing genomics for delineating conservation units. Trends in Ecology and Evolution 27, 489–496.

Futuyma, D.J., 1979. Evolutionary Biology. Sinauer, Sunderland, MA.

Futuyma, D.J., 1998. Evolutionary Biology, 3th edition. Sinauer, Sunderland, MA.

Gabriel, W., Lynch, M., Bürger, R., 1993. Muller's ratchet and mutational meltdowns. Evolution 47, 1744–1757.

Galloway, L.F., Fenster, C.B., 1999. The effect of nuclear and cytoplasmic genes on fitness and local adaptation in an annual legume, *Chamaecrista fasciculata*. Evolution, 53, 1734–1743.

Galloway, L.F., Fenster, C.B., 2001. Nuclear and cytoplasmic contributions to intraspecific divergence in an annual legume. Evolution 55, 488–497.

Gammage, B., 2011. The Greatest Estate on Earth: How Aborigines Made Australia. Allen and Unwin, Sydney, NSW, Australia.

Gao, H., Williamson, S., Bustamante, C.D., 2007. A Markov Chain Monte Carlo approach for joint inference of population structure and inbreeding rates from multilocus genotype data. Genetics 176, 1635–1651.

Garbe, J.R., Prakapenka, D., Yang, J., et al., 2016. Genomic inbreeding and relatedness in wild panda populations. PLOS ONE 11, e0160496.

Garcia-Dorado, A., 2003. Tolerant versus sensitive genomes: the impact of deleterious mutation on fitness and conservation. Conservation Genetics 4, 311–324.

Gardner, J.L., Peters, A., Kearney, M.R., et al., 2011. Declining body size: a third universal response to warming? Trends in Ecology and Evolution 26, 285–291.

Garner, B.A., Hand, B.K., Amish, S.J., et al., 2016. Genomics in conservation: case studies and bridging the gap between data and application. Trends in Ecology and Evolution 31, 81–83.

Garnett, S., Franklin, D. eds., 2014. Climate Change Adaptation Plan for Australian Birds. CSIRO Publishing, Clayton, Vic., Australia.

Garrett, L., 1994. The Coming Plague: Newly Emerging Diseases in a World out of Balance. Virago Press, London.

Garrigan, D., Hedrick, P.W., 2003. Detecting adaptive molecular polymorphism: lessons from the MHC. Evolution 57, 1707–1722.

Gavrilets, S., 2004. Fitness Landscapes and the Origin of Species. Princeton University Press, Princeton, NJ.

Geber, M.A., Griffen, L.R., 2003. Inheritance and natural selection on functional traits. International Journal of Plant Sciences 164, S21–S42.

Gemmell, N.J., Metcalf, V.J., Allendorf, F.W., 2004. Mother's curse: the effect of mtDNA on individual fitness and population viability. Trends in Ecology and Evolution 19, 238–244.

Georghiou, G.P., 1986. The magnitude of the resistance problem, In Pesticide Resistance: Strategies and Tactics for Management. ed. Committee on Strategies for the Management of Pesticide Resistant Pest Populations, pp. 14–43. National Academy Press, Washington, DC.

Gevers, D., Cohan, F.M., Lawrence, J.G., et al., 2005. Re-evaluating prokaryotic species. Nature Reviews Microbiology 3, 733–739.

Gharrett, A.J., Smoker, W.W., Reisenbichler, R.R., et al., 1999. Outbreeding depression in hybrids between odd- and even-broodyear pink salmon. Aquaculture 173, 117–129.

Gienapp, P., Teplitsky, C., Alho, J.S., et al., 2008. Climate change and evolution: disentangling environmental and genetic responses. Molecular Ecology 17, 167–178.

Giglio, R.M., Ivy, J.A., Jones, L.C., et al., 2016. Evaluation of alternative management strategies for maintenance of genetic variation in wildlife populations. Animal Conservation 19, 380–390.

Gilbert, K.J., Whitlock, M.C., 2015. Evaluating methods for estimating local effective population size with and without migration. Evolution 69, 2154–2166.

Gilbert, K.J., Andrew, R.L., Bock, D.G., et al., 2012. Recommendations for utilizing and reporting population genetic analyses: the reproducibility of genetic clustering using the program STRUCTURE. Molecular Ecology 21, 4925–4930.

Giles, B.E., Goudet, J., 1997. A case study of genetic structure in a plant metapopulation, In Metapopulation Biology: Ecology, Genetics, and Evolution. eds I.A. Hanski, M.E. Gilpin, pp. 429–454. Academic Press, San Diego.

Gillespie, J.H., 1991. The Causes of Molecular Evolution. Oxford University Press, New York.

Gilligan, D.M., Frankham, R., 2003. Dynamics of genetic adaptation to captivity. Conservation Genetics 4, 189–197.

Gilligan, D.M., Briscoe, D.A., Frankham, R., 2005. Comparative losses of quantitative and molecular genetic variation in finite populations of *Drosophila melanogaster*. Genetical Research 85, 47–55.

Gilligan, D.M., Woodworth, L.M., Montgomery, M.E., et al., 1997. Is mutation accumulation a threat to the survival of endangered populations? Conservation Biology 11, 1235–1241.

Gilpin, M., 1991. The genetic effective size of a metapopulation. Biological Journal of the Linnean Society 42, 165–176.

Gilpin, M.E., Soulé, M.E., 1986. Minimum viable populations: processes of species extinction, In Conservation Biology: The Science of Scarcity and Diversity. ed. M.E. Soulé, pp. 19–34. Sinauer, Sunderland, MA.

Gitzendanner, M.A., Weekley, C.W., Germain-Aubrey, C.C., et al., 2012. Microsatellite evidence for high clonality and limited genetic diversity in *Ziziphus celata* (Rhamnaceae), an endangered, self-incompatible shrub endemic to the Lake Wales Ridge, Florida, USA. Conservation Genetics 13, 223–234.

Glémin, S., 2003. How are deleterious mutations purged? Drift versus nonrandom mating. Evolution 57, 2678–2687.

Gogarten, J.P., Townsend, J.P., 2005. Horizontal gene transfer, genome innovation and evolution. Nature Reviews Microbiology 3, 679–687.

Goldberg, E.E., Kohn, J.R., Lande, R., et al., 2010. Species selection maintains self-incompatibility. Science 330, 493–495.

Gómez-Romano, F., Villanueva, B., de Cara, M.Á.R., et al., 2013. Maintaining genetic diversity using molecular coancestry: the effect of marker density and effective population size. Genetics Selection Evolution 45, 38.

Goodall, J., Maynard, T., Hudson, G., 2009. Hope for Animals and Their World: How Endangered Species are Being Rescued from the Brink. Icon Books, London.

Goodwillie, C., Kalisz, S., Eckert, C.G., 2005. The evolutionary enigma of mixed mating systems in plants: occurrence, theoretical explanations, and empirical evidence. Annual Review of Ecology, Evolution, and Systematics 36, 47–79.

Gottelli, D., Sillero-Zubiri, C., Appelbaum, G.D., et al., 1994. Molecular genetics of the most endangered canid: the Ethiopian wolf *Canis simensis*. Molecular Ecology 3, 301–312.

Goudet, J., 2002. FSTAT, a program to estimate and test gene diversities and fixation indices (version 2.9.3.2). http://www2.unil.ch/popgen/softwares/fstat.htm

Grant, B.R., Grant, P.R., 2008. Fission and fusion of Darwin's finches populations. Philosophical Transactions of the Royal Society of London B: Biological Sciences 363, 2821–2829.

Grant, B.S., Wiseman, L.L., 2002. Recent history of melanism in American peppered moths. Journal of Heredity 93, 86–90.

Grant, P.R., Grant, B.R., 1995. Predicting microevolutionary responses to directional selection on heritable variation. Evolution 49, 241–251.

Greeff, J.M., Jansen van Puren, G.J., Kryger, P., et al., 2009. Outbreeding and possibly inbreeding depression in a pollinating fig wasp with a mixed mating system. Heredity 102, 349–356.

Grindeland, J.M., 2008. Inbreeding depression and outbreeding depression in *Digitalis purpurea*: optimal outcrossing distance in a tetraploid. Journal of Evolutionary Biology 21, 716–726.

Groombridge, J.J., Jones, C.G., Bruford, M.W., et al., 2000. "Ghost" alleles of the Mauritius kestrel. Nature 403, 616.

Gropp, A., Winking, H., Zech, L., et al., 1972. Robertsonian chromosomal variation and identification of metacentric chromosomes in feral mice. Chromosoma 39, 265–288.

Grossman, S.R., Andersen, K.G., Shlyakhter, I., et al., 2013. Identifying recent adaptations in large-scale genomic data. Cell 152, 703–713.

Groves, C., 2004. The what, why and how of primate taxonomy. International Journal of Primatology 25, 1105–1126.

Groves, C., Grubb, P., 2011. Ungulate Taxonomy. John Hopkins University Press, Baltimore, MD.

Groves, C.P., 2001. Primate Taxonomy. Smithsonian Institution Press, Washington, DC.

Groves, C.P., Westwood, C., Shea, B.T., 1992. Unfinished business: Mahalanobis and a clockwork orang. Journal of Human Evolution 22, 327–340.

Grueber, C.E., Jamieson, I.G., 2008. Quantifying and managing the loss of genetic variation in a free-ranging population of takahe through the use of pedigrees. Conservation Genetics 9, 645–651.

Grueber, C.E., Laws, R.J., Nakagawa, S., et al., 2010. Inbreeding depression accumulation across life-history stages of the endangered takahe. Conservation Biology 24, 1617–1625.

Guillot, G., Estoup, A., Mortier, F., et al., 2005b. A spatial statistical model for landscape genetics. Genetics 170, 1261–1280.

Guillot, G., Leblois, R., Coulon, A., et al., 2009. Statistical methods in spatial genetics. Molecular Ecology 18, 4734–4756.

Guillot, G., Mortier, F., Estoup, A., 2005a. GENELAND: a computer package for landscape genetics. Molecular Ecology Notes 5, 712–715.

Guillot, G., Renaud, S., Ledevin, R., et al., 2012. A unifying model for the analysis of phenotypic, genetic, and geographic data. Systematic Biology 61, 897–911.

Gustafsson, L., 1986. Lifetime reproductive success and heritability: empirical support for Fisher's fundamental theorem. American Naturalist 128, 761–764.

Gutiérrez, J.P., Royo, L.J., Álvarez, I., et al., 2005. MolKin v2.0: a computer program for genetic analysis of populations using molecular coancestry information. Journal of Heredity 96, 718–721.

Gynther, I., Waller, N., Leung, L., 2016. Confirmation of the extinction of the Bramble Cay melomys *Melomys rubicola* on Bramble Cay, Torres Strait: results and conclusions from a comprehensive survey in August–September 2014. Unpublished report to the Department of Environment and Heritage Protection, Queensland Government, Brisbane.

Haddad, N.M., Brudvig, L.A., Damschen, E.I., et al., 2014. Potential negative ecological effects of corridors. Conservation Biology 28, 1178–1187.

Hadfield, J.D., 2010. MCMC methods for multi-response generalized linear mixed models: the MCMCglmm R package. Journal of Statistical Software 33, 1–22.

Haig, S.M., Avise, J.C., 1996. Avian conservation genetics, In Conservation Genetics: Case Histories from Nature. eds J.C. Avise, J.L. Hamrick, pp. 160–189. Chapman & Hall, New York.

Haig, S.M., Beever, E.A., Chambers, S.M., et al., 2006. Taxonomic considerations in listing subspecies under the U.S. Endangered Species Act. Conservation Biology 20, 1584–1594.

Haig, S.M., Wagner, R.S., Forsman, E.D., et al., 2001. Geographic variation and genetic structure in spotted owls. Conservation Genetics 2, 25–40.

Halbert, N.D., Derr, J.N., 2007. A comprehensive evaluation of cattle introgression into US Federal bison herds. Journal of Heredity 98, 1–12.

Haldane, J.B.S., 1924. A mathematical theory of natural and artificial selection. Part I. Transactions of the Cambridge Philosophical Society 23, 10–41.

Hall, M.C., Willis, J.H., 2006. Divergent selection on flowering time contributes to local adaptation in *Mimulus guttatus* populations. Evolution 60, 2466–2477.

Hamilton, J.A., Miller, J.M., 2016. Adaptive introgression as a resource for management and genetic conservation in a changing climate. Conservation Biology 30, 33–41.

Hamilton, N.D., 2001. Legal issues shaping society's acceptance of biotechnology and genetically modified organisms. Drake Journal of Agricultural Law 6, 81–117.

Hämmerli, A., Reusch, T.B.H., 2003. Inbreeding depression influences genet size distribution in a marine angiosperm. Molecular Ecology 12, 619–629.

Hammerly, S.C., Morrow, M.E., Johnson, J.A., 2013. A comparison of pedigree- and DNA-based measures for identifying inbreeding depression in the critically endangered Attwater's prairie-chicken. Molecular Ecology 22, 5313–5328.

Hamrick, J.L., Godt, M.J.W., 1989. Allozyme diversity in plant species, In Plant Population Genetics, Breeding, and Genetic Resources. eds A.H.D. Brown, M.T. Clegg, A.L. Kahler, B.S. Weir, pp. 43–63. Sinauer, Sunderland, MA.

Hamrick, J.L., Godt, M.J.W., 1996a. Effects of life-history traits on genetic diversity in plant species. Philosophical Transactions of the Royal Society of London B: Biological Sciences 351, 1291–1298.

Hamrick, J.L., Godt, M.J.W., 1996b. Conservation genetics of endemic plant species, In Conservation Genetics: Case Histories from Nature. eds J.C. Avise, J.L. Hamrick, pp. 281–304. Chapman & Hall, New York.

Hansen, J., Sato, M., Ruedy, R., 2012. Perception of climate change. Proceedings of the National Academy of Sciences 109, E2415–E2423.

Hansen, M.M., Limborg, M., Ferchaud, A.-L., et al., 2014. The effects of Medieval dams on genetic divergence and demographic history in brown trout populations. BMC Evolutionary Biology 14, 122.

Hanski, I., Gaggiotti, O.E. eds., 2004. Ecology, Genetics, and Evolution of Metapopulations. Elsevier, Amsterdam.

Hanski, I., Gilpin, M.E., 1997. Metapopulation Biology: Ecology, Genetics and Evolution. Academic Press, Orlando, FL.

Hanski, I., Alho, J., Moilanen, A., 2000. Estimating the parameters of survival and migration of individuals in metapopulations. Ecology 81, 239–251.

Hanski, I.A., 2011. Eco-evolutionary spatial dynamics in the Glanville fritillary butterfly. Proceedings of the National Academy of Sciences 108, 14397–14404.

Hansson, B., Bensch, S., Hasselquist, D., et al., 2004. Lifetime fitness of short-and long-distance dispersing great reed warblers. Evolution 58, 2546–2557.

Hardy, G.H., 1908. Mendelian proportions in a mixed population. Science 28, 49–50.

Hardy, O.J., Vekemans, X., 2001. Patterns of allozyme variation in diploid and tetraploid *Centaurea jacea* at different spatial scales. Evolution 55, 943–954.

Hardy, O.J., Vekemans, X., 2002. SPAGeDi: a versatile computer program to analyse spatial genetic structure at the individual or population levels. Molecular Ecology Notes 2, 618–620.

Harrisson, K.A., Pavlova, A., Gonçalves da Silva, A., et al., 2016. Scope for genetic rescue of an endangered subspecies though re-establishing natural gene flow with another subspecies. Molecular Ecology 25, 1242–1258.

Harrisson, K.A., Pavlova, A., Telonis-Scott, M., et al., 2014. Using genomics to characterize evolutionary potential for conservation of wild populations. Evolutionary Applications 7, 1008–1025.

Hartley, C.J., Newcomb, R.D., Russell, R.J., et al., 2006. Amplification of DNA from preserved specimens show blowflies were preadapted for the rapid evolution of insecticide resistance. Proceedings of the National Academy of Sciences 103, 8757–8762.

Hausdorf, B., 2011. Progress toward a general species concept. Evolution 65, 923–931.

Havird, J.C., Fitzpatrick, S.W., Kronenberger, J., et al., 2016. Sex, mitochondria, and genetic rescue. Trends in Ecology and Evolution 31, 96–99.

Hayman, R., 1974. The development of the Australian milking zebu. World Animal Review (FAO) 11, 31–35.

Hayward, M.W., Somers, M. eds., 2009. Reintroduction of Top-Order Predators. Wiley-Blackwell, Oxford, UK.

Heap, I., 2007. International Survey of Herbicide Resistant Weeds. http://www.weedscience.com.

Hebert, P.D.N., Cywinska, A., Ball, S.L., et al., 2003a. Biological identification through DNA barcodes. Proceedings of the Royal Society of London B: Biological Sciences 270, 313–321.

Hebert, P.D.N., Ratnasingham, S., de Waard, J.R., 2003b. Barcoding animal life: cytochrome *c* oxidase subunit 1 divergences among closely related species. Proceedings of the Royal Society of London B: Biological Sciences Supplement 270, S96–S99.

Hedrick, P.W., 1983. Genetics of Populations. Science Books International, Boston, MA.

Hedrick, P.W., 1995. Gene flow and genetic restoration: the Florida panther as a case study. Conservation Biology 9, 996–1007.

Hedrick, P.W., 1999. Highly variable loci and their interpretation in evolution and conservation. Evolution 53, 313–318.

Hedrick, P.W., 2005a. A standardized genetic differentiation measure. Evolution 59, 1633–1638.

Hedrick, P.W., 2005b. Genetics of Populations, 3rd edition. Jones and Bartlett, Sudbury, MA.

Hedrick, P.W., 2009. Conservation genetics and North American bison (*Bison bison*). Journal of Heredity 100, 411–420.

Hedrick, P.W., Fredrickson, R., 2008. Captive breeding and the reintroduction of Mexican and red wolves. Molecular Ecology 17, 344–350.

Hedrick, P.W., Fredrickson, R., 2010. Genetic rescue guidelines with examples from Mexican wolves and Florida panthers. Conservation Genetics 11, 615–626.

Hedrick, P.W., Parker, J.D., 1997. Evolutionary genetics of haplo-diploid organisms and X-linked chromosomes. Annual Review of Ecology and Systematics 28, 55–83.

Hedrick, P.W., Miller, P.S., Geffen, E., et al., 1997. Genetic evaluation of the three captive Mexican wolf lineages. Zoo Biology 16, 47–69.

Hedrick, P.W., Peterson, R.O., Vucetich, L.M., et al., 2014. Genetic rescue in Isle Royale wolves: genetic analysis and the collapse of the population. Conservation Genetics 15, 1111–1121.

Heiser, D.A., Shaw, R.G., 2006. The fitness effects of outcrossing in *Calylophus serrulatus*, a permanent translocation heterozygote. Evolution 60, 64–76.

Heled, J., Drummond, A.J., 2010. Bayesian inference of species trees from multilocus data. Molecular Biology and Evolution 27, 570–580.

Heller, R., Frandsen, P., Lorenzen, E., et al., 2013. Are there really twice as many bovid species as we thought? Systematic Biology 62, 490–493.

Heller, R., Siegismund, H.R., 2009. Relationship between three measures of genetic differentiation G_{ST}, D_{EST} and G'_{ST}: how wrong have we been? Molecular Ecology 18, 2080–2083.

Hendry, A.P., Kinnison, M.T., 1999. The pace of modern life: measuring rates of contemporary microevolution. Evolution 53, 1637–1653.

Hendry, A.P., Nosil, P., Rieseberg, L.H., 2007. The speed of ecological speciation. Functional Ecology 21, 455–464.

Henter, H.J., 2003. Inbreeding depression and haplodiploidy: experimental measures in a parasitoid and comparisons across diploid and haplodiploid insect taxa. Evolution 57, 1793–1803.

Hereford, J., 2009a. A quantitative survey of local adaptation and fitness trade-offs. American Naturalist 173, 579–588.

Hereford, J. 2009b. Postmating/prezygotic isolation, heterosis, and outbreeding depression in crosses within and between populations of *Diodia teres* (Rubiaceae) Walt. International Journal of Plant Sciences 170, 301–310.

Hereford, J., 2010. Does selfing or outcrossing promote local adaptation? American Journal of Botany 97, 298–302.

Hereford, J., Hansen, T.F., Houle, D., et al., 2004. Comparing strengths of directional selection: how strong is strong? Evolution 58, 2133–2143.

Herron, J.C., Freeman, S., Hodin, J., et al., 2014. Evolutionary Analysis. Pearson, Boston, MA.

Hersch-Green, E., Myburg, H., Johnson, M., 2012. Adaptive molecular evolution of a defence gene in sexual but not functionally asexual evening primroses. Journal of Evolutionary Biology 25, 1576–1586.

Heschel, M.S., Paige, K.N., 1995. Inbreeding depression, environmental stress, and population size variation in scarlet gilia (*Ipomopsis aggregata*). Conservation Biology 9, 126–133.

Hewitt, G., 2000. The genetic legacy of the Quaternary ice ages. Nature 405, 907–913.

Hey, J., 1999. The neutralist, the fly and the selectionist. Trends in Ecology and Evolution 14, 35–38.

Hey, J., 2010. Isolation with migration models for more than two populations. Molecular Biology and Evolution 27, 905–920.

Hey, J., Nielsen, R., 2004. Multilocus methods for estimating population sizes, migration rates and divergence time, with applications to the divergence of *Drosophila pseudoobscura* and *D. persimilis*. Genetics 167, 747–760.

Hey, J., Nielsen, R., 2007. Integration within the Felsenstein equation for improved Markov chain Monte Carlo methods in population genetics. Proceedings of the National Academy of Sciences 104, 2785–2790.

Hey, J., Pinho, C., 2012. Population genetics and objectivity in species diagnosis. Evolution 66, 1413–1429.

Hey, J., Waples, R.S., Arnold, M.L., et al., 2003. Understanding and confronting species uncertainty in biology and conservation. Trends in Ecology and Evolution 18, 597–603.

Hill, A.V., 1998. The immunogenetics of human infectious diseases. Annual Review of Immunology 16, 593–617.

Hill, W.G., 1982. Predictions of response to artificial selection from new mutations. Genetical Research 40, 255–278.

Hirzel, A.H., Le Lay, G., 2008. Habitat suitability modelling and niche theory. Journal of Applied Ecology 45, 1372–1381.

Hoban, S., 2014. An overview of the utility of population simulation software in molecular ecology. Molecular Ecology 23, 2383–2401.

Hoban, S., Bertorelle, G., Gaggiotti, O.E., 2012. Computer simulations: tools for population and evolutionary genetics. Nature Reviews Genetics 13, 110–122.

Hoban, S., Gaggiotti, O., Con, G.C., et al., 2013. Sample Planning Optimization Tool for conservation and population Genetics (SPOTG): a software for choosing the appropriate number of markers and samples. Methods in Ecology and Evolution 4, 299–303.

Hoban, S., Kelley, J.L., Lotterhos, K.E., et al., 2016. Finding the genomic basis of local adaptation: pitfalls, practical solutions, and future directions. American Naturalist 188, 379–397.

Hoegh-Guldberg, O., 1999. Climate change, coral bleaching and the future of the world's coral reefs. Marine and Freshwater Research 50, 839–866.

Hoegh-Guldberg, O., Hughes, L., McIntyre, S., et al., 2008. Assisted colonization and rapid climate change. Science 321, 345–346.

Hoegh-Guldberg, O., Mumby, P.J., Hooten, A.J., et al., 2007. Coral reefs under rapid climate change and ocean acidification. Science 318, 1737–1742.

Hoekstra, H.E., 2006. Genetics, development and evolution of adaptive pigmentation in vertebrates. Heredity 97, 222–234.

Hoekstra, H.E., Hoekstra, J.M., Berrigan, D., et al., 2001. Strength and tempo of directional selection in the wild. Proceedings of the National Academy of Sciences 98, 9157–9160.

Hoffman, J.I., Simpson, F., David, P., et al., 2014. High-throughput sequencing reveals inbreeding depression in a natural population. Proceedings of the National Academy of Sciences 111, 3775–3780.

Hoffmann, A., Griffin, P., Dillon, S., et al., 2015. A framework for incorporating evolutionary genomics into biodiversity conservation and management. Climate Change Responses 2, doi: 10.1186/s40665-40014-40009-x.

Hoffmann, A.A., Sgrò, C.M., 2011. Climate change and evolutionary adaptation. Nature 470, 479–485.

Hoffmann, A.A., Hallas, R.J., Dean, J.A., et al., 2003. Low potential for climatic stress adaptation in a rainforest *Drosophila* species. Science 301, 100–102.

Hoffmann, A.A., Merilä, J., Kristensen, T.N., 2016. Heritability and evolvability of fitness and nonfitness traits: lessons from livestock. Evolution 70, 1770–1779.

Hoffmann, A.A., Sgrò, C.M., Weeks, A.R., 2004. Chromosomal inversion polymorphisms and adaptation. Trends in Ecology and Evolution 19, 482–488.

Hogg, J.T., Forbes, S.H., Steele, B.M., et al., 2006. Genetic rescue of an insular population of large mammals. Proceedings of the Royal Society of London B: Biological Sciences 273, 1491–1499.

Höglund, J., Piertney, S.B., Alatalo, R.V., et al., 2002. Inbreeding depression and male fitness in black grouse. Proceedings of the Royal Society B: Biological Sciences 269, 711–715.

Hokanson, K., Hancock, J., 1998. Levels of allozymic diversity in diploid and tetraploid *Vaccinium* sect. Cyanococcus (blueberries). Canadian Journal of Plant Science 78, 327–332.

Holderegger, R., Buehler, D., Gugerli, F., et al., 2010. Landscape genetics of plants. Trends in Plant Science 15, 675–683.

Holtsford, T.P., 1996. Variation in inbreeding depression among families and populations of *Clarkia tembloriensis* (Onagraceae). Heredity 76, 83–91.

Honnay, O., Coart, E., Butaye, J., et al., 2006. Low impact of present and historical landscape configuration on the genetics of fragmented *Anthyllis vulneraria* populations. Biological Conservation 127, 411–419.

Hood, C.S., Robbins, L.W., Baker, R.J., et al., 1984. Chromosomal studies and evolutionary relationships of an endangered species, *Reithrodontomys raviventris*. Journal of Mammalogy 65, 655–667.

Hostetler, J.A., Onorato, D.P., Bolker, B.M., et al., 2012. Does genetic introgression improve female reproductive performance? A test on the endangered Florida panther. Oecologia 168, 289–300.

Hostetler, J.A., Onorato, D.P., Nichols, J.D., et al., 2010. Genetic introgression and the survival of Florida panther kittens. Biological Conservation 143, 2789–2796.

Houlden, B.A., Costello, B.H., Sharkey, D., et al., 1999. Phylogeographic differentiation in the mitochondrial control region in the koala, *Phascolarctos cinereus* (Goldfuss 1817). Molecular Ecology 8, 999–1011.

Houlden, B.A., England, P.R., Taylor, A.C., et al., 1996. Low genetic variability of the koala *Phascolarctos cinereus* in southeastern Australia. Molecular Ecology 5, 269–281.

Hu, Y., Zhan, X., Qi, D., et al., 2010. Spatial genetic structure and dispersal of giant pandas on a mountain-range scale. Conservation Genetics 11, 2145–2155.

Hudson, R.R., Coyne, J.A., 2002. Mathematical consequences of the genealogical species concept. Evolution 56, 1557–1565.

Hudson, R.R., Boos, D.D., Kaplan, N.L., 1992. A statistical test for detecting geographic subdivision. Molecular Biology and Evolution 9, 138–151.

Huey, R.B., Gilchrist, G.W., Carlson, M.L., et al., 2000. Rapid evolution of a geographic cline in size in an introduced fly. Science 287, 308–309.

Hufbauer, R.A., Szűcs, M., Kasyon, E., et al., 2015a. Three types of rescue can avert extinction in a changing environment. Proceedings of the National Academy of Sciences 112, 10557–10562.

Hufbauer, R.A., Szűcs, M., Kasyon, E., et al., 2015b. Reply to Wootton and Pfister: the search for general context should include synthesis with laboratory model systems. Proceedings of the National Academy of Sciences 112, E5904.

Hughes, A.R., Inouye, B.D., Johnson, M.T.J., et al., 2008. Ecological consequences of genetic diversity. Ecology Letters 11, 609–623.

Hughes, J.B., Daily, G.C., Ehrlich, P.R., 1997. Population diversity: its extent and extinction. Science 278, 689–692.

Hughes, W.O.H., Boomsma, J.J., 2004. Genetic diversity and disease resistance in leaf-cutting ant societies. Evolution 58, 1251–1260.

Huisman, J., Kruuk, L.E.B., Ellis, P.A., et al., 2016. Inbreeding depression across the lifespan in a wild mammal population. Proceedings of the National Academy of Sciences 113, 3585–3590.

Huntley, B., 1991. How plants respond to climate change: migration rates, individualism and the consequences for plant communities. Annals of Botany 67, 15–22.

Husband, B.C., Schemske, D.W., 1996. Evolution of the magnitude and timing of inbreeding depression in plants. Evolution 50, 54–70.

Husband, B.C., Schemske, D.W., 1997. The effect of inbreeding in diploid and tetraploid populations of *Epilobium angustifolium* (Onagraceae): implications for the genetic basis of inbreeding depression. Evolution 51, 737–746.

Husband, B.C., Baldwin, S.J., Suda, J., 2013. The incidence of polyploidy in natural plant populations: major patterns and evolutionary processes, In Plant Genome Diversity Volume 2. eds J. Greilhuber, J. Doležel, J.F. Wendel, pp. 255–276. Springer, Vienna.

IPCC, 2014. Climate Change 2014: Synthesis Report. Contribution of Working Groups I, II and III to the Fifth Assessment Report of the Intergovernmental Panel on Climate Change [Core Writing Team, R.K. Pachauri and L.A. Meyer (eds.)]. IPCC, Geneva, Switzerland.

IUCN, 2016. IUCN Red List of Threatened Species. http://www.redlist.org/.

IUCN/SSC, 2013. Guidelines for Reintroductions and Other Conservation Translocations. IUCN/SSC Reintroduction Specialist Group, Gland, Switzerland.

James, F., 1995. The status of the red-cockaded woodpecker in 1990 and the prospect for recovery, In Red-Cockaded Woodpecker: Recovery, Ecology and Management. eds D.L. Kulhavy, R.G. Hooper, R. Costa, pp. 439–451. Center for Applied Studies, Stephen F. Austin State University, Nacogdoches, TX.

James, S.H., Wylie, A.P., Johnson, M.S., et al., 1983. Complex hybridity in *Isotoma petraea*. V. Allozyme variation and the pursuit of hybridity. Heredity 51, 653–663.

Jamieson, I.G., Wallis, G.P., Briskie, J.V., 2006. Inbreeding and endangered species management: is New Zealand out of step with the rest of the world? Conservation Biology 20, 38–47.

Janczewski, D.N., Goldman, D., O'Brien, S.J., 1990. Molecular genetic divergence of orangutan (*Pongo pygmaeus*) subspecies based on isozyme and two-dimensional gel electrophoresis. Journal of Heredity 81, 375–387.

Jarne, P., Auld, J.R., 2006. Animals mix it up too: the distribution of self-fertilization among hermaphroditic animals. Evolution 60, 1816–1824.

Jenkins, D.G., Carey, M., Czerniewska, J., et al., 2010. A meta-analysis of isolation by distance: relic or reference standard for landscape genetics? Ecography 33, 315–320.

Jiménez, J.A., Hughes, K.A., Alaks, G., et al., 1994. An experimental study of inbreeding depression in a natural habitat. Science 266, 271–273.

Johansson, M.L., Raimondi, P.T., Reed, D.C., et al., 2013. Looking into the black box: simulating the role of self-fertilization and mortality in the genetic structure of *Macrocystis pyrifera*. Molecular Ecology 22, 4842–4854.

Johnson, P.M., Nolan, B.J., Schaper, D.N., 2003. Introduction of the Proserpine rock-wallaby *Petrogale persephone* from the Queensland mainland to nearby Hayman Island. Australian Mammalogy 25, 61–71.

Johnson, W.E., Onorato, D.P., Roelke, M.E., et al., 2010. Genetic restoration of the Florida panther. Science 329, 1641–1645.

Jombart, T., Devillard, S., Dufour, A., et al., 2008. Revealing cryptic spatial patterns in genetic variability by a new multivariate method. Heredity 101, 92–103.

Jones, N., Houben, A., 2003. B chromosomes in plants: escapees from the A chromosome genome? Trends in Plant Science 8, 417–423.

Joseph, L., Omland, K.E., 2009. Phylogeography: its development and impact in Australo-Papuan ornithology with special reference to paraphyly in Australian birds. Emu 109, 1–23.

Jost, L., 2008. G_{ST} and its relatives do not measure differentiation. Molecular Ecology 17, 4015–4026.

Jousimo, J., Tack, A.J.M., Ovaskainen, O., et al., 2014. Ecological and evolutionary effects of fragmentation on infectious disease dynamics. Science 344, 1289–1293.

Judson, O.P., Normark, B.B., 1996. Ancient asexual scandals. Trends in Ecology and Evolution 11, 41–46.

Jump, A.S., Peñuelas, J., 2005. Running to stand still: adaptation and the response of plants to rapid climate change. Ecology Letters 8, 1010–1020.

Jump, A.S., Peñuelas, J., 2006. Genetic effects of chronic habitat fragmentation in a wind-pollinated tree. Proceedings of the National Academy of Sciences 103, 8096–8100.

Jusaitis, M., Adams, M., 2005. Conservation implications of clonality and limited sexual reproduction in the endangered shrub *Acanthocladium dockeri* (Asteraceae). Australian Journal of Botany 53, 535–544.

Kalinowski, S.T., Hedrick, P.W., 1998. An improved method for estimating inbreeding depression in pedigrees. Zoo Biology 17, 481–497.

Kalinowski, S.T., Hedrick, P.W., 1999. Detecting inbreeding depression is difficult in captive endangered species. Animal Conservation 2, 131–136.

Kalinowski, S.T., Hedrick, P.W., Miller, P.S., 1999. No inbreeding depression observed in Mexican and red wolf captive programs. Conservation Biology 13, 1371–1377.

Kanthaswamy, S., Kurushima, J.D., Smith, D.G., 2006. Inferring *Pongo* conservation units: a perspective based on microsatellite and mitochondrial DNA analyses. Primates 47, 310–321.

Kao, K.C., Schwartz, K., Sherlock, G., 2010. A genome-wide analysis reveals no nuclear Dobzhansky–Muller pairs of determinants of speciation between *S. cerevisiae* and *S. paradoxus*, but suggests more complex incompatibilities. PLOS Genetics 6, e1001038.

Kardos, M., Luikart, G., Allendorf, F.W., 2015. Measuring individual inbreeding in the age of genomics: marker-based measures are better than pedigrees. Heredity 115, 63–72.

Kardos, M., Taylor, H.R., Ellegren, H., et al., 2016. Genomics advances the study of inbreeding depression in the wild. Evolutionary Applications 9, 1205–1218.

Kärkkäinen, K., Koski, V., Savolainen, O., 1996. Geographical variation in the inbreeding depression of Scots pine. Evolution 50, 111–119.

Karron, J.D., 1991. Patterns of genetic variation and breeding systems in rare plant species, In Genetics and Conservation of Rare Plants. eds D.A. Falk, K.E. Holsinger, pp. 87–98. Oxford University Press, New York.

Kearney, M.R., Briscoe, N.J., Karoly, D.J., et al., 2010. Early emergence in a butterfly causally linked to anthropogenic warming. Biology Letters 6, 674–677.

Keith, D.A., Akçakaya, H.R., Thuiller, W., et al., 2008. Predicting extinction risks under climate change: coupling stochastic population models with dynamic bioclimatic habitat models. Biology Letters 4, 560–563.

Keller, L.F., 1998. Inbreeding and its fitness effects in an insular population of song sparrows (*Melospiza melodia*). Evolution 52, 240–250.

Keller, L.F., Waller, D.M., 2002. Inbreeding effects in wild populations. Trends in Ecology and Evolution 17, 230–241.

Keller, M.C., Visscher, P.M., Goddard, M.E., 2011. Quantification of inbreeding due to distant ancestors and its detection using dense single nucleotide polymorphism data. Genetics 189, 237–249.

Kellermann, V.M., van Heerwaarden, B., Hoffmann, A.A., et al., 2006. Very low additive genetic variance and evolutionary potential in multiple populations of two rainforest *Drosophila* species. Evolution 60, 1104–1108.

Kennedy, E.S., Grueber, C.E., Duncan, R.P., et al., 2014. Severe inbreeding depression and no evidence of purging in an extremely inbred wild species—the Chatham Island black robin. Evolution 68, 987–995.

Kettlewell, H.B.D., 1956. A résumé of investigations on the evolution of melanism in the Lepidoptera. Proceedings of the Royal Society of London B: Biological Sciences 145, 297–303.

Kettlewell, H.B.D., 1973. The Evolution of Melanism. Clarendon Press, Oxford, UK.

Kilias, G., Alahiotis, S.N., Pelecanos, M., 1980. A multifactorial genetic investigation of speciation theory using *Drosophila melanogaster*. Evolution 34, 730–737.

Kimpton, S.K., James, E.A., Drinnan, A.N., 2002. Reproductive biology and genetic marker diversity in *Grevillea infecunda* (Proteaceae), a rare plant with no known seed production. Australian Systematic Botany 15, 485–492.

Kimura, M., 1983. The Neutral Theory of Molecular Evolution. Cambridge University Press, Cambridge, UK.

Kimura, M., Crow, J.F., 1963. On the maximum avoidance of inbreeding. Genetical Research 4, 399–415.

Kingman, J.F., 1982. On the genealogy of large populations. Journal of Applied Probability, 27–43.

Kingsolver, J.G., Pfennig, D.W., 2007. Patterns and power of phenotypic selection in nature. Bioscience 57, 561–572.

Kinnison, M.T., Hendry, A.P., 2001. The pace of modern life II: from rates of contemporary microevolution to pattern and process, In Microevolution Rate, Pattern, Process. eds A.P. Hendry, M.T. Kinnison, pp. 145–164. Kluwer Academic Publishers, Dordrecht.

Kirov, I., Divashuk, M., Van Laere, K., et al., 2014. An easy "SteamDrop" method for high quality plant chromosome preparation. Molecular Cytogenetics 7, 21, doi: 10.1186/1755-8166-7-21.

Kleiman, D.G., Rylands, A.B. eds., 2002. Lion Tamarins: Biology and Conservation. Smithsonian Institution Press, Washington, DC.

Klekowski, E.J., 1982. Genetic load and soft selection in ferns. Heredity 49, 191–197.

Knief, U., Hemmrich-Stanisak, G., Wittig, M., et al., 2015. Quantifying realized inbreeding in wild and captive animal populations. Heredity 114, 397–403.

Knowles, L.L., Carstens, B.C., 2007. Delimiting species without monophyletic gene trees. Systematic Biology 56, 887–895.

Knowlton, N., Weigt, L., 1997. Species of marine invertebrates: a comparison of the biological and phylogenetic species concepts, In Species: The Units of Biodiversity. eds M.F. Claridge, H.A. Dawah, M.R. Wilson, pp. 199–220. Chapman & Hall, London.

Koelling, V.A., Monnahan, P.J., Kelly, J.K., 2012. A Bayesian method for the joint estimation of outcrossing rate and inbreeding depression. Heredity 109, 393–400.

Koevoets, T., Van De Zande, L., Beukeboom, L.W., 2012. Temperature stress increases hybrid incompatibilities in the parasitic wasp genus *Nasonia*. Journal of Evolutionary Biology 25, 304–316.

Kolbe, J.J., Glor, R.E., Schettino, L.R., et al., 2007. Multiple sources, admixture, and genetic variation in introduced *Anolis* lizard populations. Conservation Biology 21, 1612–1625.

Kondrashov, A.S., Houle, D., 1994. Genotype–environment interactions and the estimation of the genomic mutation rate in *Drosophila melanogaster*. Proceedings of the Royal Society of London B: Biological Sciences 258, 221–227.

Kong, A., Frigge, M.L., Masson, G., et al., 2012. Rate of de novo mutations and the importance of father's age to disease risk. Nature 488, 471–475.

Kovach, R.P., Muhlfeld, C.C., Boyer, M.C., et al., 2015. Dispersal and selection mediate hybridization between a native and invasive species. Proceedings of the Royal Society of London B: Biological Sciences 282, 20142454.

Kreitman, M., Akashi, H., 1995. Molecular evidence for natural selection. Annual Review of Ecology and Systematics 26, 403–422.

Kreyling, J., Bittner, T., Jaeschke, A., et al., 2011. Assisted colonization: a question of focal units and recipient localities. Restoration Ecology 19, 433–440.

Krimbas, C., Powell, J., 2000. Inversion polymorphisms in *Drosophila*, In Evolutionary Genetics: From Molecules to Morphology. eds R.S. Singh, C.B. Krimbas, pp. 284–299. Cambridge University Press, Cambridge, UK.

Kruuk, L.E.B., 2004. Estimating genetic parameters in natural populations using the "animal model." Philosophical Transactions of the Royal Society of London B: Biological Sciences 359, 873–890.

Kruuk, L.E.B., Hadfield, J.D., 2007. How to separate genetic and environmental causes of similarity between relatives. Journal of Evolutionary Biology 20, 1890–1903.

Kruuk, L.E.B., Clutton-Brock, T.H., Slate, J., et al., 2000. Heritability of fitness in a wild mammal population. Proceedings of the National Academy of Sciences 97, 698–703.

Kruuk, L.E.B., Sheldon, B.C., Merilä, J., 2002. Severe inbreeding depression in collared flycatchers (*Ficedula albicollis*). Proceedings of the Royal Society of London B: Biological Sciences 269, 1581–1589.

Kulhavy, D.L., Hooper, R.G., Costa, R., 1995. Red-Cockaded Woodpecker: Recovery, Ecology and Management. Center for Applied Studies, Stephen F. Austin State University, Nacogdoches, TX.

Kullberg, C., Fransson, T., Hedlund, J., et al., 2015. Change in spring arrival of migratory birds under an era of climate change, Swedish data from the last 140 years. Ambio 44, 69–77.

Kuparinen, A., Hutchings, J.A., 2012. Consequences of fisheries-induced evolution for population productivity and recovery potential. Proceedings of the Royal Society of London B: Biological Sciences 279, 2571–2579.

Lachapelle, J., Bell, G., 2012. Evolutionary rescue of sexual and asexual populations in a deteriorating environment. Evolution 66, 3508–3518.

Lacy, R.C., 1987. Loss of genetic diversity from managed populations: interacting effects of drift, mutation, immigration, selection, and population subdivision. Conservation Biology 1, 143–158.

Lacy, R.C., 1997. Importance of genetic variation to the viability of mammalian populations. Journal of Mammalogy 78, 320–335.

Lacy, R.C., 1998. Partitioning additive, dominance, epistatic and maternal effects on reproductive performance in crosses between subspecies of *Peromyscus polionotes*, In 45th Annual Meeting of the Genetics Society of Australia Inc. ed. J. Sved, p. 88. Genetics Society of Australia Inc., Sydney.

Lacy, R.C., Pollak, J.P., 2014. Vortex: A Stochastic Simulation of the Extinction Process. Version 10.0. Chicago Zoological Society, Brookfield, IL.

Lacy, R.C., Alaks, G., Walsh, A., 1996. Hierarchical analysis of inbreeding depression in *Peromyscus polionotus*. Evolution 50, 2187–2200.

Lacy, R.C., Ballou, J.D., Pollak, J.P., 2012. PMx: software package for demographic and genetic analysis and management of pedigreed populations. Methods in Ecology and Evolution 3, 433–437.

Lacy, R.C., Miller, P.S., Nyhus, P.J., et al., 2013. Metamodels for transdisciplinary analysis of wildlife population dynamics. PLOS ONE 8, e84211.

Lai, Z., Nakazato, T., Salmaso, M., et al., 2005. Extensive chromosomal repatterning and the evolution of sterility barriers in hybrid sunflower species. Genetics 171, 291–303.

Laikre, L., Allendorf, F.W., Aroner, L.C., et al., 2009. Neglect of genetic diversity in implementation of the Convention on Biological Diversity. Conservation Biology 24, 86–88.

Laikre, L., Larsson, L.C., Palmé, A., et al., 2008. Potentials for monitoring gene level biodiversity: using Sweden as an example. Biodiversity and Conservation 17, 893–910.

Lamaze, F.C., Sauvage, C., Marie, A., et al., 2012. Dynamics of introgressive hybridization assessed by SNP population genomics of coding genes in stocked brook charr (*Salvelinus fontinalis*). Molecular Ecology 21, 2877–2895.

Land, D.E., Lacy, R.C., 2000. Introgression level achieved through Florida panther genetic restoration. Endangered Species Updates 17, 99–103.

Lande, R., 1979. Effective deme size during long-term evolution estimated from rates of chromosomal rearrangement. Evolution 33, 234–251.

Lande, R., 1988. Genetics and demography in biological conservation. Science 241, 1455–1460.

Lande, R., 1992. Neutral theory of quantitative genetic variance in an island model with local extinction and colonization. Evolution 46, 381–389.

Lande, R., 1995. Mutation and conservation. Conservation Biology 9, 782–791.

Lande, R., Schemske, D.W., 1985. The evolution of self-fertilization and inbreeding depression in plants. I. Genetic models. Evolution 39, 24–40.

Lande, R., Shannon, S., 1996. The role of genetic variation in adaptation and population persistence in a changing environment. Evolution 50, 434–437.

Landguth, E.L., Muhlfeld, C.C., Waples, R.S., et al., 2014. Combining demographic and genetic factors to assess population vulnerability in stream species. Ecological Applications 24, 1505–1524.

Lane, J.E., Kruuk, L.E., Charmantier, A., et al., 2012. Delayed phenology and reduced fitness associated with climate change in a wild hibernator. Nature 489, 554–557.

Lange, K., 1997. Markov chain Monte Carlo methods, In Mathematical and Statistical Methods for Genetic Analysis. ed. K. Lange, pp. 142–163. Springer, New York.

Lankau, R.A., Strauss, S.Y., 2007. Mutual feedbacks maintain both genetic and species diversity in a plant community. Science 317, 1561–1563.

Latch, E.K., Dharmarajan, G., Glaubitz, J.C., et al., 2006. Relative performance of Bayesian clustering software for inferring population substructure and individual assignment at low levels of population differentiation. Conservation Genetics 7, 295–302.

Law, J.H., Crespi, B.J., Shaw, K., 2002. Recent and ancient asexuality in *Timema* walkingsticks. Evolution 56, 1711–1717.

Leberg, P.L., Firmin, B.D., 2008. Role of inbreeding depression and purging in captive breeding and restoration programmes. Molecular Ecology 17, 334–343.

Lee, C.E., 2000. Global phylogeography of a cryptic copepod species complex and reproductive isolation between genetically proximate "populations." Evolution 54, 2014–2027.

Lee, H.-Y., Chou, J.-Y., Cheong, L., et al., 2008. Incompatibility of nuclear and mitochondrial genomes causes hybrid sterility between two yeast species. Cell 135, 1065–1073.

Lee, M., 2009. EU Regulation of GMOs: Law and Decision Making for a New Technology. Edward Elgar Publishing, Cheltenham, UK.

Leffler, E.M., Bullaughey, K., Matute, D.R., et al., 2012. Revisiting an old riddle: what determines genetic diversity levels within species? PLOS Biology 10, e1001388.

Leimu, R., Fischer, M., 2008. A meta-analysis of local adaptation in plants. PLOS ONE 3, e4010.

Leimu, R., Vergeer, P., Angeloni, F., et al., 2010. Habitat fragmentation, climate change, and inbreeding in plants. Annals of the New York Academy of Sciences 1195, 84–98.

Leinonen, T., O'Hara, R.B., Cano, J.M., et al., 2008. Comparative studies of quantitative trait and neutral marker divergence: a meta-analysis. Journal of Evolutionary Biology 21, 1–17.

Lenormand, T., 2002. Gene flow and the limits to natural selection. Trends in Ecology and Evolution 17, 183–189.

Lesica, P., Leary, R.F., Allendorf, F.W., et al., 1988. Lack of genetic diversity within and among populations of an endangered plant *Howellia aquatilis*. Conservation Biology 2, 275–282.

Letnic, M., Dickman, C.R., 2006. Boom means bust: interactions between the El Niño/Southern Oscillation (ENSO), rainfall and the processes threatening mammal species in arid Australia. Biodiversity and Conservation 15, 3847–3880.

Levi-Strauss, C., 1973. Tristes Tropiques (English translation by John and Doreen Weightman). Atheneum, New York.

Levin, D.A., Wilson, A.C., 1976. Rates of evolution in seed plants: net increase in diversity of chromosome numbers and species numbers through time. Proceedings of the National Academy of Sciences 73, 2086–2090.

Levins, R., 1969. Some demographic and genetic consequences of environmental heterogeneity for biological control. Bulletin of the Entomological Society of America 15, 237–240.

Levins, R., 1970. Extinction, In Some Mathematical Questions in Biology, Second Symposium on Mathematical Biology. ed. M. Gerstenhaber, pp. 77–107. American Mathematics Society, Providence, RI.

Levy, M., Winternheimer, P.L., 1977. Allozyme linkage disequilibria among chromosome complexes in the permanent translocation heterozygote *Oenothera biennis*. Evolution 31, 465–476.

Lewontin, R.C., 1974. The Genetic Basis of Evolutionary Change. Columbia University Press, New York.

Lewontin, R.C., Birch, L.C., 1966. Hybridization as a source of variation for adaptation to new environments. Evolution 20, 315–336.

Li, C.C., 1976. First Course in Population Genetics. Boxwood Press, Pacific Grove, CA.

Li, S., Li, B., Cheng, C., et al., 2014. Genomic signatures of near-extinction and rebirth of the crested ibis and other endangered bird species. Genome Biology 15, 557, doi: 10.1186/s13059-014-0557-1.

Liberg, O., Andrén, H., Pedersen, H.-C., et al., 2005. Severe inbreeding depression in a wild wolf (*Canis lupus*) population. Biology Letters 1, 17–20.

Lindstrom, E.W., 1941. Analysis of modern maize breeding principles and methods, In Proceeding of the VII International Congress of Genetics. ed. R.C. Punnett, pp. 151–156. Cambridge University Press, Cambridge.

Liu, K., Goodman, M., Muse, S., et al., 2003. Genetic structure and diversity among maize inbred lines as inferred from DNA microsatellites. Genetics 165, 2117–2128.

Lloyd, D.G., Webb, C., 1986. The avoidance of interference between the presentation of pollen and stigmas in angiosperms I. Dichogamy. New Zealand Journal of Botany 24, 135–162.

Lohbeck, K.T., Riebesell, U., Reusch, T.B.H., 2012. Adaptive evolution of a key phytoplankton species to ocean acidification. Nature Geoscience 5, 346–351.

Lohbeck, K.T., Riebesell, U., Reusch, T.B.H., 2014. Gene expression changes in the coccolithophore *Emiliania huxleyi* after 500 generations of selection to ocean acidification. Proceedings of the Royal Society of London B: Biological Sciences 281, 20140003.

Loiselle, B.A., Sork, V.L., Nason, J., et al., 1995. Spatial genetic structure of a tropical understory shrub, *Psychotria officinalis* (Rubiaceae). American Journal of Botany 82, 1420–1425.

Losdat, S., Chang, S.M., Reid, J.M., 2014. Inbreeding depression in male gametic performance. Journal of Evolutionary Biology 27, 992–1011.

Losos, J.B., Warheit, K., Schoener, T., 1997. Adaptive differentiation following experimental island colonization in *Anolis* lizards. Nature 387, 70–73.

Loveless, M.D., Hamrick, J.L., 1984. Ecological determinants of genetic structure in plant populations. Annual Review of Ecology and Systematics 15, 65–95.

Lu, Z., Johnson, W.E., Menotti-Raymond, M., et al., 2001. Patterns of genetic diversity in remaining giant panda populations. Conservation Biology 15, 1596–1607.

Luijten, S.H., Kéry, M., Oostermeijer, J.G.B., et al., 2002. Demographic consequences of inbreeding and outbreeding in *Arnica montana*: a field experiment. Journal of Ecology 90, 593–603.

Luikart, G., Ryman, N., Tallmon, D.A., et al., 2010. Estimation of census and effective population sizes: the increasing usefulness of DNA-based approaches. Conservation Genetics 11, 355–373.

Luna, M.G., Hawkins, B.A., 2004. Effects of inbreeding versus outbreeding in *Nasonia vitripennis* (Hymenoptera: Pteromalidae). Environmental Entomology 33, 765–775.

Luttikhuizen, P.C., Stift, M., Kuperus, P., et al., 2007. Genetic diversity in diploid vs. tetraploid *Rorippa amphibia* (Brassicaceae). Molecular Ecology 16, 3544–3553.

Lynch, A.J.J., Barnes, R.W., Cambecedes, J., et al., 1998. Genetic evidence that *Lomatia tasmanica* (Proteaceae) is an ancient clone. Australian Journal of Botany 46, 25–33.

Lynch, M., Lande, R., 1993. Evolution and extinction in response to environmental change, In Biotic Interactions and Global Change. eds P.M. Kareiva, J.G. Kingslover, R.B. Huey, pp. 234–250. Sinauer, Sunderland, MA.

Lynch, M., Walsh, B., 1998. Genetics and Analysis of Quantitative Traits. Sinauer, Sunderland, MA.

Lynch, M., Bürger, R., Butcher, D., et al., 1993. The mutational meltdown in asexual populations. Journal of Heredity 84, 339–344.

Lynch, M., Conery, J., Bürger, R., 1995a. Mutational meltdowns in sexual populations. Evolution 49, 1067–1080.

Lynch, M., Conery, J., Bürger, R., 1995b. Mutation accumulation and the extinction of small populations. American Naturalist 146, 489–518.

Lynch, M., Koskella, B., Schaack, S., 2006. Mutation pressure and the evolution of organelle genomic architecture. Science 311, 1727–1730.

Lysak, M.A., Mandáková, T., 2013. Analysis of plant meiotic chromosomes by chromosome painting, In Plant Meiosis: Methods and Protocols. eds P.W. Pawlowski, M. Grelon, S. Armstrong, pp. 13–24. Humana Press, Totowa, NJ.

Mace, G.M., 2004. The role of taxonomy in species conservation. Philosophical Transactions of the Royal Society B: Biological Sciences 359, 711–719.

Mackay, T.F.C., 1986. A quantitative genetic analysis of fitness and its components in *Drosophila melanogaster*. Genetics Research 47, 59–70.

Mackay, T.F.C., Fry, J.D., Lyman, R.F., et al., 1994. Polygenic mutation in *Drosophila melanogaster*: estimates from response to selection of inbred strains. Genetics 136, 937–951.

Madlung, A., 2013. Polyploidy and its effect on evolutionary success: old questions revisited with new tools. Heredity 110, 99–104.

Madsen, T., Shine, R., Olsson, M., et al., 1999. Restoration of an inbred adder population. Nature 402, 34–35.

Madsen, T., Stille, B., Shine, R., 1996. Inbreeding depression in an isolated population of adders *Vipera berus*. Biological Conservation 75, 113–118.

Madsen, T., Ujvari, B., Olsson, M., 2004. Novel genes continue to enhance population growth in adders (*Vipera berus*). Biological Conservation 120, 145–147.

Magoulick, D.D., Kobza, R.M., 2003. The role of refugia for fishes during drought: a review and synthesis. Freshwater Biology 48, 1186–1198.

Maheshwari, S., Wang, J., Barbash, D.A., 2008. Recurrent positive selection of the *Drosophila* hybrid incompatibilty gene *Hmr*. Molecular Biology and Evolution 25, 2421–2430.

Mahy, G., Bruederle, L.P., Connors, B., et al., 2000. Allozyme evidence for genetic autopolyploidy and high genetic diversity in tetraploid cranberry, *Vaccinium oxycoccos* (Ericaceae). American Journal of Botany 87, 1882–1889.

Mailund, T., Dutheil, J.Y., Hobolth, A., et al., 2011. Estimating divergence time and ancestral effective population size of Bornean and Sumatran orangutan subspecies using a coalescent hidden Markov model. PLOS Genetics 7, e1001319.

Majerus, M.E.N., 1998. Melanism. Evolution in Action. Oxford University Press, Oxford, UK.

Major, R.E., Sladek, J., 2012. Is an island reserve enough? The decline and fall of the white-fronted chat (Aves: Meliphagidae) in southern Sydney. Proceedings of the Linnean Society of New South Wales 134, B205–B214.

Major, R.E., Johnson, R.N., King, A.G., et al., 2014. Genetic isolation of endangered bird populations inhabiting salt marsh remnants surrounded by intensive urbanization. Animal Conservation 17, 419–429.

Malécot, G., 1955. The decrease of relationship with distance. Cold Spring Harbor Symposium on Quantitative Biology 20, 52–53.

Malécot, G., 1969. The Mathematics of Heredity. W.H. Freeman and Company, San Francisco, CA.

Manel, S., Gaggiotti, O.E., Waples, R.S., 2005. Assignment methods: matching biological questions with appropriate techniques. Trends in Ecology and Evolution 20, 136–142.

Manel, S., Schwartz, M.K., Luikart, G., et al., 2003. Landscape genetics: combining landscape ecology and population genetics. Trends in Ecology and Evolution 18, 189–197.

Manichaikul, A., Mychaleckyj, J.C., Rich, S.S., et al., 2010. Robust relationship inference in genome-wide association studies. Bioinformatics 26, 2867–2873.

Manly, B.F.J., 1985. The Statistics of Natural Selection on Animal Populations. Chapman & Hall, London.

Mansergh, I., Heinze, D., Weeks, A., et al., 2013. Gene-pool mixing: lessons from the high country. Wildlife Australia 50, 32.

Mansfield, K.G., Land, E.D., 2002. Cryptorchidism in Florida panthers: prevalence, features, and influence of genetic restoration. Journal of Wildlife Diseases 38, 693–698.

Mantel, N., 1967. The detection of disease clustering and a generalized regression approach. Cancer Research 27, 209–220.

Margan, S.H., Nurthen, R.K., Montgomery, M.E., et al., 1998. Single large or several small? Population fragmentation in the captive management of endangered species. Zoo Biology 17, 467–480.

Markham, A., 2016. Australia's iconic Great Barrier Reef World Heritage Site at risk from global warming, In Union of Concerned Sciences. Blog May 26, 2016.

Marko, P.B., Hart, M.W., 2011. The complex analytical landscape of gene flow inference. Trends in Ecology and Evolution 26, 448–456.

Maschinski, J., Wright, S.J., Koptur, S., et al., 2013. When is local the best paradigm? Breeding history influences conservation reintroduction survival and population trajectories in times of extreme climate events. Biological Conservation 159, 277–284.

Matthies, D., Bräuer, I., Maibom, W., et al., 2004. Population size and the risk of local extinction: empirical evidence from rare plants. Oikos 105, 481–488.

Mattick, J.S., Taft, R.J., Faulkner, G.J., 2010. A global view of genomic information—moving beyond the gene and the master regulator. Trends in Genetics 26, 21–28.

Mattila, A.L.K., Duplouy, A., Kirjokangas, M., et al., 2012. High genetic load in an old isolated butterfly population. Proceedings of the National Academy of Sciences 109, E2496–E2505.

Mavárez, J., Salazar, C.A., Bermingham, E., et al., 2006. Speciation by hybridization in *Heliconius* butterflies. Nature 441, 868–871.

Mayden, R.L., 1997. A hierarchy of species concepts: the denouncement in the saga of the species problem, In Species: The Units of Biodiversity. eds M.F. Claridge, H.A. Dawah, M.R. Wilson, pp. 381–424. Chapman & Hall, London.

Maynard Smith, J., 1978. The Evolution of Sex. Cambridge University Press, Cambridge, UK.

Mayr, E., 1942. Systematics and the Origin of Species. Columbia University Press, New York.

Mayr, E., 1963. Animal Species and Evolution. Harvard University Press, Cambridge, MA.

McCartney, M.A., Lessios, H.A., 2004. Adaptive evolution of sperm bindin tracks egg incompatibility in neotropical sea urchins of the genus *Echinometra*. Molecular Biology and Evolution 21, 732–745.

McCauley, D.E., 1994. Contrasting the distribution of chloroplast DNA and allozyme polymorphism among local populations of *Silene alba*: implications for studies of gene flow in plants. Proceedings of the National Academy of Sciences 91, 8127–8131.

McCleery, R.H., Pettifor, R.A., Armbruster, P., et al., 2004. Components of variance underlying fitness in a natural population of the great tit *Parus major*. American Naturalist 164, E62–E72.

McClelland, E.K., Naish, K.A., 2007. What is the fitness outcome of crossing unrelated fish populations? A meta-analysis and an evaluation of future research directions. Conservation Genetics 8, 397–416.

McDade, L.A., 1995. Species concepts and problems in practice: insight from botanical monographs. Systematic Botany, 20, 606–622.

McDevitt, A.D., Mariani, S., Hebblewhite, M., et al., 2009. Survival in the Rockies of an endangered hybrid swarm from diverged caribou (*Rangifer tarandus*) lineages. Molecular Ecology 18, 665–679.

McFarlane, S.E., Gorrell, J.C., Coltman, D.W., et al., 2014. Very low levels of direct additive genetic variance in fitness and fitness components in a red squirrel population. Ecology and Evolution 4, 1729–1738.

McKenzie, J.A., 1996. Ecological and Evolutionary Aspects of Insecticide Resistance. Academic Press and R.G. Lands Company, Austin, TX.

McKenzie, J.A., Batterham, P., 1994. The genetic, molecular and phenotypic consequences of selection for insecticide resistance. Trends in Ecology and Evolution 9, 166–169.

McLachlan, J.S., Hellmann, J.J., Schwartz, M.W., 2007. A framework for debate of assisted migration in an era of climate change. Conservation Biology 21, 297–302.

McLaughlin, J.F., Hellmann, J.J., Bloggs, C.L., et al., 2002. Climate change hastens population extinctions. Proceedings of the National Academy of Sciences 99, 6070–6074.

McNeely, J.A., Miller, K.R., Reid, W.V., et al., 1990. Conserving the World's Biological Diversity. IUCN, World Resources Institute, Conservation International, WWF-US and the World Bank, Washington, DC.

McVean, G., 2009. A genealogical interpretation of principal components analysis. PLOS Genetics 5, e1000686.

Meagher, S., Penn, D.J., Potts, W.K., 2000. Male–male competition magnifies inbreeding depression in wild house mice. Proceedings of the National Academy of Sciences 97, 3324–3329.

Meffe, G.K., Carroll, C.R., 1997. Principles of Conservation Biology, 2nd edition. Sinauer, Sunderland, MA.

Meirmans, P.G., 2014. Nonconvergence in Bayesian estimation of migration rates. Molecular Ecology Resources 14, 726–733.

Meirmans, P.G., Hedrick, P.W., 2011. Assessing population structure: F_{ST} and related measures. Molecular Ecology Resources 11, 5–18.

Meirmans, P.G., Van Tienderen, P.H., 2004. GENOTYPE and GENODIVE: two programs for the analysis of genetic diversity of asexual organisms. Molecular Ecology Notes 4, 792–794.

Mengel, R.M., Jackson, J.A., 1977. Geographic variation of the red-cockaded woodpecker. Condor 79, 349–355.

Menzel, A., Sparks, T.H., Estrella, N., et al., 2006. European phenological response to climate change matches the warming pattern. Global Change Biology 12, 1969–1976.

Merilä, J., Sheldon, B.C., Kruuk, L.E.B., 2001. Explaining stasis: microevolutionary studies in natural populations. Genetica 112, 199–222.

Metcalfe, C.J., Eldridge, M.D.B., Johnston, P.G., 2002. Mapping the distribution of the telomeric sequence (T2AG3) in rock-wallabies, *Petrogale* (Marsupialia: Macropodidae) by fluorescence in situ hybridization. II. The lateralis complex. Cytogenetic and Genome Research 96, 169–175.

Meyer, K., 2007. WOMBAT—a tool for mixed model analyses in quantitative genetics by restricted maximum likelihood (REML). Journal of Zhejiang University Science B 8, 815–821.

Mickelberg, J.L., 2011. Understanding and managing isolation in a fragmented population of golden lion tamarins, *Leontopithecus rosalia*. Ph.D. thesis, George Mason University, Fairfax, VA.

Miller, J., Poissant, J., Hogg, J., et al., 2012. Genomic consequences of genetic rescue in an insular population of bighorn sheep (*Ovis canadensis*). Molecular Ecology 21, 1583–1596.

Miller-Rushing, A.J., Primack, R.B., 2008. Global warming and flowering times in Thoreau's Concord: a community perspective. Ecology 89, 332–341.

Mills, L.S., Allendorf, F.W., 1996. The one-migrant-per-generation rule in conservation and management. Conservation Biology 10, 1509–1518.

Minteer, B.A., Collins, J.P., 2010. Move it or lose it? The ecological ethics of relocating species under climate change. Ecological Applications 20, 1801–1804.

Miskelly, C.M., Taylor, G.A., Gummer, H., et al., 2009. Translocations of eight species of burrow-nesting seabirds (genera *Pterodroma*, *Pelecanoides*, *Pachyptila* and *Puffinus*: Family Procellariidae). Biological Conservation 142, 1965–1980.

Mitrovski, P., Heinze, D.A., Broome, L., et al., 2007. High levels of variation despite genetic fragmentation in populations of the endangered mountain pygmy-possum, *Burramys parvus*, in alpine Australia. Molecular Ecology 16, 75–87.

Mitrovski, P., Hoffmann, A.A., Heinze, D., et al., 2008. Rapid loss of genetic variation in an endangered possum. Biology Letters 4, 134–138.

Mittell, E.A., Nakagawa, S., Hadfield, J.D., 2015. Are molecular markers useful predictors of adaptive potential? Ecology Letters 18, 772–778.

Mlot, C., 2015. Inbred wolf population on Isle Royale collapses. Science 348, 383.

Mlot, C., 2016. Extreme inbreeding likely spells doom for Isle Royale wolves. Science doi: 10.1126/science.aaf9930.

Moll, R.H., Lonnquist, J.H., Fortuno, J.V., et al., 1965. The relationship of heterosis and genetic divergence in maize. Genetics 52, 139–144.

Møller, A.P., Nielsen, J.T., 2015. Genetic variation in birds in relation to predation risk by hawks: a comparative analysis. Current Zoology 61, 1–9.

Montgomery, M.E., Ballou, J.D., Nurthen, R.K., et al., 1997. Minimizing kinship in captive breeding programs. Zoo Biology 16, 377–389.

Montgomery, M.E., Woodworth, L.M., England, P.R., et al., 2010. Widespread selective sweeps affecting microsatellites in *Drosophila* populations adapting to captivity: implications for captive breeding programs. Biological Conservation 143, 1842–1849.

Montgomery, M.E., Woodworth, L.M., Nurthen, R.K., et al., 2000. Relationships between population size and genetic diversity: comparisons of experimental results with theoretical predictions. Conservation Genetics 1, 33–43.

Moody, M.E., Mueller, L.D., Soltis, D.E., 1993. Genetic variation and random drift in autotetraploid populations. Genetics 134, 649–657.

Mora, C., Metzger, R., Rollo, A., et al., 2007. Experimental simulations about the effects of overexploitation and habitat fragmentation on populations facing environmental warming. Proceedings of the Royal Society of London B: Biological Sciences 274, 1023–1028.

Mora, C., Tittensor, D.P., Adl, S., et al., 2011. How many species are there on earth and in the ocean? PLOS Biology 9, e1001127.

Morgan, J.W., Meyer, M.J., Young, A.G., 2013. Severe habitat fragmentation leads to declines in genetic variation, mate availability, and reproductive success in small populations of a once-common Australian grassland daisy. International Journal of Plant Sciences 174, 1209–1218.

Moritz, C., 1994. Defining "Evolutionarily Significant Units" for conservation. Trends in Ecology and Evolution 9, 373–375.

Moritz, C., 1999. Conservation units and translocations: strategies for conserving evolutionary processes. Hereditas 130, 217–228.

Morton, N.E., Crow, J.F., Muller, H.J., 1956. An estimate of the mutational damage in man from data on consanguineous marriages. Proceedings of the National Academy of Sciences 42, 855–863.

Mousseau, T.A., Sinervo, B., Endler, J.A., 2000. Adaptive Genetic Variation in the Wild. Oxford University Press, New York.

Moyle, L.C., Stinchcombe, J.R., Hudgens, B.R., et al., 2003. Conservation genetics in the recovery of endangered animal species: a review of US endangered species recovery plans (1977–1998). Animal Biodiversity and Conservation 26, 85–95.

Muir, C.C., Galdikas, B.M.F., Beckenbach, A.T., 1998. Is there sufficient evidence to elevate the orangutan of Borneo and Sumatra to separate species. Journal of Molecular Evolution 46, 378–381.

Muller, H.J., 1940. Bearings of the "Drosophila" work on systematics, In The New Systematics. ed. J.S. Huxley, pp. 195–268. Clarendon Press, Oxford, UK.

Murphy, W.J., Larkin, D.M., der Wind, A.E.-v., et al., 2005. Dynamics of mammalian chromosome evolution inferred from multispecies comparative maps. Science 309, 613–617.

Murray, B.G., Young, A.G., 2001. Widespread chromosomal variation in the endangered grassland forb *Rutidosis leptorrhynchoides* F. Muell. (Asteraceae: Gnaphalieae). Annals of Botany 87, 83–90.

Nason, J.D., Ellstrand, N.C., 1995. Lifetime estimates of biparental inbreeding depression in the self-incompatible annual plant *Raphanus sativus*. Evolution 49, 307–316.

Neaves, L.E., Zenger, K.R., Prince, R.I.T., et al., 2012. Impact of Pleistocene aridity oscillations on the population history of a widespread, vagile Australian mammal, *Macropus fuliginosus*. Journal of Biogeography 39, 1545–1563.

Nei, M., 1972. Genetic distance between populations. American Naturalist, 106, 283–292.

Nei, M., 1973. Analysis of gene diversity in subdivided populations. Proceedings of the National Academy of Sciences 70, 3321–3323.

Nei, M., 1987. Molecular Evolutionary Genetics. Columbia University Press, New York.

Neigel, J.E., 1996. Estimation of effective population size and migration parameters from genetic data, In Molecular Genetic Approaches in Conservation. eds T.B. Smith, R.K. Wayne, pp. 329–346. Oxford University Press, New York.

Neigel, J.E., 2002. Is F_{ST} obsolete? Conservation Genetics 3, 167–173.

Neiman, M., Sharbel, T., Schwander, T., 2014. Genetic causes of transitions from sexual reproduction to asexuality in plants and animals. Journal of Evolutionary Biology 27, 1346–1359.

Nelis, M., Esko, T., Mägi, R., et al., 2009. Genetic structure of Europeans: a view from the north-east. PLOS ONE 4, e5472.

Nevo, E., Fu, Y.-B., Pavlicek, T., et al., 2012. Evolution of wild cereals during 28 years of global warming in Israel. Proceedings of the National Academy of Sciences 109, 3412–3415.

Newhouse, A.E., Polin-McGuigan, L.D., Baier, K.A., et al., 2014. Transgenic American chestnuts show enhanced blight resistance and transmit the trait to T1 progeny. Plant Science 228, 88–97.

Newman, D., Pilson, D., 1997. Increased probability of extinction due to decreased genetic effective population size: experimental populations of *Clarkia pulchella*. Evolution 51, 354–362.

Newman, D., Tallmon, D.A., 2001. Experimental evidence for beneficial fitness effects of gene flow in recently isolated populations. Conservation Biology 15, 1054–1063.

Ng, K.K.S., Lee, S.L., Koh, C.L., 2004. Spatial structure and genetic diversity of two tropical tree species with contrasting breeding systems and different ploidy levels. Molecular Ecology 13, 657–669.

Nichols, R., 2001. Gene trees and species trees are not the same. Trends in Ecology and Evolution 16, 358–364.

Nicotra, A.B., Atkin, O.K., Bonser, S.P., et al., 2010. Plant phenotypic plasticity in a changing climate. Trends in Plant Science 15, 684–692.

Nieminen, M., Singer, M.C., Fortelius, W., et al., 2001. Experimental confirmation that inbreeding depression increases extinction risk in butterfly populations. American Naturalist 157, 237–244.

NOAA, 2016. National Centers for Environmental Information. http://www.ncdc.noaa.gov/cag/.

Normark, B.B., Judson, O.P., Moran, N.A., 2003. Genomic signatures of ancient asexual lineages. Biological Journal of the Linnean Society 79, 69–84.

Nosil, P., 2007. Divergent host plant adaptation and reproductive isolation between ecotypes of *Timema cristinae* walking sticks. American Naturalist 169, 151–162.

Nosil, P., Schluter, D., 2011. The genes underlying the process of speciation. Trends in Ecology and Evolution 26, 160–167.

Nosil, P., Crespi, B.J., Sandoval, C.P., 2002. Host-plant adaptation drives the parallel evolution of reproductive isolation. Nature 417, 440–443.

Nowak, R., 2004. Forty years of rabbit rage. New Scientist 182(2445), 50–51.

NSW National Parks and Wildlife Service, 1998. Wollemi Pine (*Wollemi nobilis*) Recovery Plan. NSW National Parks and Wildlife Service, Hurstville, NSW, Australia.

Nunes, A.C., Catalan, J., Lopez, J., et al., 2011. Fertility assessment in hybrids between monobrachially homologous Rb races of the house mouse from the island of Madeira: implications for modes of chromosomal evolution. Heredity 106, 348–356.

O'Brien, S.J., Mayr, E., 1991. Bureaucratic mischief: recognizing endangered species and subspecies. Science 251, 1187–1188.

O'Brien, S.J., Menninger, J.C., Nash, W.G. eds., 2006. Atlas of Mammalian Chromosomes. Wiley, Hoboken, NJ.

O'Grady, J.J., Brook, B.W., Reed, D.H., et al., 2006. Realistic levels of inbreeding depression strongly affect extinction risk in wild populations. Biological Conservation 133, 42–51.

Oakley, C.G., Winn, A.A., 2012. Effects of population size and isolation on heterosis, mean fitness, and inbreeding depression in a perennial plant. New Phytologist 196, 261–270.

Oedekoven, K., 1980. The vanishing forest. Environmental Policy and Law 6, 184–185.

Oneal, E., Knowles, L.L., 2013. Ecological selection as the cause and sexual differentiation as the consequence of species divergence? Proceedings of the Royal Society B: Biological Sciences 280, 20122236.

Oostermeijer, J.G.B., 2000. Population viability analysis of the rare *Gentiana pneumonanthe*; the importance of genetics, demography and reproductive biology, In Genetics, Demography and Viability of Fragmented Populations. eds A.G. Young, G.M. Clarke, pp. 313–334. Cambridge University Press, Cambridge, UK.

Orr, H.A., Unckless, R.L., 2014. The population genetics of evolutionary rescue. PLOS Genetics 10, e1004551.

Orr, H.A., Masly, J.P., Phadnis, N., 2007. Speciation in *Drosophila*: from phenotypes to molecules. Journal of Heredity 98, 103–110.

Ortego, J., Aguirre, M.P., Noguerales, V., et al., 2015. Consequences of extensive habitat fragmentation in landscape-level patterns of genetic diversity and structure in the Mediterranean esparto grasshopper. Evolutionary Applications 8, 621–632.

Osborne, P.E., Seddon, P.J., 2012. Selecting suitable habitats for reintroductions: variation, change and the role of species distribution modelling, In Reintroduction Biology: Integrating Science and Management. eds J.G. Ewen, D.P. Armstrong, K.A. Parker, P.J. Seddon, pp. 87–104. Wiley-Blackwell, Oxford, UK.

Osborne, W.S., Zentelis, R.A., Lau, M., 1996. Geographic variation in corroboree frogs, *Pseudophryne corroboree* Moore (Anura: Myobatrachidae): a reappraisal supports recognition of *P. pengilleyi* Wells and Wellington. Australian Journal of Zoology 44, 569–587.

Ottewell, K.M., Bickerton, D.C., Byrne, M., et al., 2016. Bridging the gap: a genetic assessment framework for population-level threatened plant conservation prioritization and decision-making. Diversity and Distributions 22, 174–188.

Otto, S.P., 2007. The evolutionary consequences of polyploidy. Cell 131, 452–462.

Ozimec, B., Husband, B.C., 2011. Effect of recurrent selfing on inbreeding depression and mating system evolution in an autopolyploid plant. Evolution 65, 2038–2049.

Pacifici, M., Foden, W.B., Visconti, P., et al., 2015. Assessing species vulnerability to climate change. Nature Climate Change 5, 215–224.

Paetkau, D., Calvert, W., Stirling, I., et al., 1995. Microsatellite analysis of population structure in Canadian polar bears. Molecular Ecology 4, 347–354.

Paland, S., Lynch, M., 2006. Transitions to asexuality result in excess amino acid substitutions. Science 311, 990–992.

Paland, S., Schmid, B., 2003. Population size and the nature of genetic load in *Gentianella germanica*. Evolution 57, 2242–2251.

Paland, S., Colbourne, J.K., Lynch, M., 2005. Evolutionary history of contagious asexuality in *Daphnia pulex*. Evolution 59, 800–813.

Palstra, F.P., Fraser, D.J., 2012. Effective/census population size ratio estimation: a compendium and appraisal. Ecology and Evolution 2, 2357–2365.

Palstra, F.P., Ruzzante, D.E., 2008. Genetic estimates of contemporary effective population size: what can they tell us about the importance of genetic stochasitcity for wild population persistence? Molecular Ecology 17, 3428–3447.

Palumbi, S.R., 2001. The Evolution Explosion: How Humans Cause Rapid Evolutionary Change. W. W. Norton & Company, New York.

Pamilo, P., Crozier, R.H., 1997. Population biology of social insect conservation. Memoirs of the Museum of Victoria 56, 411–419.

Pannell, J.R., Charlesworth, B., 1999. Neutral genetic diversity in a metapopulation with recurrent local extinction and recolonization. Evolution, 53, 664–676.

Papadopulos, A.S., Baker, W.J., Crayn, D., et al., 2011. Speciation with gene flow on Lord Howe Island. Proceedings of the National Academy of Sciences 108, 13188–13193.

Park, C.B., Larsson, N-G., 2011. Mitochondrial DNA mutations in disease and aging. The Journal of Cell Biology 193, 809–818.

Parmesan, C., Yohe, G., 2003. A globally coherent fingerprint of climate change impacts across natural systems. Nature 421, 37–42.

Patterson, N., Price, A.L., Reich, D., 2006. Population structure and eigenanalysis. PLOS Genetics 2, e190.

Pavlova, A., Amos, J.N., Joseph, L., et al., 2013. Perched at the mito-nuclear crossroads: divergent mitochondrial lineages correlate with environment in the face of ongoing nuclear gene flow in an Australian bird. Evolution 67, 3412–3428.

Peakall, R., Smouse, P.E., 2006. GENALEX 6: genetic analysis in Excel. Population genetic software for teaching and research. Molecular Ecology Notes 6, 288–295.

Peakall, R., Ebert, D., Scott, L.J., et al., 2003. Comparative genetic study confirms exceptionally low genetic variation in the ancient and endangered relictual conifer, *Wollemia nobilis* (Araucariaceae). Molecular Ecology 12, 2331–2343.

Pearson, D., 2013. Recovery plan for five species of rock wallabies: black-footed rock wallaby (*Petrogale lateralis*), Rothschild rock wallaby (*Petrogale rothschildi*), short-eared rock wallaby (*Petrogale brachyotis*), monjon (*Petrogale burbidgei*) and nabarlek (*Petrogale concinna*). Department of Parks and Wildlife, Perth, WA.

Pearson, R.G., Dawson, T.P., 2003. Predicting the impacts of climate change on the distribution of species: are bioclimate envelope models useful? Global Ecology and Biogeography 12, 361–371.

Peer, K., Taborsky, M., 2005. Outbreeding depression, but no inbreeding depression in haplodiploid ambrosia beetles with regular sib mating. Evolution 59, 317–323.

Peery, M.Z., Hall, L.A., Sellas, A., et al., 2010. Genetic analyses of historic and modern marbled murrelets suggest decoupling of migration and gene flow after habitat fragmentation. Proceedings of the Royal Society of London B: Biological Sciences 277, 697–706.

Peichel, C.L., Nereng, K.S., Ohgi, K.A., et al., 2001. The genetic architecture of divergence between threespine stickleback species. Nature 414, 901–905.

Pelikan, S., Rogstad, S.H., 2013. NEWGARDEN: a computer program to model the population dynamics and genetics of establishing and fragmented plant populations. Conservation Genetics Resources 5, 857–862.

Pelz, H-J., Rost, S., Hünerberg, M., et al., 2005. The genetic basis of resistance to anticoagulants in rodents. Genetics 170, 1839–1847.

Penn, D.J., Damjanovich, K., Potts, W.K., 2002. MHC heterozygosity confers a selective advantage against multiple-strain infections. Proceedings of the National Academy of Sciences 99, 11260–11264.

Perelman, P., Johnson, W.E., Roos, C., et al., 2011. A molecular phylogeny of living primates. PLOS Genetics 7, e1001342.

Pérez-Figueroa, A., Saura, M., Fernández, J., et al., 2009. METAPOP—a software for the management and analysis of subdivided populations in conservation programs. Conservation Genetics 10, 1097–1099.

Pertoldi, C., Bijlsma, R., Loeschcke, V., 2007. Conservation genetics in a globally changing environment: present problems, paradoxes and future challenges. Biodiversity and Conservation 16, 4147–4163.

Pespeni, M.H., Sanford, E., Gaylord, B., et al., 2013. Evolutionary change during experimental ocean acidification. Proceedings of the National Academy of Sciences 110, 6937–6942.

Peterson, A.T., Ortega-Huerta, M.A., Bartley, J., et al., 2002. Future projections for Mexican faunas under global climate change scenarios. Nature 416, 626–629.

Petit, R.J., Hampe, A., 2006. Some evolutionary consequences of being a tree. Annual Review of Ecology, Evolution, and Systematics 37, 187–214.

Phillips, R.B., Kapuscinski, A.R., 1988. High frequency of translocation heterozygotes in odd year populations of pink salmon (*Onchrhynchus gorbuscha*). Cytogenetics and Cell Genetics 48, 178–182.

Phillips, S.J., Dudík, M., 2008. Modeling of species distributions with Maxent: new extensions and a comprehensive evaluation. Ecography 31, 161–175.

Phillipsen, I.C., Kirk, E.H., Bogan, M.T., et al., 2015. Dispersal ability and habitat requirements determine landscape-level genetic patterns in desert aquatic insects. Molecular Ecology 24, 54–69.

Pickup, M., 2008. Local Adaptation and Outbreeding Depression in Fragmented Populations of *Rutidosis leptorrhynchoides* (Asteraceae). Ph.D. thesis, Australian National University, Canberra, ACT.

Pickup, M., Young, A.G., 2008. Population size, self-incompatibility and genetic rescue in diploid and tetraploid races of *Rutidosis leptorrhynchoides* (Asteraceae). Heredity 100, 268–274.

Pickup, M., Field, D.L., Rowell, D.M., et al., 2013. Source population characteristics affect heterosis following genetic rescue of fragmented plant populations. Proceedings of the Royal Society B: Biological Sciences 280, 20122058.

Piel, W.H., Chan, L., Dominus, M.J., et al., 2009. TreeBASE v. 2: a database of phylogenetic knowledge. https://treebase.org/treebase-web

Pierson, J.C., Beissinger, S.R., Bragg, J.G., et al., 2015. Incorporating evolutionary processes into population viability models. Conservation Biology 29, 755–764.

Pierson, J.C., Coates, D.J., Oostermeijer, J.G.B., et al., 2016. Consideration of genetic factors in threatened species recovery plans on three continents. Frontiers in Ecology and Environment 14, 433–440.

Piertney, S.B., Oliver, M.K., 2006. The evolutionary ecology of the major histocompatibility complex. Heredity 96, 7–21.

Pinto-Torres, E., Koptur, S., 2009. Hanging by a coastal strand: breeding system of a federally endangered morning-glory of the south-eastern Florida coast, *Jacquemontia reclinata*. Annals of Botany 104, 1301–1311.

Piou, C., Prévost, E., 2012. A demo-genetic individual-based model for Atlantic salmon populations: model structure, parameterization and sensitivity. Ecological Modelling 231, 37–52.

Pistevos, J.C.A., Calosi, P., Widdicombe, S., et al., 2011. Will variation among genetic individuals influence species responses to global climate change? Oikos 120, 675–689.

Platt, A., Horton, M., Huang, Y.S., et al., 2010. The scale of population structure in *Arabidopsis thaliana*. PLOS Genetics 6, e1000843.

Polans, N.O., Allard, R.W., 1989. An experimental evaluation of the recovery potential of ryegrass populations from genetic stress resulting from restriction of population size. Evolution 43, 1320–1324.

Poloczanska, E.S., Brown, C.J., Sydeman, W.J., et al., 2013. Global imprint of climate change on marine life. Nature Climate Change 3, 919–925.

Porcher, E., Lande, R., 2005. Loss of gametophytic self-incompatibility with evolution of inbreeding depression. Evolution 59, 46–60.

Postma, E., 2014. Four decades of estimating heritabilities in wild vertebrate populations: improved methods, more data, better estimates, In Quantitative Genetics in the Wild. eds A. Charmentier, D. Garant, L.E.B. Kruuk, pp. 16–33. Oxford University Press, Oxford, UK.

Potter, S., Cooper, S.J., Metcalfe, C.J., et al., 2012a. Phylogenetic relationships of rock-wallabies, *Petrogale* (Marsupialia: Macropodidae) and their biogeographic history within Australia. Molecular Phylogenetics and Evolution 62, 640–652.

Potter, S., Eldridge, M.D., Taggart, D.A., et al., 2012b. Multiple biogeographical barriers identified across the monsoon tropics of northern Australia: phylogeographic analysis of the *brachyotis* group of rock-wallabies. Molecular Ecology 21, 2254–2269.

Potter, S., Moritz, C., Eldridge, M.D.B., 2015. Gene flow despite complex Robertsonian fusions among rock-wallaby (*Petrogale*) species. Biology Letters 11, 20150731.

Potvin, C., Tousignant, D., 1996. Evolutionary consequences of simulated global change: genetic adaptation or adaptive phenotypic plasticity. Oecologia 108, 683–693.

Powers, D.A., Lauerman, T., Crawford, D., et al., 1991. Genetic mechanisms for adapting to a changing environment. Annual Review of Genetics 25, 629–659.

Prado, J.R., Segers, G., Voelker, T., et al., 2014. Genetically engineered crops: from idea to product. Annual Review of Plant Biology 65, 769–790.

Presgraves, D.C., 2010. The molecular evolutionary basis of species formation. Nature Reviews Genetics 11, 175–180.

Primack, R.B., 2014. Walden Warming: Climate Change Comes to Thoreau's Woods. University of Chicago Press, Chicago, IL.

Princee, F.P.G., 1995. Overcoming the constraints of social structure and incomplete pedigree data through low-intensity genetic management, In Population Management for Survival & Recovery: Analytical Methods and Strategies in Small Population Conservation. eds J.D. Ballou, M. Gilpin, T.J. Foose, pp. 124–154. Columbia University Press, New York.

Pritchard, J.K., Stephens, M., Donnelly, P., 2000. Inference of population structure using multilocus genotype data. Genetics 155, 945–959.

Prober, S.M., Byrne, M., McLean, E.H., et al., 2015. Climate-adjusted provenancing: a strategy for climate-resilient ecological restoration. Frontiers in Ecology and Evolution 3, 65.

Puechmaille, S.J., 2016. The program STRUCTURE does not reliably recover the correct population structure when sampling is uneven: subsampling and new estimators alleviate the problem. Molecular Ecology Resources 16, 608–627.

Pulsford, I., Worboys, G., Gough, J., et al., 2003. A potential new continental-scale conservation corridor for Australia: combining the Australian Alps and the Great Escarpment of eastern Australia conservation corridors. Mountain Research and Development 23, 291–293.

Pursel, V., Pinkert, C., Miller, K., et al., 1989. Genetic engineering of livestock. Science 244, 1281–1288.

Pusey, A., Wolf, M., 1996. Inbreeding avoidance in animals. Trends in Ecology and Evolution 11, 201–206.

Raduski, A.R., Haney, E.B., Igić, B., 2012. The expression of self-incompatibility in angiosperms is bimodal. Evolution 66, 1275–1283.

Rakotonirina, L., Rajaonson, A., Ratolojanahary, T., et al., 2011. New distributional records and conservation implications for the critically endangered greater bamboo lemur *Prolemur simus*. Folia Primatologica 82, 118–129.

Ralls, K., Ballou, J., 1983. Extinction: lessons from zoos, In Genetics and Conservation: A Reference for Managing Wild Animal and Plant Populations. eds C.M. Schonewald-Cox, S.M. Chambers, B. MacBryde, L. Thomas, pp. 164–184. Benjamin/Cummings, Menlo Park, CA.

Ralls, K., Ballou, J.D., 2004. Genetic status and management of California condors. Condor 106, 215–228.

Ralls, K., Ballou, J.D., Templeton, A., 1988. Estimates of lethal equivalents and the cost of inbreeding in mammals. Conservation Biology 2, 185–193.

Ralls, K., Harvey, P.H., Lyles, A.M., 1986. Inbreeding in natural populations of birds and mammals, In Conservation Biology: The Science of Scarcity and Diversity. ed. M.E. Soule, pp. 35–56. Sinauer, Sunderland, MA.

Ramsey, J., Schemske, D.W., 2002. Neoployploidy in flowering plants. Annual Review of Ecology and Systematics 33, 589–639.

Randi, E., 2008. Detecting hybridization between wild species and their domesticated relatives. Molecular Ecology 17, 285–293.

Rannala, B., Mountain, J.L., 1997. Detecting immigration by using multilocus genotypes. Proceedings of the National Academy of Sciences 94, 9197–9201.

Rathbun, G.B., Hatfield, B.B., Murphey, T.G., 2000. Status of translocated sea otters at San Nicolas Island, California. Southwestern Naturalist 45, 322–328.

Réale, D., McAdam, A.G., Boutin, S., et al., 2003. Genetic and plastic responses of a northern mammal to climate change. Proceedings of the Royal Society of London B: Biological Sciences 270, 591–596.

Reed, D.H., Bryant, E.H., 2000. Experimental test of minimum viable population size. Animal Conservation 3, 7–14.

Reed, D.H., Frankham, R., 2001. How closely correlated are molecular and quantitative measures of genetic variation? A meta-analysis. Evolution 55, 1095–1103.

Reed, D.H., Briscoe, D.A., Frankham, R., 2002. Inbreeding and extinction: effects of environmental stress and lineages. Conservation Genetics 3, 301–307.

Reed, D.H., Fox, C.W., Enders, L.S., et al., 2012. Inbreeding–stress interactions: evolutionary and conservation consequences. Annals of the New York Academy of Sciences 1256, 33–48.

Reed, D.H., Lowe, E., Briscoe, D.A., et al., 2003a. Inbreeding and extinction: effects of rate of inbreeding. Conservation Genetics 4, 405–410.

Reed, D.H., Lowe, E.H., Briscoe, D.A., et al., 2003b. Fitness and adaptation in a novel environment: effect of inbreeding, prior environment, and lineage. Evolution 57, 1822–1828.

Reed, T.E., Schindler, D.E., Hague, M.J., et al., 2011. Time to evolve? Potential evolutionary responses of Fraser River sockeye salmon to climate change and effects on persistence. PLOS ONE 6, e20380.

Reid, A.L., 1996. Review of the Peripatopsidae (Onychophora) in Australia, with comments on peripatopsid relationships. Invertebrate Systematics 10, 663–936.

Reid, J.M., Arcese, P., Keller, L.F., et al., 2015. Quantifying inbreeding avoidance through extra-pair reproduction. Evolution 69, 59–74.

Reid, J.M., Keller, L.F., Marr, A.B., et al., 2014. Pedigree error due to extra-pair reproduction substantially biases estimates of inbreeding depression. Evolution 68, 802–815.

Remold, S.K., Lenski, R.E., 2001. Contribution of individual random mutations to genotype-by-environment interactions in *Escherichia coli*. Proceedings of the National Academy of Sciences 98, 11388–11393.

Renner, S.S., 2014. The relative and absolute frequencies of angiosperm sexual systems: dioecy, monoecy, gynodioecy, and an updated online database. American Journal of Botany 101, 1588–1596.

Reusch, T.B.H., Wood, T.E., 2007. Molecular ecology of global change. Molecular Ecology 16, 3973–3992.

Reusch, T.B.H., Ehlers, A., Hämmerli, A., et al., 2005. Ecosystem recovery after climatic extremes enhanced by genotypic diversity. Proceedings of the National Academy of Sciences 102, 2826–2831.

Reznick, D.N., Ghalambor, C.K., 2001. The population ecology of contemporary adaptations: what empirical studies reveal about the conditions that promote adaptive evolution. Genetica 112, 183–198.

Reznick, D.N., Shaw, F.H., Rodd, F.H., et al., 1997. Evaluation of the rate of evolution in natural populations of guppies (*Poecilia reticulata*). Science 275, 1934–1937.

Rhymer, J.M., Simberloff, D., 1996. Extinction by hybridization and introgression. Annual Review of Ecology and Systematics 27, 83–109.

Ricciardi, A., Simberloff, D., 2009. Assisted colonization is not a viable conservation strategy. Trends in Ecology and Evolution 24, 248–253.

Rice, W.R., Hostert, E.E., 1993. Laboratory experiments on speciation: what have we learned in 40 years? Evolution 47, 1637–1653.

Rich, S.S., Bell, A.E., Wilson, S.P., 1979. Genetic drift in small populations of *Tribolium*. Evolution 33, 579–583.

Richards, A.J., 1997. Plant Breeding Systems, 2nd edition. Chapman & Hall, London.

Richman, A.D., Kohn, J.R., 1996. Learning from rejection: the evolutionary biology of single-locus incompatibility. Trends in Ecology and Evolution 11, 497–502.

Rieseberg, L.H., 2001. Chromosomal rearrangements and speciation. Trends in Ecology and Evolution 16, 351–357.

Rieseberg, L.H., Willis, J.H., 2007. Plant speciation. Science 317, 910–914.

Rieseberg, L.H., Raymond, O., Rosenthal, D.M., et al., 2003. Major ecological transitions in wild sunflowers facilitated by hybridization. Science 301, 1211–1216.

Rieseberg, L.H., Sinervo, B., Linder, C.R., et al., 1996. Role of gene interactions in hybrid speciation: evidence from ancient and experimental hybrids. Science 272, 741–745.

Ritland, K., 1989. Correlated matings in the partial selfer *Mimulus guttatus*. Evolution 43, 848–859.

Ritland, K., 2002. Extensions of models for the estimation of mating systems using *n* independent loci. Heredity 88, 221–228.

Roach, D.A., Wulff, R.D., 1987. Maternal effects in plants. Annual Review of Ecology and Systematics 18, 209–235.

Roberts, D.G., Baker, J., Perrin, C., 2011. Population genetic structure of the endangered eastern bristlebird, *Dasyornis brachypterus*; implications for conservation. Conservation Genetics 12, 1075–1085.

Roberts, R.C., 1960. The effects on litter size of crossing lines of mice inbred without selection. Genetics Research 1, 239–252.

Roberts, R.C., 1966. The limits to artificial selection for body weight in the mouse. I. The limits attained in earlier experiment. Genetical Research 8, 347–360.

Robertson, A., 1960. A theory of limits in artificial selection. Proceedings of the Royal Society of London B: Biological Sciences 153, 234–249.

Robertson, A., 1962. Selection for heterozygotes in small populations. Genetics 47, 1291–1300.

Robertson, A., 1964. The effect of non-random mating within inbred lines on the rate of inbreeding. Genetical Research 5, 164–167.

Robertson, A., 1978. The time of detection of recessive visible genes in small populations. Genetical Research 31, 255–264.

Robertson, D.E., 1969. Selection in small populations. Ph.D. thesis, University of New South Wales, Sydney.

Robichaux, R.H., Friar, E.A., Mount, D.W., 1997. Molecular genetic consequences of a population bottleneck associated with reintroduction of the Mauna Kea silversword (*Argyroxiphium sandwicense* ssp. *sandwicense* [Asteraceae]). Conservation Biology 11, 1140–1146.

Rockman, M.V., Rowell, D.M., 2002. Episodic chromosomal evolution in *Planipapillus* (Onychophora: Peripatopsidae): a phylogenetic approach to evolutionary dynamics and speciation. Evolution 56, 58–69.

Rockwood, L.L., 2015. Introduction to Population Ecology, 2nd edition. Wiley-Blackwell, Hoboken, NJ.

Roelke, M.E., Martenson, J., O'Brien, S.J., 1993. The consequences of demographic reduction and genetic depletion in the endangered Florida panther. Current Biology 3, 340–350.

Rohland, N., Reich, D., Mallick, S., et al., 2010. Genomic DNA sequences from mastodon and woolly mammoth reveal deep speciation of forest and savanna elephants. PLOS Biology 8, e1000564.

Rollinson, N., Keith, D.M., Houde, A.L.S., et al., 2014. Risk assessment of inbreeding and outbreeding depression in a captive-breeding program. Conservation Biology 28, 529–540.

Romiguier, J., Gayral, P., Ballenghien, M., et al., 2014. Comparative population genomics in animals uncovers the determinants of genetic diversity. Nature 515, 261–263.

Ronquist, F., Huelsenbeck, J.P., 2003. MrBayes 3: Bayesian phylogenetic inference under mixed models. Bioinformatics 19, 1572–1574.

Root, T.L., Price, J.T., Hall, K.R., et al., 2003. Fingerprints of global warming on wild animals and plants. Nature 421, 57–60.

Rosenberg, N.A., Nordborg, M., 2002. Genealogical trees, coalescent theory and the analysis of genetic polymorphisms. Nature Reviews Genetics 3, 380–390.

Rossetto, M., Jezierski, G., Hopper, S.J., et al., 1999. Conservation genetics and clonality in two critically endangered eucalypts from the highly endemic south-western Australian flora. Biological Conservation 88, 321–331.

Rousset, F., 1997. Genetic differentiation and estimation of gene flow from *F*-statistics under isolation by distance. Genetics 145, 1219–1228.

Rousset, F., 2004. Genetic Structure and Selection in Subdivided Populations. Princeton University Press, Princeton, NJ.

Rousset, F., 2008. GENEPOP'007: a complete re-implementation of the GENEPOP software for Windows and Linux. Molecular Ecology Resources 8, 103–106.

Rowell, D.M., Higgins, A.V., Briscoe, D.A., et al., 1995. The use of chromosomal data in the systematics of viviparous onychophorans from Australia (Onychophora: Peripatopsidae). Zoological Journal of the Linnean Society 114, 139–153.

Rowell, D.M., Lim, S.L., Grutzner, F., 2011. Chromosome analysis in invertebrates and vertebrates, In Molecular Methods for Evolutionary Genetics. eds V. Orgogozo, M.V. Rockman, pp. 13–35. Humana Press, New York.

Rowell, D.M., Rockman, M.V., Tait, N.N., 2002. Extensive Robertsonian rearrangement: implications for the radiation and biogeography of *Planipapillus* Reid (Onychophora: Peripatopsidae). Journal of Zoology 257, 171–179.

Rubinoff, D., Cameron, S., Will, K., 2006. A genomic perspective on the shortcomings of mitochondrial DNA for "barcoding" identification. Journal of Heredity 97, 581–594.

Rundle, H.D., 2002. A test of ecologically dependent postmating isolation between sympatric sticklebacks. Evolution 56, 322–329.

Rundle, H.D., Whitlock, M.C., 2001. A genetic interpretation of ecologically dependent isolation. Evolution 55, 198–201.

Rundle, H.D., Nagel, L., Boughman, J.W., et al., 2000. Natural selection and parallel speciation in sympatric sticklebacks. Science 287, 306–308.

Russello, M.A., Amato, G., 2004. Ex situ population management in the absence of pedigree information. Molecular Ecology 13, 2829–2840.

Russello, M.A., Amato, G., 2014. Operationalism matters in conservation: comments on Frankham et al. (2012). Biological Conservation 170, 332–333.

Rutter, M.T., Fenster, C.B., 2007. Testing for adaptation to climate in *Arabidopsis thaliana*: a calibrated common garden approach. Annals of Botany 99, 529–536.

Ryder, O.A., 1986. Species conservation and systematics: the dilemma of subspecies. Trends in Ecology and Evolution 1, 9–10.

Ryder, O.A., Chemnick, L.G., 1993. Chromosomal and mitochondrial DNA variation in orangutans. Journal of Heredity 84, 405–409.

Ryder, O.A., Kumamoto, A.T., Durrant, B.S., et al., 1989. Chromosomal divergence and reproductive isolation in dik-diks, In Speciation and its Consequences. eds D. Otte, J.A. Endler, pp. 208–225. Sinauer, Sunderland, MA.

Ryman, N., Palm, S., 2006. POWSIM: a computer program for assessing statistical power when testing for genetic differentiation. Molecular Ecology Notes 6, 600–602.

Ryman, N., Utter, F., Hindar, K., 1995. Introgression, supportive breeding, and genetic conservation, In Population Management for Survival & Recovery: Analytical Methods and Strategies in Small Population Conservation. eds J.D. Ballou, M. Gilpin, T.J. Foose, pp. 341–365. Columbia University Press, New York.

Saccheri, I.J., Brakefield, P.M., 2002. Rapid spread of immigrant genomes into inbred populations. Proceedings of the Royal Society of London B: Biological Sciences 269, 1073–1078.

Saccheri, I., Kuussaari, M., Kankare, M., et al., 1998. Inbreeding and extinction in a butterfly metapopulation. Nature 392, 491–494.

Sackton, T.B., Haney, R.A., Rand, D.M., 2003. Cytonuclear coadaptation in *Drosophila*: disruption of cytochrome c oxidase activity in backcross genotypes. Evolution 57, 2315–2325.

Safner, T., Miller, M.P., McRae, B.H., et al., 2011. Comparison of Bayesian clustering and edge detection methods for inferring boundaries in landscape genetics. International Journal of Molecular Sciences 12, 865–889.

Sakai, A.K., Allendorf, F.W., Holt, J.S., et al., 2001. The population biology of invasive species. Annual Review of Ecology and Systematics 32, 305–332.

Sánchez-Molano, E., Caballero, A., Fernández, J., 2013. Efficiency of conservation management methods for subdivided populations under local adaptation. Journal of Heredity 104, 554–564.

Sardos, J., Perrier, X., Doležel, J., et al., 2016. DArT whole genome profiling provides insights on the evolution and taxonomy of edible banana (*Musa* spp.). Annals of Botany. doi: 10.1093/aob/mcw170.

Sarre, S.D., Aitken, N., Adamack, A.T., et al., 2014. Creating new evolutionary pathways through bioinvasion: the population genetics of brushtail possums in New Zealand. Molecular Ecology 23, 3419–3433.

Sattler, T., Bontadina, F., Hirzel, A.H., et al., 2007. Ecological niche modelling of two cryptic bat species calls for a reassessment of their conservation status. Journal of Applied Ecology 44, 1188–1199.

Savage, A.E., Zamudio, K.R., 2011. MHC genotypes associate with resistance to a frog-killing fungus. Proceedings of the National Academy of Sciences 108, 16705–16710.

Savolainen, O., Lascoux, M., Merilä, J., 2013. Ecological genomics of local adaptation. Nature Reviews Genetics 14, 807–820.

Savolainen, V., Anstett, M.-C., Lexer, C., et al., 2006. Sympatric speciation in palms on an oceanic island. Nature 441, 210–213.

Sawada, H., Morita, M., Iwano, M., 2014. Self/non-self recognition mechanisms in sexual reproduction: new insight into the self-incompatibility system shared by flowering plants and hermaphroditic animals. Biochemical and Biophysical Research Communications 450, 1142–1148.

Sawaya, M.A., Kalinowski, S.T., Clevenger, A.P., 2014. Genetic connectivity for two bear species at wildlife crossing structures in Banff National Park. Proceedings of the Royal Society B: Biological Sciences 281, 20131705.

Schemske, D.W., Lande, R., 1985. The evolution of self-fertilization and inbreeding depression in plants. II. Empirical observations. Evolution 39, 41–52.

Schiegg, K., Daniels, S.J., Walters, J.R., et al., 2006. Inbreeding in red-cockaded woodpeckers: effects of natal dispersal distance and territory location. Biological Conservation 131, 544–552.

Schierenbeck, K., Ellstrand, N., 2009. Hybridization and the evolution of invasiveness in plants and other organisms. Biological Invasions 11, 1093–1105.

Schlick-Steiner, B.C., Steiner, F.M., Seifert, B., et al., 2010. Integrative taxonomy: a multisource approach to exploring biodiversity. Annual Review of Entomology 55, 421–438.

Schluter, D., 1988. Estimating the form of natural selection on a quantitative trait. Evolution, 42, 849–861.

Schou, M.F., Kristensen, T.N., Kellermann, V., et al., 2014. A *Drosophila* laboratory evolution experiment points to low evolutionary potential under increased temperatures likely to be experienced in the future. Journal of Evolutionary Biology 27, 1859–1868.

Schwander, T., Henry, L., Crespi, B.J., 2011. Molecular evidence for ancient asexuality in *Timema* stick insects. Current Biology 21, 1129–1134.

Schwartz, M.W., 1993. Allozyme variation of the endangered Florida torreya (*Torreya taxifolia*). Canadian Journal of Forest Research 23, 2598–2602.

Schwartz, M.W., Hellmann, J.J., McLachlan, J.M., et al., 2012. Managed relocation: integrating the scientific, regulatory, and ethical challenges. Bioscience 62, 732–743.

Schwartz, M.W., Iverson, L.R., Prasad, A.M., et al., 2006. Predicting extinctions as a result of climate change. Ecology 87, 1611–1615.

Schwarz, D., Matta, B.M., Shakir-Botteri, N.L., et al., 2005. Host shift to an invasive plant triggers rapid animal hybrid speciation. Nature 436, 546–549.

Scriven, P.N., 1992. Robertsonian translocations introduced into an island population of house mice. Journal of Zoology London 227, 493–502.

Scobie, A.R., Wilcock, C.C., 2009. Limited mate availability decreases reproductive success of fragmented populations of *Linnaea borealis*, a rare, clonal self-incompatible plant. Annals of Botany 103, 835–846.

Seddon, P.J., Armstrong, D.P., Soorae, P., et al., 2009. The risks of assisted colonization. Conservation Biology 23, 788–789.

Segarra-Moragues, J., Palo-Esteban, M., González-Candelas, F., et al., 2005. On the verge of extinction: genetics of the critically endangered Iberian plant species, *Borderea chouardii* (Dioscoreaceae) and implications for conservation management. Molecular Ecology 14, 969–982.

Segelbacher, G., Cushman, S.A., Epperson, B.K., et al., 2010. Applications of landscape genetics in conservation biology: concepts and challenges. Conservation Genetics 11, 375–385.

Selwood, K.E., McGeoch, M.A., Mac Nally, R., 2015. The effects of climate change and land-use change on demographic rates and population viability. Biological Reviews 90, 837–853.

Severns, P.M., Liston, A., 2008. Intraspecific chromosome number variation: a neglected threat to the conservation of rare plants. Conservation Biology 22, 1641–1647.

Sexton, J.P., Hangartner, S.B., Hoffmann, A.A., 2014. Genetic isolation by environment or distance: which pattern of gene flow is most common? Evolution 68, 1–15.

Sexton, J.P., Strauss, S.Y., Rice, K.J., 2011. Gene flow increases fitness at the warm edge of a species' range. Proceedings of the National Academy of Sciences 108, 11704–11709.

Seymour, A.M., Montgomery, M.E., Costello, B.H., et al., 2001. High effective inbreeding coefficients correlate with morphological abnormalities in populations of South Australian koalas (*Phascolarctos cinereus*). Animal Conservation 4, 211–219.

Shaffer, M., Hood, L., Snape, W.J. III, et al., 2000. Population viability analysis and conservation policy, In Population Viability Analysis. eds S.R. Beissinger, D.R. McCullough, pp. 123–142. University of Chicago Press, Chicago, IL.

Shaffer, M.K., 1981. Minimum viable populations size for species conservation. Bioscience 31, 131–134.

Sharbel, T.F., Haubold, B., Mitchell-Olds, T., 2000. Genetic isolation by distance in *Arabidopsis thaliana*: biogeography and postglacial colonization of Europe. Molecular Ecology 9, 2109–2118.

Sharman, G., Close, R., Maynes, G., 1990. Chromosome evolution, phylogeny and speciation of rock wallabies (*Petrogale*, Macropodidae). Australian Journal of Zoology 37, 351–363.

Sherwin, W.B., Timms, P., Wilcken, J., et al., 2000. Analysis and conservation implications of koala genetics. Conservation Biology 14, 639–649.

Shoffner, R.N., Narayanan, S., Pilkey, A.M., 1966. The potential of the advanced generation cross, In Proceedings of the 13th World's Poultry Congress. pp. 36–41, Kiev.

Sillero-Zubiri, C., Aungerbjorn, A., 2009. Arctic foxes and climate change, In IUCN Red List. IUCN, Gland, Switzerland.

Silvertown, J., 2008. The evolutionary maintenance of sexual reproduction: evidence from the ecological distribution of asexual reproduction in clonal plants. International Journal of Plant Sciences 169, 157–168.

Simmons, M.J., Crow, J.F., 1983. The mutation load in *Drosophila*, In The Genetics and Biology of *Drosophila*, Vol 3c. eds M. Ashburner, H.L. Carson, J.N.J. Thompson, pp. 1–35. Academic Press, London.

Simons, A.M., Roff, D.A., 1994. The effect of environmental variability on the heritabilities of traits of a field cricket. Evolution 48, 1637–1649.

Simpson, G.G., 1951. The species concept. Evolution 5, 285–298.

Simpson, G.G., 1961. Principles of Animal Taxonomy. Columbia University Press, New York.

Sinha, S.K., Khanna, R., 1975. Physiological, biochemical, and genetic basis of heterosis. Advances in Agronomy 27, 123–174.

Slate, J., Kruuk, L.E.B., Marshall, T.C., et al., 2000. Inbreeding depression influences lifetime breeding success in a wild population of red deer (*Cervus elaphus*). Proceedings of the Royal Society of London B: Biological Sciences 267, 1657–1662.

Slatkin, M., 1985. Rare alleles as indicators of gene flow. Evolution 39, 53–65.

Slatkin, M., 1995. A measure of population subdivision based on microsatellite allele frequencies. Genetics 139, 457–462.

Sletvold, N., Mousset, M., Hagenblad, J., et al., 2013. Strong inbreeding depression in two Scandinavian populations of the self-incompatible perennial herb *Arabidopsis lyrata*. Evolution 67, 2876–2888.

Smith, S., Hughes, J., 2008. Microsatellite and mitochondrial DNA variation defines island genetic reservoirs for reintroductions of an endangered Australian marsupial, *Perameles bougainville*. Conservation Genetics 9, 547–557.

Soanes, K., Lobo, M.C., Vesk, P.A., et al., 2013. Movement re-established but not restored: inferring the effectiveness of road-crossing mitigation for a gliding mammal by monitoring use. Biological Conservation 159, 434–441.

Sobel, J.M., Chen, G.F., Watt, L.R., et al., 2010. The biology of speciation. Evolution 64, 295–315.

Soderquist, T., 2011. What we don't know and haven't learned about cost-benefit prioritisation of rock-wallaby management. Australian Mammalogy 33, 202–213.

Soltis, D.E., Rieseberg, L.H., 1986. Autopolyploidy in *Tolmiea menziesii* (Saxifragaceae): genetic insights from enzyme electrophoresis. American Journal of Botany 73, 310–318.

Sommer, S., 2005. The importance of immune gene variability (MHC) in evolutionary ecology and conservation. Frontiers of Zoology 2, 16.

Soulé, M.E., 1985. What is conservation biology? Bioscience 35, 727–734.

Soulé, M.E., Terborgh, J., 1999. Conserving nature at regional and continental scales—a scientific program for North America. Bioscience 49, 809–817.

Souza, R.O., Del Lama, M.A., Cervini, M., et al., 2010. Conservation genetics of neotropical pollinators revisited: microsatellite analysis suggests that diploid males are rare in orchid bees. Evolution 64, 3318–3326.

Spielman, D., Frankham, R., 1992. Modeling problems in conservation genetics using captive *Drosophila* populations: improvement in reproductive fitness due to immigration of one individual into small partially inbred populations. Zoo Biology 11, 343–351.

Spielman, D., Brook, B.W., Frankham, R., 2004a. Most species are not driven to extinction before genetic factors impact them. Proceedings of the National Academy of Sciences 101, 15261–15264.

Spielman, D., Brook, B.W., Briscoe, D.A., et al., 2004b. Does inbreeding and loss of genetic diversity reduce disease resistance? Conservation Genetics 5, 439–448.

Stangel, P.W., Lennartz, M.R., Smith, M.H., 1992. Genetic variation and population structure of red-cockaded woodpeckers. Conservation Biology 6, 283–292.

Stathos, A., Fishman, L., 2014. Chromosomal rearrangements directly cause underdominant F1 pollen sterility in *Mimulus lewisii–Mimulus cardinalis* hybrids. Evolution 68, 3109–3119.

Steane, D.A., Potts, B.M., McLean, E., et al., 2014. Genome-wide scans detect adaptation to aridity in a widespread forest tree species. Molecular Ecology 23, 2500–2513.

Stearns, F.W., Fenster, C.B., 2016. Fisher's geometric model predicts the effects of random mutations when tested in the wild. Evolution 70, 495–501.

Steinberg, E.K., Jordan, C.E., 1998. Using molecular genetics to learn about ecology of threatened species: the allure and the illusion of measuring genetic structure in natural populations, In Conservation Biology: For the Coming Decade, 2nd edition. eds P.L. Fiedler, P.M. Karieva, pp. 440–460. Chapman & Hall, New York.

Stelkens, R.B., Seehausen, O., 2009. Phenotypic divergence but not genetic distance predicts assortative mating among species of a cichlid fish radiation. Journal of Evolutionary Biology 22, 1679–1694.

Stelkens, R.B., Brockhurst, M.A., Hurst, G.D.D., et al., 2014. Hybridization facilitates evolutionary rescue. Evolutionary Applications 7, 1209–1217.

Stephen, C.L., Devos, J.C., Lee, T.E., et al., 2005. Population genetic analysis of sonoran pronghorn (*Antilocapra americana sonoriensis*). Journal of Mammalogy 86, 782–792.

Stoddart, J.A., Babcock, R.C., Heyward, A.J., 1988. Self-fertilization and maternal enzymes in the planulae of the coral *Goniastrea favulus*. Marine Biology 99, 489–494.

Stoffel, M.A., Esser, M., Kardos, M., et al., 2016. inbreedR: an R package for the analysis of inbreeding based on genetic markers. Methods in Ecology and Evolution 7, 1331–1339.

Stokes, H.W., Gillings, M.R., 2011. Gene flow, mobile genetic elements and the recruitment of antibiotic resistance genes into Gram-negative pathogens. FEMS Microbiology Reviews 35, 790–819.

Stow, A.J., Sunnucks, P., 2004. Inbreeding avoidance in Cunningham's skinks (*Ergenia cunninghami*) in natural and fragmented habitats. Molecular Ecology 13, 443–447.

Stringfield, G.H., 1950. Heterozygosis and hybrid vigor in maize. Agronomy Journal 42, 145–152.

Sunday, J.M., Calosi, P., Dupont, S., et al., 2014. Evolution in an acidifying ocean. Trends in Ecology and Evolution 29, 117–125.

Sunday, J.M., Crim, R.N., Harley, C.D.G., et al., 2011. Quantifying rates of evolutionary adaptation in response to ocean acidification. PLOS ONE 6, e22881.

Sunnucks, P., England, P.R., Taylor, A.C., et al., 1996. Microsatellite and chromosome evolution of parthenogenetic sitobion aphids in Australia. Genetics 144, 747–756.

Sutherland, G.D., Harestad, A.S., Price, K., et al., 2000. Scaling of natal dispersal distances in terrestrial birds and mammals. Conservation Ecology 4, [online] http://www.consecol.org/vol4/iss1/art16/

Sutton, J.T., Nakagawa, S., Robertson, B.C., et al., 2011. Disentangling the roles of natural selection and genetic drift in shaping variation at MHC immunity genes. Molecular Ecology 20, 4408–4420.

Swanson-Wagner, R.A., Jia, Y., DeCook, R., et al., 2006. All possible modes of gene action are observed in a global comparison of gene expression in a maize F_1 hybrid and its inbred parents. Proceedings of the National Academy of Sciences 103, 6805–6810.

Swindell, W., Bouzat, J., 2006. Gene flow and adaptive potential in *Drosophila melanogaster*. Conservation Genetics 7, 79–89.

Szulkin, M., Sheldon, B.C., 2007. The environmental dependence of inbreeding depression in a wild bird population. PLOS ONE 2, e1027.

Szulkin, M., Garant, D., McCleery, R.H., et al., 2007. Inbreeding depression along a life-history continuum in the great tit. Journal of Evolutionary Biology 20, 1531–1543.

Tallmon, D.A., Gregovich, D., Waples, R.S., et al., 2010. When are genetic methods useful for estimating contemporary abundance and detecting population trends? Molecular Ecology Resources 10, 684–692.

Tallmon, D.A., Luikart, G., Waples, R.S., 2004. The alluring simplicity and the complex reality of genetic rescue. Trends in Ecology and Evolution 19, 489–496.

Tang, S., Dai, W., Li, M., et al., 2008. Genetic diversity of relictual and endangered plant *Abies ziyuanensis* (Pinaceae) revealed by AFLP and SSR markers. Genetica 133, 21–30.

Taylor, A.C., Sherwin, W.B., Wayne, R.K., 1994. Genetic variation of microsatellite loci in a bottlenecked species: the northern hairy-nosed wombat *Lasiorhinus krefftii*. Molecular Ecology 3, 277–290.

Taylor, H.R., 2015. The use and abuse of genetic marker-based estimates of relatedness and inbreeding. Ecology and Evolution 5, 3140–3150.

Taylor, H.R., Kardos, M.D., Ramstad, K.M., et al., 2015. Valid estimates of individual inbreeding coefficients from marker-based pedigrees are not feasible in wild populations with low allelic diversity. Conservation Genetics 16, 901–913.

Taylor, P.J., Eppley, S.M., Jesson, L.K., 2007. Sporophytic inbreeding depression in mosses occurs in a species with separate sexes but not in a species with combined sexes. American Journal of Botany 94, 1853–1859.

Templeton, A.R., 1986. Coadaptation and outbreeding depression, In Conservation Biology: The Science of Scarcity and Diversity. ed. M.E. Soulé, pp. 105–116. Sinauer, Sunderland, MA.

Templeton, A.R., 2008. The reality and importance of founder speciation in evolution. Bioessays 30, 470–479.

Tenaillon, O., 2014. The utility of Fisher's geometric model in evolutionary genetics. Annual Review of Ecology, Evolution, and Systematics 45, 179–201.

Thomas, C.D., Cameron, A., Green, R.E., et al., 2004. Extinction risk from climate change. Nature 427, 145–148.

Thomas, D.A., Barber, H.N., 1974. Studies on leaf characteristics of a cline of *Eucalyptus urnigera* from Mount Wellington, Tasmania. I. Water repellency and the freezing of leaves. Australian Journal of Biological Sciences 22, 501–512.

Thompson, J., Charpentier, A., Bouguet, G., et al., 2013. Evolution of a genetic polymorphism with climate change in a Mediterranean landscape. Proceedings of the National Academy of Sciences 110, 2893–2897.

Thompson, J.N., 1998. Rapid evolution as an ecological process. Trends in Ecology and Evolution 13, 329–332.

Thompson, J.N., 2013. Relentless Evolution. University of Chicago Press, Chicago, IL.

Thorn, J.S., Nijman, V., Smith, D., et al., 2009. Ecological niche modelling as a technique for assessing threats and setting conservation priorities for Asian slow lorises (Primates: *Nycticebus*). Diversity and Distributions 15, 289–298.

Thornhill, N.W., 1993. The Natural History of Inbreeding and Outbreeding: Theoretical and Empirical Perspectives. University of Chicago Press, Chicago, IL.

Thorpe, R.S., Surget-Groba, Y., Johansson, H., 2010. Genetic tests for ecological and allopatric speciation in anoles on an island archipelago. PLOS Genetics 6, e1000929.

Tobias, J.A., Seddon, N., Spottiswoode, C.N., et al., 2010. Quantitative criteria for species delimitation. IBIS 152, 724–746.

Todesco, M., Pascual, M.A., Owens, G.L., et al., 2016. Hybridization and extinction. Evolutionary Applications 9, 892–908.

Toro, M.A., Villanueva, B., Fernández, J., 2014. Genomics applied to management strategies in conservation programmes. Livestock Science 166, 48–53.

Tortajada, A.M., Carmona, M.J., Serra, M., 2009. Does haplodiploidy purge inbreeding depression in rotifer populations? PLOS ONE 4, e8195.

Tortajada, A.M., Carmona, M.J., Serra, M., 2010. Effects of population outcrossing on rotifer fitness. BMC Evolutionary Biology 10, 1.

Trinkel, M., Ferguson, N., Reid, A., et al., 2008. Translocating lions into an inbred lion population in the Hluhluwe-iMfolozi Park, South Africa. Animal Conservation 11, 138–143.

Tsitrone, A., Jarne, P., David, P., 2003. Delayed selfing and resource reallocations in relation to mate availability in the freshwater snail *Physa acuta*. American Naturalist 162, 474–488.

Tsuchimatsu, T., Suwabe, K., Shimizu-Inatsugi, R., et al., 2010. Evolution of self-compatibility in *Arabidopsis* by a mutation in the male specificity gene. Nature 464, 1342–1346.

Tucker, A.E., Ackerman, M.S., Eads, B.D., et al., 2013. Population-genomic insights into the evolutionary origin and fate of obligately asexual *Daphnia pulex*. Proceedings of the National Academy of Sciences 110, 15740–15745.

Turček, F., Hickey, J., 1951. Effect of introductions on two game populations in Czechoslovakia. Journal of Wildlife Management 15, 113–114.

Turesson, G., 1922. The genotypical response of the plant species to the habitat. Hereditas 3, 211–350.

UNEP, 1992. Convention on Biological Diversity. www.cbd.int

UNEP, 2007. The Global Environment Outlook 4 (GEO-4). UNEP, Nairobi.

United States Census Bureau, 2014. Historical estimates of world population. http://www.census.gov/ipc/www/worldhis.html

Urban, M.C., 2015. Accelerating extinction risk from climate change. Science 348, 571–573.

US Fish and Wildlife Service, 2003. Recovery Plan for the Red-Cockaded Woodpecker (*Picoides borealis*): Second Revision. US Fish and Wildlife Service, Atlanta, GA.

Vale, P.F., Choisy, M., Froissart, R., et al., 2012. The distribution of mutational fitness effects of phage φx174 on different hosts. Evolution 66, 3495–3507.

Vallejo-Marín, M., Dorken, M.E., Barrett, S.C., 2010. The ecological and evolutionary consequences of clonality for plant mating. Annual Review of Ecology, Evolution, and Systematics 41, 193–213.

van der Werf, J., Graser, H.-U., Frankham, R., et al. eds., 2009. Adaptation and Fitness in Animal Populations: Evolutionary and Breeding Perspectives on Genetic Resource Management. Springer, Dordrecht.

van Heerwaarden, B., Sgrò, C.M., 2014. Is adaptation to climate change really constrained in niche specialists? Proceedings of the Royal Society of London B: Biological Sciences 281, 20140396.

van Oppen, M.J.H., Oliver, J.K., Putnam, H.M., et al., 2015. Building coral reef resilience through assisted evolution. Proceedings of the National Academy of Sciences 112, 2307–2313.

Van Valen, L., 1973. A new evolutionary law. Evolutionary Theory 1, 1–30.

van't Hof, A.E., Campagne, P., Rigden, D.J., et al., 2016. The industrial melanism mutation in British peppered moths is a transposable element. Nature 534, 102–105.

van't Hof, A.E., Edmonds, N., Dalíková, M., et al., 2011. Industrial melanism in British peppered moths has a singular and recent mutational origin. Science 332, 958–960.

Vasey, N., Tattersall, I., 2002. Do ruffed lemurs form a hybrid zone? Distribution and discovery of *Varecia*, with systematic and conservation implications. American Museum Novitates, 3376, 1–26.

Vayssade, C., De Fazio, C.l., Quaglietti, B., et al., 2014. Inbreeding depression in a parasitoid wasp with single-locus complementary sex determination. PLOS ONE 9, e97733.

Vekemans, X., Poux, C., Goubet, P.M., et al., 2014. The evolution of selfing from outcrossing ancestors in Brassicaceae: what have we learned from variation at the S-locus? Journal of Evolutionary Biology 27, 1372–1385.

Vilà, C., Sundqvist, A.-K., Flagstad, Ö., et al., 2003. Rescue of a severely bottlenecked wolf (*Canis lupus*) population by a single immigrant. Proceedings of the Royal Society of London B: Biological Sciences 270, 91–97.

Vilas, C., García, C., 2006. The role of genetic mechanisms of sex determination in the survival of small populations of *Silene littorea*: a reintroduction experiment. Biological Conservation 129, 124–133.

Vilas, C., Miguel, E.S., Amaro, R., et al., 2006. Relative contribution of inbreeding depression and eroded adaptive diversity to extinction risk in small populations of shore campion. Conservation Biology 20, 229–238.

Visser, M.E., 2008. Keeping up with a warming world; assessing the rate of adaptation to climate change. Proceedings of the Royal Society of London B: Biological Sciences 275, 649–659.

Vitti, J.J., Grossman, S.R., Sabeti, P.C., 2013. Detecting natural selection in genomic data. Annual Review of Genetics 47, 97–120.

Vogler, A.P., DeSalle, R., 1994. Diagnosing units of conservation management. Conservation Biology 8, 354–363.

Vrijenhoek, R.C., 1994. Genetic diversity and fitness in small populations, In Conservation Genetics. eds V. Loeschcke, J. Tomiuk, S.K. Jain, pp. 37–53. Birkhauser, Basel, Switzerland.

Vrijenhoek, R.C., Parker Jr, E.D., 2009. Geographical parthenogenesis: general purpose genotypes and frozen niche variation, In Lost Sex: The Evolutionary Biology of Parthenogenesis. eds I. Schön, K. Martens, P. van Dijk, pp. 99–131. Springer, Dordrecht.

Vucetich, J.A., Waite, T.A., 2000. Is one migrant per generation sufficient for the genetic management of fluctuating populations? Animal Conservation 3, 261–266.

Wahlund, S., 1928. Zusammensetzung von populationen und korrelationserscheinungen vom standpunkt der vererbungslehre aus betrachtet. Hereditas 11, 65–101.

Waits, L., Taberlet, P., Swenson, J.E., et al., 2000. Nuclear DNA microsatellite analysis of genetic diversity and gene flow in the Scandinavian brown bear (*Ursus arctos*). Molecular Ecology 9, 421–431.

Wakeley, J., 2007. Coalescent Theory: An Introduction. Roberts & Co., Greenwood Village, CO.

Wallace, A.R., 1865. On the phenomena of variation and geographical distribution as illustrated by the Papilionidæ of the Malayan Region. Transactions of the Linnean Society of London 25, 1–71.

Waller, D.M., 2015. Genetic rescue: a safe or risky bet? Molecular Ecology 24, 2595–2597.

Walling, C., Nussey, D., Morris, A., et al., 2011. Inbreeding depression in red deer calves. BMC Evolutionary Biology 11, 318.

Wandeler, P., Hoeck, P.E., Keller, L.F., 2007. Back to the future: museum specimens in population genetics. Trends in Ecology and Evolution 22, 634–642.

Wang, I.J., Bradburd, G.S., 2014. Isolation by environment. Molecular Ecology 23, 5649–5662.

Wang, I.J., Summers, K., 2010. Genetic structure is correlated with phenotypic divergence rather than geographic isolation in the highly polymorphic strawberry poison-dart frog. Molecular Ecology 19, 447–458.

Wang, J., 2004a. Application of the one-migrant-per-generation rule to conservation and management. Conservation Biology 18, 332–343.

Wang, J., 2004b. Monitoring and managing genetic variation in group breeding populations without individual pedigrees. Conservation Genetics 5, 813–825.

Wang, J., 2005. Estimation of effective population size from data on genetic markers. Philosophical Transactions of the Royal Society of London B: Biological Sciences 360, 1395–1409.

Wang, J., 2011. COANCESTRY: a program for simulating, estimating and analysing relatedness and inbreeding coefficients. Molecular Ecology Resources 11, 141–145.

Wang, J., 2014a. Marker-based estimates of relatedness and inbreeding coefficients: an assessment of current methods. Journal of Evolutionary Biology 27, 518–530.

Wang, J., 2014b. Estimation of migration rates from marker-based parentage analysis. Molecular Ecology 23, 3191–3213.

Wang, J., 2016. Pedigrees or markers: which are better in estimating relatedness and inbreeding coefficient? Theoretical Population Biology 107, 4–13.

Wang, J., Caballero, A., 1999. Developments in predicting the effective size of subdivided populations. Heredity 82, 212–226.

Waples, R.S., 1991. Definition of "Species" Under the Endangered Species Act: Application to Pacific Salmon. NOAA Technical Memorandum NMFS F/NWC-194, National Marine Fisheries Service, Seattle, WA.

Waples, R.S., Gaggiotti, O., 2006. What is a population? An empirical evaluation of some genetic methods for identifying the number of gene pools and their degree of connectivity. Molecular Ecology 15, 1419–1439.

Waples, R.S., Luikart, G., Faulkner, J.R., et al., 2013. Simple life-history traits explain key effective population size ratios across diverse taxa. Proceedings of the Royal Society B: Biological Sciences 280, 20131339.

Warburton, C.L., James, E.A., Fripp, Y.J., et al., 2000. Clonality and sexual reproductive failure in remnant populations of *Santalum lanceolatum* (Santalaceae). Biological Conservation 96, 45–54.

Ward, O.G., Graphodatsky, A.S., Wurster-Hill, D.H., et al., 1991. Cytogenetics of beavers: a case of speciation by monobrachial centric fusions. Genome 34, 324–328.

Waser, N.M., Price, M.V., Shaw, R.G., 2000. Outbreeding depression varies among cohorts of *Ipomopsis aggregata* planted in nature. Evolution 54, 485–491.

Wayne, R.K., Lehman, N., Girman, D., et al., 1991. Conservation genetics of the endangered Isle Royale gray wolf. Conservation Biology 5, 41–51.

Webb, C., Lloyd, D.G., 1986. The avoidance of interference between the presentation of pollen and stigmas in angiosperms II. Herkogamy. New Zealand Journal of Botany 24, 163–178.

Weber, K., 2004. Population size and long-term selection. Plant Breeding Reviews 24, 249–268.

Weber, K.E., 1996. Large genetic change at small fitness cost in large populations of *Drosophila melanogaster* selected for wind tunnel flight: rethinking fitness surfaces. Genetics 144, 205–213.

Weekley, C.W., Kubisiak, T.L., Race, T.M., 2002. Genetic impoverishment and cross-incompatibility in remnant genotypes of *Ziziphus celata* (Rhamnaceae), a rare shrub endemic to the Lake Wales Ridge, Florida. Biodiversity and Conservation 11, 2027–2046.

Weeks, A.R., Moro, D., Thavornkanlapachai, R., et al., 2015. Conserving and enhancing genetic diversity in translocation programs, In Advances in Reintroduction Biology of Australian and New Zealand Fauna. eds D. Armstrong, M. Hayward, D. Moro, P. Seddon, pp. 127–140. CSIRO Publishing, Melbourne, Vic., Australia.

Weeks, A.R., Sgrò, C.M., Young, A.G., et al., 2011. Assessing the benefits and risks of translocations in changing environments: a genetic perspective. Evolutionary Applications 4, 709–725.

Weeks, A.R., Stoklosa, J., Hoffmann, A.A., 2016. Conservation of genetic uniqueness of populations may increase extinction likelihood of endangered species: the case of Australian mammals. Frontiers in Zoology 13, 1.

Weeks, A.R., van Rooyen, A., Mitrovski, P., et al., 2013. A species in decline: genetic diversity and conservation of the Victorian eastern barred bandicoot, *Perameles gunnii*. Conservation Genetics 14, 1243–1254.

Weinberg, W., 1908. Über den nachweis der vererbung beim menschen. Jahreshefte des Vereins für Vaterlandische Naturkunde in Württemberg 64, 369–382.

Welbergen, J.A., Klose, S.M., Markus, N., et al., 2008. Climate change and the effects of temperature extremes on Australian flying-foxes. Proceedings of the Royal Society B: Biological Sciences 275, 419–425.

Weller, S.G., Sakai, A.K., Thai, D.A., et al., 2005. Inbreeding depression and heterosis in populations of *Schiedea viscosa*, a highly selfing species. Journal of Evolutionary Biology 18, 1434–1444.

Werren, J.H., 1993. The evolution of inbreeding in haplodiploid organisms, In The Natural History of Inbreeding and Outbreeding: Theoretical and Empirical Perspectives. ed. N.W. Thornhill, pp. 42–59. University of Chicago Press, Chicago, IL.

Westemeier, R.L., Brawn, J.D., Simpson, S.A., et al., 1998. Tracking the long-term decline and recovery of an isolated population. Science 282, 1695–1698.

White, M.J.D., 1973. Animal Cytology and Evolution, 3rd edition. Cambridge University Press, Cambridge, UK.

White, M.J.D., 1978. Modes of Speciation. W.H. Freeman, San Francisco, CA.

Whitehorn, P.R., Tinsley, M.C., Brown, M.J.F., et al., 2009. Impacts of inbreeding on bumblebee colony fitness under field conditions. BMC Evolutionary Biology 9, 152.

Whiteley, A.R., Fitzpatrick, S.W., Funk, W.C., et al., 2015. Genetic rescue to the rescue. Trends in Ecology and Evolution 30, 42–49.

Whiteman, N.K., Matson, K.D., Bollmer, J.L., et al., 2006. Disease ecology in the Galápagos hawk (*Buteo galapagoensis*): host genetic diversity, parasite load and natural antibodies. Proceedings of the Royal Society of London B: Biological Sciences 273, 797–804.

Whitham, T.G., Bailey, J.K., Schweitzer, J.A., et al., 2006. A framework for community and ecosystem genetics: from genes to ecosystems. Nature Reviews Genetics 7, 510–523.

Whitlock, M.C., 2000. Fixation of new alleles and the extinction of small populations: drift load, beneficial alleles, and sexual selection. Evolution 54, 1855–1861.

Whitlock, M.C., 2004. Selection and drift in metapopulations, In Ecology, Genetics, and Evolution of Metapopulations. eds I. Hanski, O.E. Gaggiotti, pp. 153–173. Elsevier, Burlington, MA.

Whitlock, M.C., McCauley, D.E., 1999. Indirect measures of gene flow and migration: $F_{ST} \neq 1/(4Nm+1)$. Heredity 82, 117–125.

Whitlock, M.C., Phillips, P.C., Moore, F.B.G., et al., 1995. Multiple fitness peaks and epistasis. Annual Review of Ecology and Systematics 26, 601–629.

Whitlock, R., Stewart, G., Goodman, S., et al., 2013. A systematic review of phenotypic responses to between-population outbreeding. Environmental Evidence 2, 13.

Wilcken, J., 2002. The Cost of Inbreeding Revisited. M.Sc. thesis, Macquarie University, Sydney, NSW, Australia.

Wilczek, A.M., Cooper, M.D., Korves, T.M., et al., 2014. Lagging adaptation to warming climate in *Arabidopsis thaliana*. Proceedings of the National Academy of Sciences 111, 7906–7913.

Wiley, E.O., 1978. The evolutionary species concept reconsidered. Systematic Biology 27, 17–26.

Wilkins, J.S., 2009. Species: A History of the Idea. University of California Press, Berkeley, CA.

Willi, Y., Fischer, M., 2005. Genetic rescue in interconnected populations of small and large size of the self-incompatible *Ranunculus reptans*. Heredity 95, 437–443.

Williams, J.G., Armstrong, G., Katopodis, C., et al., 2012. Thinking like a fish: a key ingredient for development of effective fish passage facilities at river obstructions. River Research and Applications 28, 407–417.

Williams, W., Gilbert, N., 1960. Heterosis and the inheritance of yield in the tomato. Heredity 14, 133–145.

Willis, J.H., 1992. Genetic analysis of inbreeding depression caused by chlorophyll-deficient lethals in *Mimulus guttatus*. Heredity 69, 562–572.

Willis, S.G., Hill, J.K., Thomas, C.D., et al., 2009. Assisted colonization in a changing climate: a test-study using two U.K. butterflies. Conservation Letters 2, 45–51.

Wilson, A.J., Réale, D., Clements, M.N., et al., 2010. An ecologist's guide to the animal model. Journal of Animal Ecology 79, 13–26.

Wilson, D.E., Reeder, D.M. eds., 2005. Mammal Species of the World, 3rd edition: A Taxonomic and Geographic Reference. Johns Hopkins University Press, Baltimore, MD.

Wilson, G.A., Rannala, B., 2003. Bayesian inference of recent migration rates using multilocus genotypes. Genetics 163, 1177–1191.

Winker, K., 2010. Is it a species? IBIS 152, 679–682.

Winn, A.A., Elle, E., Kalisz, S., et al., 2011. Analysis of inbreeding depression in mixed-mating plants provides evidence for selective interference and stable mixed mating. Evolution 65, 3339–3359.

Wolf, C.M., Griffith, B., Reed, C., et al., 1996. Avian and mammalian translocations: update and reanalysis of 1987 survey data. Conservation Biology 10, 1142–1154.

Wood, J.L.A., Yates, M.C., Fraser, D.J., 2016. Are heritability and selection related to population size in nature? Meta-analysis and conservation implications. Evolutionary Applications 9, 640–657.

Wood, T.E., Takebayashi, N., Barker, M.S., et al., 2009. The frequency of polyploid speciation in vascular plants. Proceedings of the National Academy of Sciences 106, 13875–13879.

Woodruff, D., 1989. The problems of conserving genes and species, In Conservation for the Twenty-First Century. eds D. Western, M.C. Pearl, pp. 76–88. Oxford University Press, New York.

Woodworth, L.M., Montgomery, M.E., Briscoe, D.A., et al., 2002. Rapid genetic deterioration in captivity: causes and conservation implications. Conservation Genetics 3, 277–288.

Wootton, J.T., Pfister, C.A., 2013. Experimental separation of genetic and demographic factors on extinction risk in wild populations. Ecology 94, 2117–2123.

Wootton, J.T., Pfister, C.A., 2015. Processes affecting extinction risk in the laboratory and in nature. Proceedings of the National Academy of Sciences 112, E5903.

Wright, S., 1922. Coefficients of inbreeding and relationship. American Naturalist 56, 330–338.

Wright, S., 1931. Evolution in Mendelian populations. Genetics 16, 97–159.

Wright, S., 1939. The distribution of self-sterility alleles in populations. Genetics 24, 538–552.

Wright, S., 1940. Breeding structure of populations in relation to speciation. American Naturalist 74, 232–248.

Wright, S., 1943. Isolation by distance. Genetics 28, 114–138.

Wright, S., 1951. The genetical structure of populations. Annals of Eugenics 15, 322–354.

Wright, S., 1969. Evolution and the Genetics of Populations. 2. The Theory of Gene Frequencies. University of Chicago Press, Chicago, IL.

Wright, S., 1977. Evolution and the Genetics of Populations. 3. Experimental Results and Evolutionary Deductions. University of Chicago Press, Chicago, IL.

Wyner, Y.M., Amato, G., DeSalle, R., 1999. Captive breeding, reintroduction, and the conservation genetics of black and white ruffed lemurs, *Varecia variegata variegata*. Molecular Ecology 8, S107–S115.

Xu, X., Arnason, U., 1996. The mitochondrial DNA molecule of Sumatran orangutan and a molecular proposal for two (Bornean and Sumatran) species of orangutan. Journal of Molecular Evolution 43, 431–437.

Xue, Y., Prado-Martinez, J., Sudmant, P.H., et al., 2015. Mountain gorilla genomes reveal the impact of long-term population decline and inbreeding. Science 348, 242–245.

Yang, F., Fu, B., O'Brien, P.M., et al., 2004. Refined genome-wide comparative map of the domestic horse, donkey and human based on cross-species chromosome painting: insight into the occasional fertility of mules. Chromosome Research 12, 65–76.

Yang, Z., 2015. The BPP program for species tree estimation and species delimitation. Current Zoology 61, 854–865.

Young, A.G., Murray, B.G., 2000. Genetic bottlenecks and dysgenic gene flow into re-established populations of the grassland daisy *Rutidosis leptorrynchoides*. Australian Journal of Botany 48, 409–416.

Young, A.G., Brown, A.H.D., Murray, B.G., et al., 2000. Genetic erosion, restricted mating and reduced viability in fragmented populations of the endangered grassland herb *Rutidosis leptorrhynchoides*, In Genetics, Demography and Viability of Fragmented Populations. eds A.G. Young, G.M. Clarke, pp. 335–359. Cambridge University Press, Cambridge, UK.

Yuan, N., Comes, H.P., Mao, Y.-R., et al., 2012. Genetic effects of recent habitat fragmentation in the Thousand-Island Lake region of southeast China on the distylous herb *Hedyotis chrysotricha* (Rubiaceae). American Journal of Botany 99, 1715–1725.

Zachos, F.E., Apollonio, M., Bärmann, E.V., et al., 2013. Species inflation and taxonomic artefacts—A critical comment on recent trends in mammalian classification. Mammalian Biology 78, 1–6.

Zapata, L., Ding, J., Willing, E.-M., et al., 2016. Chromosome-level assembly of *Arabidopsis thaliana* Ler reveals the extent of translocation and inversion polymorphisms. Proceedings of the National Academy of Sciences 113, E4052–E4060.

Zayed, A., Packer, L., 2005. Complementary sex determination substantially increases extinction proneness of haplodiploid populations. Proceedings of the National Academy of Sciences 102, 10742–10746.

Zayed, A., Roubik, D.W., Packer, L., 2004. Use of diploid male frequency data as an indicator of pollinator decline. Proceedings of the Royal Society of London B: Biological Sciences 271, S9–S12.

Zeyl, C., Mizesko, M., Arjan, J., et al., 2001. Mutational meltdown in laboratory yeast populations. Evolution 55, 909–917.

Zhang, B., Li, M., Zhang, Z., et al., 2007. Genetic viability and population history of the giant panda, putting an end to the "evolutionary dead end"? Molecular Biology and Evolution 24, 1801–1810.

Zhang, C., Zhang, D.-X., Zhu, T., et al., 2011. Evaluation of a Bayesian coalescent method of species delimitation. Systematic Biology 60, 747–761.

Zhi, L., Karesh, W.B., Janczewski, D.N., et al., 1996. Genomic differentiation among natural populations of orang-utan (*Pongo pygmaeus*). Current Biology 6, 1326–1336.

Zhu, L., Zhang, S., Gu, X., et al., 2011. Significant genetic boundaries and spatial dynamics of giant pandas occupying fragmented habitat across southwest China. Molecular Ecology 20, 1122–1132.

Zigler, K.S., McCartney, M.A., Levitan, D.R., et al., 2005. Sea urchin bindin divergence predicts gamete compatibility. Evolution 59, 2399–2404.

Zima, J., Fedyk, S., Fredga, K., et al., 1996. The list of the chromosome races of the common shrew (*Sorex araneus*). Hereditas 125, 97–107.

Index

Note: Tables, figures, and boxes are indicated by an italic *t*, *f*, and *b* following the page number.

C

Q

R

S

The manufacturer's authorised representative in the EU for product
safety is Oxford University Press España S.A. of El Parque Empresarial
San Fernando de Henares, Avenida de Castilla, 2 - 28830 Madrid
(www.oup.es/en or product.safety@oup.com). OUP España S.A. also acts
as importer into Spain of products made by the manufacturer.
Printed and bound by CPI Group (UK) Ltd, Croydon, CR0 4YY
14/09/2026
02216286-0018